Advances in freshwater decapod systematics and biology

CRUSTACEANA MONOGRAPHS constitutes a series of books on carcinology in its widest sense. Contributions are handled by the Series Editor(s) and may be submitted through the office of KONINKLIJKE BRILL Academic Publishers N.V., P.O. Box 9000, NL-2300 PA Leiden, The Netherlands.

Published in this series:

CRM 001 - *Stephan G. Bullard*	Larvae of anomuran and brachyuran crabs of North Carolina
CRM 002 - *Spyros Sfenthourakis et al. (eds.)*	The biology of terrestrial isopods, V
CRM 003 - *Tomislav Karanovic*	Subterranean Copepoda from arid Western Australia
CRM 004 - *Katsushi Sakai*	Callianassoidea of the world (Decapoda, Thalassinidea)
CRM 005 - *Kim Larsen*	Deep-sea Tanaidacea from the Gulf of Mexico
CRM 006 - *Katsushi Sakai*	Upogebiidae of the world (Decapoda, Thalassinidea)
CRM 007 - *Ivana Karanovic*	Candoninae (Ostracoda) from the Pilbara region in Western Australia
CRM 008 - *Frank D. Ferrari & Hans-Uwe Dahms*	Post-embryonic development of the Copepoda
CRM 009 - *Tomislav Karanovic*	Marine interstitial Poecilostomatoida and Cyclopoida (Copepoda) of Australia
CRM 010 - *Carrie E. Schweitzer et al.*	Systematic list of fossil decapod crustacean species
CRM 011 - *Peter Castro et al. (eds.)*	Studies on Brachyura: a homage to Danièle Guinot
CRM 012 - *Patricio R. De los Ríos-Escalante*	Crustacean zooplankton communities in Chilean inland waters
CRM 013 - *Katsushi Sakai*	Axioidea of the world and a reconsideration of the Callianassoidea (Decapoda, Thalassinidea, Callianassida)
CRM 014 - *Charles H.J.M. Fransen et al. (eds.)*	Studies on Malacostraca: Lipke Bijdeley Holthuis Memorial Volume
CRM 015 - *Akira Asakura et al. (eds.)*	New Frontiers in Crustacean Biology: Proceedings of the TCS Summer Meeting, Tokyo, 20-24 September 2009
CRM 016 - *Danielle Defaye et al. (eds.)*	Studies on Freshwater Copepoda: a Volume in Honour of Bernard Dussart
CRM 017 - *Hironori Komatsu et al. (eds.)*	Studies on Eumalacostraca: a homage to Masatsune Takeda
CRM 018 - *Masahiro Dojiri & Ju-Shey Ho*	Systematics of the Caligidae, copepods parasitic on fishes

Editors' addresses:
Darren C.J. Yeo, Department of Biological Sciences, National University of Singapore (NUS), 14 Science Drive 4, Singapore 117543, Republic of Singapore; e-mail: dbsyeod@nus.edu.sg
Neil Cumberlidge, Department of Biology, Northern Michigan University, 1401 Presque Isle Avenue, Marquette, MI 49855, U.S.A.; e-mail: ncumberl@nmu.edu
Sebastian Klaus, Department of Ecology and Evolution, J.W. Goethe-Universität, Max-von-Laue-Straße 13, 60438 Frankfurt am Main, Germany; e-mail: klaus@bio.uni-frankfurt.de

Cover: Cherax preissii (Erichson, 1846) by Ahyong; see p. 256, fig. 1.

Advances in freshwater decapod systematics and biology

By
Darren C.J. Yeo, Neil Cumberlidge and Sebastian Klaus
(Editors)

CRUSTACEANA MONOGRAPHS, 19

BRILL

LEIDEN • BOSTON

This book is printed on acid-free paper.

Library of Congress Cataloging-in-Publication Data

Advances in freshwater decapod systematics and biology / by Darren C.J. Yeo, Neil Cumberlidge, and Sebastian Klaus (editors).
pages cm. — (Crustaceana monographs ; 19)
Includes bibliographical references.
ISBN 978-90-04-20760-8 (hardback : alk. paper) — ISBN 978-90-04-20761-5 (e-book)
1. Decapoda (Crustacea) 2. Decapoda (Crustacea)—Classification. 3. Freshwater invertebrates. I. Yeo, Darren C. J. (Darren Chong Jinn) II. Cumberlidge, Neil. III. Klaus, Sebastian.
QL444.M33A33 2014
595.3'8—dc23

2014016758

ISBN13: 978 90 04 20760 8
E-ISBN: 978 90 04 20761 5

PRINTED IN THE NETHERLANDS

CONTENTS

PREFACE — FRESHWATER DECAPOD BIOLOGY IN THE 21ST CENTURY

BY

DARREN C. J. YEO[1,4]), NEIL CUMBERLIDGE[2,5]) and SEBASTIAN KLAUS[1,3,6])

[1]) Department of Biological Sciences, National University of Singapore, 14 Science Drive 4, Singapore 117543, Republic of Singapore

[2]) Department of Biology, Northern Michigan University, Marquette, MI 49855, U.S.A.

[3]) Department of Ecology and Evolution, J.W. Goethe-Universität, Max-von-Laue-Straße 13, 60438 Frankfurt am Main, Germany

Among an estimated 15 000 described decapod crustaceans, about 3000 species (ca. 20%) can be considered freshwater species (De Grave et al., 2008; Yeo et al., 2008) that depend on intact freshwater habitats for their survival. The inland waters of continents and islands comprise less than 3% of the total water on earth and are highly discontinuous habitats. Despite this, the high diversity of ecological niches in freshwater habitats and the frequent occurrence of genetic isolation have resulted in the evolution of an impressively rich assemblage of species.

Most of the recent studies on freshwater decapods are the result of international collaborative efforts, bringing together taxonomic and methodological expertise and access to a broad selection of species from most parts of the world. An initial Freshwater Crab Symposium in 2008 held at the National University of Singapore brought together, for the first time, a multi-national group of specialists working on the primary freshwater crabs in all major regions of the animals' global distribution. This was a follow-up to informal collaborations that began in 2005 between several of the participants, which resulted in the publication of the first global checklist and assessment of freshwater crab diversity (Yeo et al., 2008a). The overall objective of the Freshwater Crab Symposium in Singapore was to expand collaboration on the production of more synthetic works that build on the collective expertise of primary freshwater crabs throughout their range. One of the key topics discussed was the

[4]) Corresponding author; e-mail: dbsyeod@nus.edu.sg

[5]) e-mail: ncumberl@nmu.edu

[6]) e-mail: klaus@bio.uni-frankfurt.de

conservation of freshwater crabs. Up to this point there had only been conservation assessments of regional faunas using the IUCN Red List criteria as guidelines (e.g., Ng & Yeo, 2005; Yeo et al., 2008b), and there was a clear need for a worldwide conservation assessment of the freshwater crabs now that a global checklist was available. This resulted in the first global IUCN Red List assessment of freshwater crabs (Cumberlidge et al., 2009). By then it also became evident that more collaborative works would be fruitful and indeed necessary. Consequently, the idea came up for a broader platform to discuss future approaches and to initiate a wider network of collaboration especially for tackling molecular and morphological phylogenetics of freshwater decapod taxa. Michael Türkay from the Senckenberg Research Institute, Frankfurt am Main, generously offered to organise and host this meeting (together with Peter K. L. Ng and Neil Cumberlidge), which took place in December 2010. The present volume was initiated at that meeting, and neither claims to be a synthesis of current freshwater decapod biology in general, nor an exhaustive review on the current state of systematics and phylogeny of freshwater decapods. Rather, it is a selection of 14 papers that represents a sample of the current state of research in various fields for the main freshwater decapod taxa.

The present volume includes two general overviews of freshwater decapods as a whole (Vogt — longevity and mortality; Cumberlidge — conservation); two dealing with caridean prawns (Guerao et al. — phylogenetics of Xiphocaridae; Cai — atyids of Hainan Island); one on freshwater anomurans (Santos et al. — aeglids of Uruguay); two on crayfish (Ahyong — diversity of parastacids; Furse — conservation of parastacids); and seven dealing with freshwater brachyurans (Klaus & Türkay — systematics and biogeography/legacy of Richard Bott; Cumberlidge — overview of Afrotropical crabs; Magalhães et al. — new species of *Fredius* (Pseudothelphusidae); Keikhosravi & Schubart — new species of *Potamon* (Potamidae); Klaus & Prieto — new potamid fossils; Mendoza & Yeo — new species of *Isolapotamon*; Schubart & Santl — new subspecies of *Sesarma*). The freshwater brachyurans covered in this volume include primary freshwater crabs as well as secondary freshwater crabs. The utilitarian term "primary freshwater crabs" is preferred here as it is more descriptive and apt for referring to brachyurans in the "wholly freshwater" families of Pseudothelphusidae, Potamidae, Potamonautidae, Gecarcinucidae, and Trichodactylidae (Yeo et al., 2008; Cumberlidge & Ng, 2009), which do not necessarily represent a single monophyletic grouping, but commonly lack any marine members. Additionally the immediate marine ancestors or sister groups of the wholly freshwater crabs remain unclear. The use of "true freshwater crabs" as the collective term for these taxa in earlier publications (e.g., Ng, 1988) could potentially cause confusion because it was not meant to refer to "partly freshwater" or secondary freshwater families, e.g., Sesarmidae, Varunidae, Gecarcinidae, Hymenosomatidae (see Yeo et al., 2008). This latter group comprises mainly marine families where although most of the members are marine or brackish water species a significant number are completely freshwater or terrestrial in habit (although many of these species found in freshwater habitats are amphidromous and their larvae need a marine or brackish water phase for development).

What is the status of freshwater decapod biology at the beginning of the 21st century? A simple literature search on the Web of Science® data base shows a rapid increase in publications on decapods in general since the 1990s, but most deal with marine species and only a small proportion of these (9%) dealt with freshwater decapods (fig. 1). Although the informative value of such statistics has to be treated with caution, it is nevertheless obvious that an exhaustive knowledge of freshwater decapod crustaceans is a goal that still needs to be realised. Basic alpha-taxonomic research has been built upon

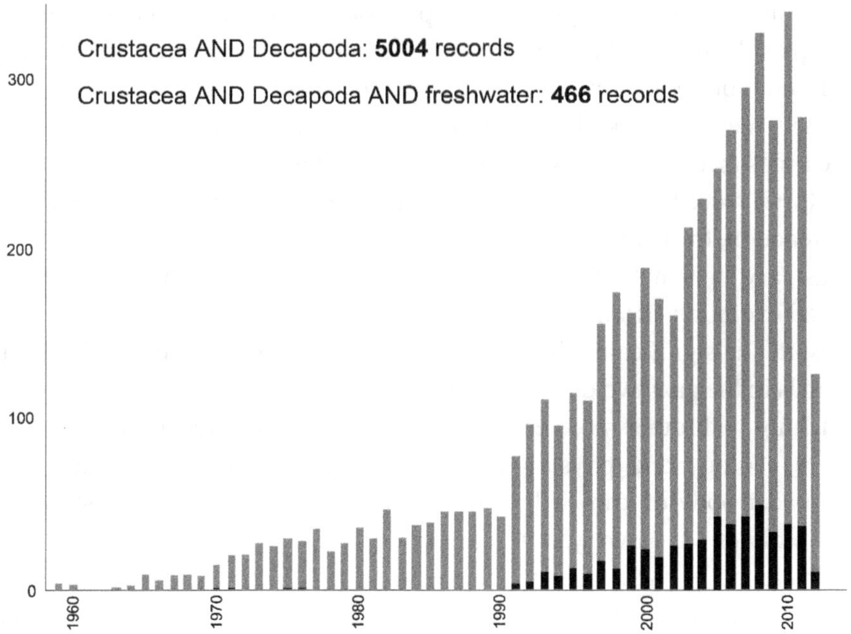

Fig. 1. Increase in publications on freshwater decapod Crustacea (black bars) in relation to articles on decapod Crustacea in general (grey bars) during the last 50 years (citations retrieved from Web of Science®, 1899 to 2012; records prior to 1959 not shown).

the work of earlier researchers (see article by Klaus & Türkay, present volume). A total of seven new taxa are described in this volume: two new atyid species from Hainan, China (see Cai, present volume), one new aeglid species from Uruguay (see Santos et al., present volume); one new pseudothelphusid species from Brazil (see Magalhães et al., present volume); two new potamid species — one from Iran (Keikhosravi & Schubart, present volume) and one from Philippines (Mendoza & Yeo, present volume); and one new sesarmid subspecies (Schubart & Santl, present volume). Conventional morphological approaches (see articles by Cai, Mendoza & Yeo, Santos et al., all present volume) still play a major role alongside diverse new methodological approaches that employ a combination of molecular, morphometric, and histological tools (see articles by Keikhosravi & Schubart, Magalhães et al., Schubart & Santl, all present volume).

Since the mid-1990s to mid-2000s, molecular phylogenetic and phylogeographic studies have been conducted on various groups of freshwater decapods (e.g., crayfish: Crandall & Fitzpatrick, 1996; Crandall et al., 1999, 2000; crabs: Segawa, 2000; Daniels et al., 2002, 2006; Shih et al., 2004; Klaus et al., 2006; aeglids: Pérez-Losada et al., 2004; prawns: Murphy & Austin, 2005). This trend is unlikely to slow given the rapid pace of development of molecular techniques, laboratory procedures, and data analysis (see article by Guerao et al., present volume). However, these advances also require a critical review of these new methods, e.g., the calibration of molecular clocks to estimate divergence times (see article by Klaus & Prieto, present volume). Less well studied is the fossil record of freshwater decapods, reflecting the much more restricted occurrence of freshwater sediments. Fossils are the only direct window into the past, and thus make a vital contribution to our understanding of freshwater decapod evolution (Klaus & Prieto, present volume).

Despite the significant advances in systematics and taxonomy of freshwater decapods in the past two decades (see above), these fields have some way to go to catch up with other more established areas of freshwater decapod research such as ecology and physiology, which have accumulated a much larger and broader body of knowledge (e.g., Warner, 1977; Burggren & McMahon, 1988; Holdich & Lowery, 1988; Holdich, 2002; Thorp & Covich, 2009). The challenge in such cases would be to consolidate and synthesise various disparate studies in order to develop a comprehensive picture of aspects of the life history of these animals (see article by Vogt, present volume).

In our rapidly changing world, perhaps the most urgent topic in freshwater decapod research is the growing realization that there are large numbers of species of freshwater crabs, crayfish, freshwater shrimp, and freshwater aeglids that are threatened with extinction, and that there is an urgent need for the effective implementation of conservation measures (see articles by Cumberlidge, Furse, present volume). The fact that all groups of freshwater decapods include high numbers of endemic species that have a restricted distribution makes them more likely to be strongly affected by a range of human activities that are seriously impacting vulnerable freshwater habitats throughout the inland waters in most parts of the world.

REFERENCES

BURGGREN, W. W. & B. R. MCMAHON (eds.), 1988. Biology of the land crabs: i-xii, 1-479. (Press Syndicate of the University of Cambridge, Cambridge).

CRANDALL, K. A., J. W. FETZNER JR., C. G. JARA & L. BUCKUP, 2000. On the phylogenetic positioning of the South American freshwater crayfish genera (Decapoda: Parastacidae). Journal of Crustacean Biology, **20**(3): 530-540.

CRANDALL, K. A., J. W. FETZNER JR., S. H. LAWLER, M. KINNERSLEY & C. M. AUSTIN, 1999. Phylogenetic relationships among the Australian and New Zealand genera of freshwater crayfishes (Decapoda: Parastacidae). Australian Journal of Zoology, **47**(2): 199-214.

CRANDALL, K. A. & J. F. FITZPATRICK JR., 1996. Crayfish molecular systematics: using a combination of procedures to estimate phylogeny. Systematic Biology, **45**: 1-26.

CUMBERLIDGE, N. & P. K. L. NG, 2009. Systematics, evolution, and biogeography of the freshwater crabs. In: J. W. MARTIN, K. A. CRANDALL & D. L. FELDER (eds.), Decapod crustacean phylogenetics. Crustacean Issues, **18**: 491-504. (Taylor & Francis/CRC Press, Boca Raton, Florida).

CUMBERLIDGE, N., P. K. L. NG, D. C. J. YEO, C. MAGALHAES, M. R. CAMPOS, F. ALVAREZ, T. NARUSE, S. R. DANIELS, L. J. ESSER, F. Y. K. ATTIPOE, F.-L. CLOTILDE-BA, W. R. T. DARWALL, A. MCIVOR, M. RAM & B. COLLEN, 2009. Freshwater crabs and the biodiversity crisis: importance, threats, status, and conservation challenges. Biological Conservation, **142**: 1665-1673.

DANIELS, S. R., N. CUMBERLIDGE, M. PEREZ-LOSADA, S. A. E. MARIJNISSEN & K. A. CRANDALL, 2006. Evolution of Afrotropical freshwater crab lineages obscured by morphological convergence. Molecular Phylogenetics and Evolution **40**: 227-235.

DANIELS, S. R., B. A. STEWART, G. GOUWS, M. CUNNINGHAM & C. A. MATTHEE, 2002. Phylogenetic relationships of the southern African freshwater crab fauna derived from multiple data sets reveal biogeographic patterning. Molecular Phylogenetics and Evolution, **25**: 511-523.

DE GRAVE, S., Y. CAI & A. ANKER, 2008. Global diversity of shrimps (Caridea) in freshwater. Hydrobiologia, **595**: 287-293.

HOLDICH, D. M. (ed.), 2002. Biology of freshwater crayfish: 1-702. (Blackwell Science, Oxford).

HOLDICH, D. M. & R. S. LOWERY (eds.), 1988. Freshwater crayfish: biology, management and exploitation: 1-498. (Timber Press, Portland, Oregon).

KLAUS, S., C. D. SCHUBART & D. BRANDIS, 2006. Phylogeny, biogeography and a new taxonomy for the Gecarcinucoidea Rathbun, 1904 (Decapoda: Brachyura). Organisms, Diversity and Evolution, **6**: 199-217.

MURPHY, N. P. & C. M. AUSTIN, 2005. Phylogenetic relationships of the globally distributed freshwater prawn genus *Macrobrachium* (Crustacea: Decapoda: Palaemonidae): biogeography, taxonomy and the convergent evolution of abbreviated larval development. Zoologica Scripta, **34**: 187-197.

NG, P. K. L., 1988. The freshwater crabs of peninsular Malaysia and Singapore: i-viii, 1-156, figs. 1-63, 4 colour plates. (Department of Zoology, National University of Singapore, Shinglee Press, Singapore).

NG, P. K. L. & D. C. J. YEO, 2007. Malaysian freshwater crabs: conservation prospects and challenges. In: L. S. L. CHUA, L. G. KIRTON & L. G. SHAW (eds.), Status of biological diversity in Malaysia and threat assessment of plant species in Malaysia. Proceedings of the Seminar and Workshop, 28-30 June 2005: 95-120.

PÉREZ-LOSADA, M., G. BOND-BUCKUP, C. G. JARA & K. A. CRANDALL, 2004. Molecular systematics and biogeography of the Southern South American freshwater "crabs" *Aegla* (Decapoda: Anomura: Aeglidae) using multiple heuristic tree search approaches. Systematic Biology, **53**: 767-780.

SEGAWA, R., 2000. Molecular phylogenetic study of potamoid crabs in Ryukyu Islands. Kaiyo Monthly, **32**: 241-245.

SHIH, H.-T., P. K. L. NG & H.-W. CHANG, 2004. The systematics of the genus *Geothelphusa* (Crustacea, Decapoda, Brachyura, Potamidae) from southern Taiwan: a molecular appraisal. Zoological Studies, **43**: 561-570.

THORP, J. H. & A. P. COVICH (eds.), 2009. Ecology and classification of North American freshwater invertebrates (3rd edition): 1-1036. (Academic Press, Boston).

WARNER, G. F., 1977. The biology of crabs: 1-202. (Elek Science, London).

YEO, D. C. J., P. K. L. NG, N. CUMBERLIDGE, C. MAGALHÃES, S. R. DANIELS & M. R. CAMPOS, 2008a. Global diversity of crabs (Crustacea: Decapoda: Brachyura) in freshwater. Hydrobiologia, **595**: 275-286.

YEO, D. C. J., S. H. TAN & P. K. L. NG, 2008b. Horseshoe crabs (phylum Arthropoda: subphylum Chelicerata: class Merostomata) decapod crustaceans (phylum Arthropoda: subphylum Crustacea: order Decapoda). In: G. H. W. DAVISON, P. K. L. NG & H. C. HO (eds.), The Singapore Red Data Book: threatened plants and animals of Singapore (2nd edition): 110-128. (Nature Society (Singapore), Singapore).

FRESHWATER CRAB SYSTEMATICS AND BIOGEOGRAPHY: THE LEGACY OF RICHARD BOTT (*1902-†1974)

BY

SEBASTIAN KLAUS[1,3]) and MICHAEL TÜRKAY[2,4])

[1]) Department of Ecology and Evolution, J.W. Goethe-Universität, Max-von-Laue-Straße 13, 60438 Frankfurt am Main, Germany

[2]) Senckenberg Forschungsinstitut und Naturmuseum, Senckenberganlage 25, 60325 Frankfurt am Main, Germany

There are three good reasons to commemorate and discuss the scientific achievements of Richard Bott in this Crustacean Monographs volume on freshwater decapod biology. First, 2012 was the 110th anniversary of the birth of Richard Bott, and 2014 will mark the 40th anniversary of his death. Second, the first international Freshwater Decapod meeting, which is the incentive for the present volume, was held in 2010 in Frankfurt am Main, Germany. This is where Bott conducted his carcinological studies over several decades (1948-1974) as the honorary head of the Crustacean section at the Naturmuseum Senckenberg (fig. 1) (for orbituaries see Türkay, 1974a, b, 1975a). Bott contributed substantially to the taxonomy and systematics of all groups of primary freshwater crabs, laying the foundations for any later work in this field. Although Bott's biogeographic work is less well known (possibly because Bott published exclusively in German), his ideas on the origins and dispersal of the freshwater crabs were novel at that time and emphasised the importance of using geographic and palaeogeographic data to understand taxonomic and systematic relationships.

REPRODUCTIVE MORPHOLOGY AND TAXONOMY

Bott earned his doctorate with a dissertation on coleopteran eye morphology in 1927 (published in Bott, 1928). His interest in freshwater crabs arose when

[3]) Corresponding author; e-mail: klaus@bio.uni-frankfurt.de

[4]) e-mail: Michael.Tuerkay@senckenberg.de

 Advances in freshwater decapod systematics and biology: 7-16

Fig. 1. Richard Bott in his office at the Senckenberg Forschungsinstitut und Naturmuseum, Frankfurt am Main.

he worked as a teacher in Istanbul, Turkey, from 1940 to 1944, which resulted in his first publication on the ecology and behaviour of *Potamon fluviatile* (see Bott, 1944) that included figures of the gonopods (though the material for that work was collected in Istanbul, and so his specimens were most likely *P. ibericum* rather than *P. fluviatile*). Bott (1944) made observations on copulation in freshwater crabs that later influenced his approach to freshwater crab taxonomy.

Bott recognised that the number of morphological characters that could be used to delimit species of primary freshwater crabs was limited. He also realised that carapace characters (that were widely used at the time) were prone to convergent evolution, whereby species of freshwater crabs from different parts of the world that live in a similar habitat may have evolved similar-looking morphological adaptations. A reliance on evolutionary convergent morphologies can lead to taxonomic errors when trying to group taxa on the basis of a shared common ancestry (Bott, 1970a). These difficulties are reflected by the fact that (apart from genera with highly aberrant external morphology, such as *Gecarcinucus*, *Platythelphusa*, *Trichodactylus*, and *Erimetopus*) very few of the hundreds of other genera and sub-genera of freshwater crabs that are known today had been described by the first half of the twentieth century.

Instead of using carapace characters Bott was convinced that the male reproductive organs, especially the male first pleopods or first gonopods (G1), represented the key to recognising natural groups within the freshwater crabs. Although G1 morphology had previously been used as an additional diagnostic character by Rathbun (1904), Colosi (1920), and Balss (1937), it was Bott who consistently applied G1 characters to resolve taxonomic relationships. He also argued that at the species level, carapace characters were often subject to environmentally driven variability, and could therefore be inconsistent, even within a species. The male gonopods are shielded from environmental selection because they are covered by the tightly fitting abdomen, and their structural differences therefore represented adaptations for successful fertilisation (Türkay, 1975b). Bott did not propose a lock and key mechanism for the interaction of the male pleopods (the intromittant organs) with the female reproductive openings of freshwater crabs. Instead, Bott argued that differences between the G1 of different species of freshwater crabs were good indications of a species boundary, and that similarities between the G1 characters shared by different taxa were good indications of a natural group with a shared ancestry (Bott, 1970a).

It should be noted that Bott's work preceded the modern understanding of evolutionary and phylogenetic concepts. When Bott started his work on freshwater crabs in the 1950's synthetic evolutionary theory and phylogenetic systematics were just being developed, and were rarely incorporated into taxonomic works. This is reflected in Bott's biogeographic hypotheses that allowed paraphyletic relationships between taxa that are not recognised today.

An example of Bott's use of gonopod characters at the genus level can be seen in the freshwater crabs of the genus *Platythelphusa* from Lake Tanganyika. The highly unusual combination of characters of the carapace and mouthparts of this genus resulted in a number of authors assigning it to either a unique subfamily or family (Colosi, 1920; see Cumberlidge, 1999). The opposite view of the taxonomic position of *Platythelphusa* was taken by Bott (1955) who included this taxon within an existing family (the Potamonautidae) based on his assignment of a higher taxonomic weight to the characters of the G1. Bott (1955) treated *Platythelphusa* as a subgenus of *Potamonautes*, arguing that its G1 morphology was not sufficiently different from that of *Potamonautes* to warrant its assignment to a separate genus. Subsequent morphological studies of second gonopods (G2) (Klaus et al., 2006), and molecular phylogenetic studies of the Afrotropical freshwater crabs (Daniels et al., 2006) both support the placement of *Platythelphusa* within the genus *Potamonautes*.

The case of *Platythelphusa* also highlights the risk that valid species will not be recognised when G1 morphology is more conservative. Bott (1955) recognised only one species of *Platythelphusa* (out of three taxa that had been described at that time); today nine species of this genus are recognised (Capart, 1954; Cumberlidge et al., 1999; Marijnissen et al., 2004). As a taxonomist Bott was a 'lumper' of taxa rather than a 'splitter' of taxa. This approach resulted in his suppression of a number of taxa — as either junior synonyms or subspecies — that have since proved to be valid species.

Today a description of a new species of freshwater crab from any part of the world would be considered incomplete without including characters of the G1 and G2 of an adult male specimen. The recent addition of histological characters of the G1 and G2 have further refined the use of gonopod characters to define the boundaries of species and genera (fig. 2; Brandis et al., 1999). Bott's three monographs on the freshwater crabs of the world (Bott, 1955, 1969a, 1970b) are still important reference works, not least because of the photographs of the carapaces and gonopods of many type specimens.

Bott's works profoundly influenced the systematics and higher taxonomy of freshwater crabs. Before Bott all primary freshwater crabs of the Old World were assigned to one of two families, either the Eurasian Potamidae Ortmann, 1896, or the African Deckeniidae Ortmann, 1897, the latter comprising only two species. Bott (1970a) recognised that primary freshwater crabs as a group shared two major characteristics: direct development from egg to hatchling crab, and a restriction to the freshwater habitat. However, neither one of these characteristics is unique to the primary freshwater crabs, and are shared with species of secondary freshwater crabs (e.g., within the family Sesarmidae).

Bott (1969a, 1970a) recognized three superfamilies and 11 families of primary freshwater crabs and nine subfamilies, using characters of the G1, G2, mandible, and medial frontal region. Although not all of Bott's higher level taxonomic assignments were accepted, his system was in widespread use for the last 30 years of the 20th Century. The currently accepted systematics of the primary freshwater crabs now recognises five families: the Potamidae,

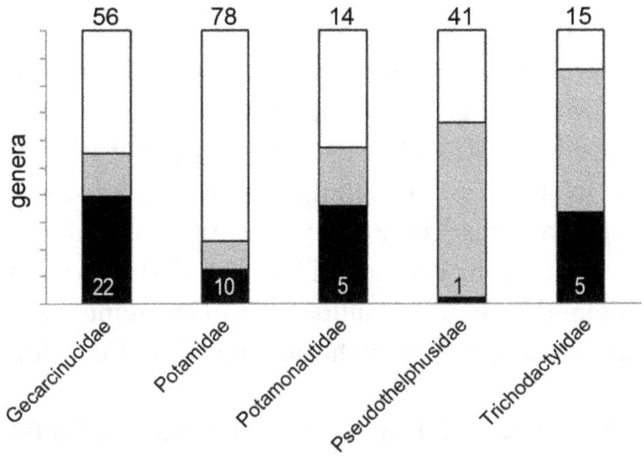

Fig. 2. Number of genera of the five primary freshwater crab families described by Bott (in black) in proportion to the genera described until 1974 by other researchers (in grey) and described after Bott (in white). Values inside the bar graphs indicate absolute genus numbers described by Bott, numbers above the total number of currently recognized genera. Based on the data given by Ng et al. (2008) and Yeo et al. (2008).

Potamonautidae, Gecarcinucidae, Pseudothelphusidae, and Trichodactylidae (see Cumberlidge & Ng, 2009; Klaus et al., 2011). This taxonomic system combines additional morphological characters of the gonopods, mandibles, and sternum (Yeo & Ng, 2004; Klaus et al., 2006; Cumberlidge et al., 2008) with the results of molecular studies (see Daniels et al., 2006; Klaus et al., 2009; Shih et al., 2009).

TRANSITIONS INTO FRESHWATER

Bott (1955, 1970a, b, 1972) proposed that the colonisation of freshwater habitats by the marine ancestors of the freshwater crabs during the Late Cretaceous/Tertiary was driven by the loss of their shallow marine littoral habitat when the epicontinental seas regressed, being influenced in his thinking by the ideas of Beuerlen (1931). Bott (1955, 1969a, b, 1970a, b, 1972) assumed that the ancestors of freshwater crabs became pre-adapted to the low salinities and changing temperatures of freshwater by their life in the estuarine conditions around river mouths. Bott hypothesised that the flat, wide, spiny-edged carapace of many freshwater crabs represented an adaptation to prevent their ancestors from sinking into the soft mud typical of estuaries. Bott (1969a, 1970a) assumed that a number of different marine crab ancestors of freshwater crabs made multiple transitions into freshwater, and thereby gave rise to the different families and genera of freshwater crabs. The reasoning behind this idea was the hypothesis that the G1 of the marine ancestors of freshwater crabs was 3-segmented, straight, and conical, and that the G2 morphologies developed later after entering in fresh water habitats. Two of today's primary freshwater crab families have a three-segmented G1: the Pseudothelphusidae and the Trichodactylidae, both of which are endemic to the Neotropical regions. In contrast, all of the Old World freshwater crab families (the Potamidae, Potamonautidae, and Gecarcinucidae) have a four-part G1, although the morphology of the gonopod differs considerably between these families.

Bott argued that different G1 morphologies represented different colonisation events, and that different degrees of complexity in G1 morphology reflected the length of time the taxon had spent in freshwater since its initial colonization (i.e., a complex G1 indicated an ancient invasion of inland waters). Bott's linkage of time spent in a freshwater habitat with G1 morphology is difficult to explain in terms of functional morphology because it contradicts his assumption that the sheltering of the male reproductive organs of freshwater crabs under the abdomen rendered them independent of ecological constraints.

Bott illustrated his ideas with maps that indicated idealized centers of origin and dispersal paths of the different families of freshwater crab (Bott, 1970a, 1972; reproduced in fig. 3). Bott conceived of the ancestors of freshwater crabs as basal groups ("Basisgruppen"), rather than as single species. According to Bott, the Potamonautidae colonised Africa from the southern shores of the former Tethys Ocean, while the ancestors of the Potamidae, the sister group of the Potamonautidae according to Bott, colonised Eurasia from the northern shores of the Tethys Sea (fig. 3B, C). Bott hypothesised that the marine ancestors of the Gecarcinucidae were widely distributed in the Indian Ocean;

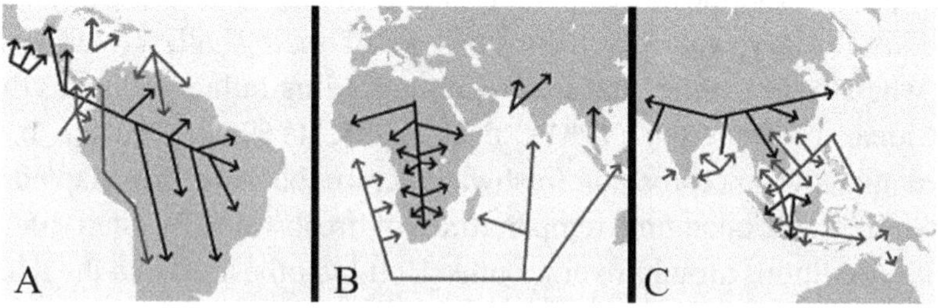

Fig. 3. Multiple independent colonisations of the freshwater habitat by the marine ancestors of the primary freshwater crabs according to Bott (1970a, 1972). A, Pseudothelphusidae and Trichodactylidae; B, Potamonautidae and Gecarcinucoidea; C, Potamoidea and Gecarcinucoidea. Adapted from Bott (1970a, 1972).

while the ancestors of the Parathelphusidae sensu Bott (now recognised as part of the family Gecarcinucidae) originated in what is now the South China Sea (fig. 3C). In the Neotropics, Bott assumed a Pacific Ocean origin for both the Pseudothelphusidae and Trichodactylidae, with two independent dispersal pathways for the pseudothelphusids into South America, east and west of the Cordilleras in Venezuela and Colombia.

Bott was the first author to realize that the Potamonautidae were found exclusively in the Afrotropcial region, and that the other Old World families were either found exclusively in the Eurasian-Oriental region (Potamidae), or exclusively in the Oriental-Australasian region (Gecarcinucidae). Before this, the Old World freshwater crabs were grouped by similar carapace morphologies regardless of their continent of origin, which led to cases where African, Madagascan, and Asian species were included in the same genus or sub-genus (Rathbun, 1904).

Bott's hypotheses on the biogeography and evolutionary origins of the freshwater crabs are difficult to test because the marine sister-group(s) of primary freshwater crabs still remain elusive (Sternberg et al., 1999), and the impact of marine regressions on the colonization of freshwater during the late Cretaceous/early Tertiary is poorly understood. However, current molecular clock estimates support at least Bott's assumption of a post-cretaceous evolution of freshwater crabs (Klaus et al., 2011; Tsang et al., 2014). It is possible that the same ecological conditions that we see today would have existed at earlier times when the ancestors of today's freshwater crabs first colonised freshwaters, perhaps in response to dramatic falls in sea levels in the past. In fact, eustatic regressions took place many times since the Late Cretaceous, most remarkably since the current icehouse climate established during the Oligocene (Miller et al., 2005). Bott's hypotheses concerning freshwater crab origins could be tested by focusing on species of secondary freshwater crabs that exhibit different levels of adaptation to freshwater, because these are present-day examples of ongoing multiple invasions of inland waters by species from a number of marine crab families (e.g., Sesarmidae, Varunidae, or Hymenosomatidae) (Yeo et al., 2008).

CONCLUSION

Bott's pioneering work on the freshwater crabs of the world laid the foundations of the present highly active field of freshwater crab biology. Bott introduced a new taxonomic system that consistently used differences in G1 morphology to delimit genera and species, and his biogeographic hypotheses for freshwater crab origins and dispersal have stimulated a number of subsequent authors to enter this field. Although Bott's taxonomic system has been profoundly changed during the last decade in the light of new molecular studies, his contributions to the higher taxonomy of the group have nevertheless set the stage for the present explosion of interest in the taxonomy and phylogeny of the primary freshwater crabs. These contributions have in turn driven further studies on the biogeography and conservation of these important and interesting crustaceans.

ACKNOWLEDGEMENTS

We thank Neil Cumberlidge and Darren Yeo for their helpful suggestions.

REFERENCES

BALSS, H., 1937. Potamoniden (Dekapoda Brachyura) der Philippinen und des Malayischen Archipels. Internationale Revue der gesamten Hydrobiologie und Hydrographie, **34**(3/5): 143-187.

BEURLEN, K., 1931. Die Besiedlung der Tiefsee. Natur und Volk, **61**: 269-278.

BOTT, R., 1928. Beiträge zur Kenntnis von *Gyrinus natator substriatus* Steph. I. Lebensweise und Entwicklung. II. Der Sehapparat. Zeitschrift für Morphologie und Ökologie der Tiere, **10**: 207-306.

— —, 1944. Die Süßwasserkrabbe (*Potamon fluviatile*). Natur und Volk, **74**: 136-142.

— —, 1955. Die Süßwasserkrabben von Afrika (Crust., Decap.) und ihre Stammesgeschichte. Annalen van het Koninklijk Museum van Belgisch Congo, C, Dierkunde, Reeks III, **1**: 209-352.

— —, 1969a. Die Süßwasserkrabben Süd-Amerikas und ihre Stammesgeschichte. Eine Revision der Trichodactylidae und der Pseudothelphusidae östlich der Anden (Crustacea, Decapoda). Abhandlungen der Senckenbergischen Naturforschenden Gesellschaft, **518**: 1-94.

— —, 1969b. Präadaptation, Evolution und Besiedlungsgeschichte der Süßwasserkrabben der Erde. Natur und Museum, **99**: 266-275.

— —, 1970a. Betrachtungen über die Entwicklungsgeschichte und Verbreitung der Süßwasserkrabben nach der Sammlung des Naturhistorischen Museums in Genf/Schweiz. Revue Suisse de Zoologie, **77**: 327-344.

— —, 1970b. Die Süßwasserkrabben von Europa, Asien, Australien und ihre Stammesgeschichte. Eine Revision der Potamoidea und der Parathelphusoidea (Crustacea, Decapoda). Abhandlungen der Senckenbergischen Naturforschenden Gesellschaft, **526**: 1-338.

— —, 1972. Stammesgeschichte und geographische Verbreitung der Süßwasserkrabben. Natur und Museum, **102**: 63-77.

BRANDIS, D., V. STORCH & M. TÜRKAY, 1999. Morphology and function of the copulatory system in freshwater crabs of the genus *Potamon*. Journal of Morphology, **239**(2): 157-166.

COLOSI, G., 1920. I Potamonidi conservati del R. Museo Zoologico di Torino. Bollettino dei Musei di Zoologia ed Anatomia comparata della R. Università di Torino, **35**(734): 1-39.

CUMBERLIDGE, N., 1999. The freshwater crabs of West Africa. Faune et flore tropicales, **35**: 1-382. (IRD, Paris).

CUMBERLIDGE, N. & P. K. L. NG, 2009. Systematics, evolution, and biogeography of freshwater crabs. In: J. W. MARTIN, K. A. CRANDALL & D. L. FELDER (eds.), Decapod crustacean phylogenetics. Crustacean Issues, **18**: 491-508. (CRC Press, Taylor & Francis Group, Boca Raton, London, New York).

CUMBERLIDGE, N., P. K. L. NG, D. C. J. YEO, C. MAGALHAES, M. R. CAMPOS, F. ALVAREZ, T. NARUSE, S. R. DANIELS, L. J. ESSER, F. Y. K. ATTIPOE, F.-L. CLOTILDE-BA, W. DARWALL, A. MCIVOR, M. RAM & B. COLLEN, 2009. Freshwater crabs and the biodiversity crisis: importance, threats, status, and conservation challenges. Biological Conservation, **142**: 1665-1673.

CUMBERLIDGE, N., R. V. STERNBERG, I. R. BILLS & H. A. MARTIN, 1999. A revision of the genus *Platythelphusa* A. Milne-Edwards, 1887 from Lake Tanganyika, East Africa (Decapoda: Potamoidea: Platythelphusidae). Journal of Natural History, **33**: 1487-1512.

CUMBERLIDGE, N., R. V. STERNBERG & S. R. DANIELS, 2008. A revision of the higher taxonomy of the Afrotropical freshwater crabs (Decapoda: Brachyura) with a discussion of their biogeography. Biological Journal of the Linnean Society, **93**: 399-413.

DANIELS, S. R., N. CUMBERLIDGE, M. PÉREZ-LOSADA, S. A. E. MARIJNISSEN & K. A. CRANDALL, 2006. Evolution of Afrotropical freshwater crab lineages obscured by morphological convergence. Molecular Phylogenetics and Evolution, **40**: 227-235.

KLAUS, S., D. BRANDIS, P. K. L. NG, D. C. J. YEO & C. D. SCHUBART, 2009. Phylogeny and biogeography of Asian freshwater crabs of the family Gecarcinucidae (Brachyura: Potamoidea). In: J. W. MARTIN, K. A. CRANDALL & D. L. FELDER (eds.), Decapod crustacean phylogenetics. Crustacean Issues, **18**: 509-531. (CRC Press, Taylor & Francis Group, Boca Raton, London, New York).

KLAUS, S., C. D. SCHUBART & D. BRANDIS, 2006. Phylogeny, biogeography and a new taxonomy for the Gecarcinucoidea Rathbun, 1904 (Decapoda: Brachyura). Organisms, Diversity & Evolution, **6**: 199-217.

KLAUS, S., C. D. SCHUBART, B. STREIT & M. PFENNINGER, 2010. When Indian crabs were not yet Asian — evidence for Eocene proximity of India and Southeast Asia from freshwater crab biogeography. BMC Evolutionary Biology, **10**: 287.

KLAUS, S., D. C. J. YEO & S. T. AHYONG, 2011. Freshwater crab origins — laying Gondwana to rest. Zoologischer Anzeiger, **250**: 449-456.

MARIJNISSEN, S. A. E., F. R. SCHRAM, N. CUMBERLIDGE & E. MICHEL, 2004. Two new species of *Platythelphusa* A. Milne-Edwards, 1887 (Decapoda, Potamoidea, Platythelphusidae) and comments on the taxonomic position of *P. denticulata* Capart, 1952 from Lake Tanganyika, East Africa. Crustaceana, **77**(5): 513-532.

NG, P. K. L., D. GUINOT & P. J. F. DAVIE, 2008. Systema Brachyurorum, Part I. An annotated checklist of extant brachyuran crabs of the world. Raffles Bulletin of Zoology, (Supplement) **17**: 1-286.

RATHBUN, M. J., 1904. Les crabes d'eau douce (Potamonidae). Nouvelles archives du muséum d'histoire naturelle, Paris, (4) **6**: 225-312.

SHIH, H.-T., D. C. J. YEO & P. K. L. NG, 2009. The collision of the Indian plate with Asia: molecular evidence for its impact on the phylogeny of freshwater crabs (Brachyura: Potamidae). Journal of Biogeography, **36**: 703-719.

TSANG, M. L., C. D. SCHUBART, S. T. AHYONG, J. C. Y. LAI, E. Y. C. AU, T.-Y. CHAN, P. K. L. NG & K. H. CHU, 2014. Evolutionary history of true crabs (Crustacea: Decapoda: Brachyura) and the origin of freshwater crabs. Molecular Biology and Evolution, **31**(5): 1173-1187.

TÜRKAY, M., 1974a. Dr. Richard Bott (1902-1974). Natur und Museum, **104**(4): 135-136.

— —, 1974b. Richard Bott (1902-1974). Senckenbergiana biologica, **55**(4/6): 409-414.

— —, 1975a. Dr. phil. nat. Richard Bott (1902-1974). Leben und carcinologisches Werk. Crustaceana, **28**(3): 228-302.

— —, 1975b. Statement: Die Bedeutung des Gonopodenaufbaus für die Aufklärung von Verwandtschaftsverhältnissen bei dekapoden Crustaceen. Aufsätze und Reden der senckenbergischen naturforschenden Gesellschaft, **27**: 114-115.

YEO, D. C. J. & P. K. L. NG, 2004. Recognition of two subfamilies in the Potamidae Ortmann, 1896 (Brachyura, Potamidae) with a note on the genus *Potamon* Savigny, 1816. Crustaceana, **76**: 1219-1235.

YEO, D. C. J., P. K. L. NG, N. CUMBERLIDGE, C. MAGALHÃES, S. R. DANIELS & M. R. CAMPOS, 2008. Global diversity of crabs (Crustacea: Decapoda: Brachyura) in freshwater. Hydrobiologia, **595**: 275-286.

First received 4 October 2012.
Final version accepted 29 July 2013.

LIFE SPAN, EARLY LIFE STAGE PROTECTION, MORTALITY, AND SENESCENCE IN FRESHWATER DECAPODA

BY

GÜNTER VOGT[1])

Faculty of Biosciences, Centre for Organismal Studies (COS), University of Heidelberg, Im Neuenheimer Feld 230, 69120 Heidelberg, Germany

ABSTRACT

This article reviews the present knowledge on ageing and longevity in the freshwater Decapoda and examines the impact of abbreviated development and postembryonic brood care, two major adaptations to fresh water, on life expectancy. Life span data are available for only 4% of freshwater decapods. Reliably determined maximum life spans in freshwater shrimps, crayfish, crabs, and aeglids vary from 8 months to 38 years and may be underestimated in slow-growing species. Decapods that live at high latitudes and high altitudes tend to live longer, which may reflect life history adaptations to cool water temperatures. Particularly long-lived species are found among crayfish and in subterranean habitats. Abbreviated and direct development and postembryonic brood care reduce mortality of the early life stages of freshwater decapods and are associated with an increase of individual life expectancy but the longevity of freshwater decapod species is not extended when compared to marine decapods. Long-lived freshwater decapods maintain structural and functional integrity into old age and possess several effective anti-ageing mechanisms including life-long stem cell activity. The most obvious anti-ageing mechanisms are moulting and the regeneration of damaged appendages, both of which obviate mechanical senescence. Some species of freshwater decapods are suitable models for the investigation of general topics of biogerontology.

INTRODUCTION

Approximately 20% of the Decapoda are freshwater species, including caridean shrimps, freshwater crayfish, and brachyuran and anomuran freshwater crabs. Caridean freshwater species belong to the Alpheidae, Atyidae, Desmocarididae, Kakaducarididae, Palaemonidae, Typhlocarididae and

[1]) e-mail: gunter.vogt@web.de

 Advances in freshwater decapod systematics and biology: 17-51

Xiphocarididae (De Grave et al., 2008). Freshwater carideans are dominated by the nearly exclusively freshwater Atyidae and the Palaemonidae, but these families also include many brackish water and marine representatives. Freshwater crayfish include three families, the Astacidae, Cambaridae, and Parastacidae (Crandall & Buhay, 2008; Gherardi et al., 2010). The anomuran freshwater species belong to the Aeglidae (Bond-Buckup et al., 2008), with the exception of one representative of the Diogenidae (McLaughlin & Murray, 1990). The adults of freshwater shrimps, crayfish, and aeglids are confined to freshwater bodies for their entire life cycle.

The freshwater crabs are less uniform with regard to their relationship to freshwater. They can be subdivided into primary freshwater crabs that belong to the Gecarcinucidae, Potamidae, Potamonautidae, Pseudothelphusidae, and Trichodactylidae, and the secondary freshwater crabs that have representatives in the Gecarcinidae, Goneplacidae, Hymenosomatidae, Ocypodidae, Sesarmidae, Portunidae, Xanthidae, and Varunidae (Yeo et al., 2008; De Grave et al., 2009; Klaus et al., 2011). The primary freshwater crabs live their entire lives in freshwater and are independent of salt water. Secondary freshwater crabs belong to primarily marine brachyuran families (Yeo et al., 2008) and live most of their lives in freshwater but most require saltwater for their eggs and larvae to develop (Burggren & McMahon, 1988; Diesel, 1992; Diesel & Horst, 1995). Some of them are semiterrestrial or terrestrial.

Longevity is of fundamental importance to a broad range of disciplines and is central to population biology (Carey, 2003). Despite the high number of publications on freshwater decapods every year, very few focus on longevity and ageing, possibly because of the lack of convenient methods to determine age. Precise information on the age of a decapod crustacean can only be obtained either in captivity or by the long-term marking of free-living animals with internal tags, but data from such studies are still rare. Most studies on longevity employ growth models based on mark and recapture, length-frequency distribution and the analysis of moult increment, intermoult duration and reproduction parameters. An alternative approach relies on measurement of the pigment lipofuscin, which is continuously accumulated with age in some areas of the brain and the eyestalks. In all of these growth models the mean age prediction error is small at a young age and large at older ages. A detailed discussion of the advantages and disadvantages of the various ageing techniques usually applied in decapods is provided by Hartnoll (2001), Vogt (2011), and Chang et al. (2012).

The present review provides an overview of life span, early life stage protection, mortality and senescence in Decapoda that either live in freshwater (including groundwater, temporary pools and water filled leaf axils of plants) or use freshwater for moulting and/or reproduction. The article begins with a comparison of the life spans of freshwater decapods and possible correlations between longevity and taxonomic affiliation, geographical latitude, altitude, and habitat. It then addresses strategies to reduce the mortality of the early life stages and compares individual life expectancy and life spans of freshwater and marine species. Finally, anti-ageing strategies, the relationship of growth format and pattern of senescence, and the advantages of using freshwater decapods for ageing research are discussed.

The photographs and micrographs in figs. 1 and 3 were retouched (background, dirt spots and some light reflexes) using Adobe Photoshop 6.0 to emphasize the relevant parts.

LIFE SPAN IN FRESHWATER DECAPODA

There are approximately 2833 species of freshwater decapods (Bond-Buckup et al., 2008; Crandall & Buhay, 2008; De Grave et al., 2008; Yeo et al., 2008; Bond-Buckup, pers. comm.). Longevity data have been collected for 109 species corresponding to 3.8% of decapod diversity. This list includes mostly primary freshwater species, with a few species of semi-terrestrial and terrestrial brachyuran crabs that use freshwater for moulting and reproduction. The highest percentage of species with a known age is found in the aeglids (7.4% of 68 species), followed by crayfish (7.3% of 634 species), freshwater shrimps (3.7% of 655 species), and brachyuran freshwater crabs (2.3% of 1476 species). The longevity data used here were obtained only from reliable literature sources but not from aquarist and aquarium trader websites that tend to rely on anecdotal accounts.

The life span data published in the literature and presented in tables I and II are not all of the same type. Depending on the ageing methods applied the data indicate the oldest individual (laboratory rearing), the oldest cohort (size-frequency distribution) or the estimated minimum and maximum ages (growth models). Longevity is also sometimes expressed as the mean age of the reproducing adults or the mean age of the oldest 10% of the population. Of course, these parameters differ greatly from the maximum life span. For example, in the parthenogenetic marbled crayfish, *Procambarus fallax* (Hagen, 1870) *f. virginalis*, the mean life span of 49 reproducing adults was 718.8 days, the mean life span of the longest-lived 10% was 1154 days, and the maximum

TABLE I
Life spans of freshwater shrimps and crabs

Life span	Species	Family	Method	Reference
Caridea				
1 yr	*Caridina simoni* Bouvier, 1904	Atyidae	GM-RP	De Silva, 1988
1 yr	*Palaemonetes paludosus* (Gibbes, 1850)	Palaemonidae	GM-RP	Beck & Cowell, 1976
1.3 yr	*Exopalaemon modestus* (Heller, 1862)	Palaemonidae	GM-RP	Oh et al., 2002
1.3 yr	*Atyaephyra desmarestii* (Millet, 1831)	Atyidae	GM-RP	Dhaouadi-Hassen & Boumaïza, 2009
1.8 yr	*Caridina cantonensis* Yu, 1938	Atyidae	GM-RP	Yam & Dudgeon, 2005
3 yr	*Macrobrachium rosenbergii* (De Man, 1879)	Palaemonidae	C	Brown et al., 2010
8 yr	*Macrobrachium carcinus* (Linnaeus, 1758)	Palaemonidae	C	Brown et al., 2010
8 yr	*Atya lanipes* Holthuis, 1963	Atyidae	GM-RP	Cross et al., 2008
15 yr[a]	*Palaemonias ganteri* Hay, 1901	Atyidae	C, LR	U.S. Fish and Wildlife Service, 2010
18 yr	*Xiphocaris elongata* (Guérin-Méneville, 1855)	Xiphocarididae	GM-RP, MR	Cross et al., 2008
Brachyura				
0.8 yr[b]	*Elamenopsis kempi* (Chopra & Das, 1930)	Hymenosomatidae	GM-RP, LR	Ali et al., 1995
4 yr	*Oziothelphusa ceylonensis* (Fernando, 1960)	Gecarcinucidae	LR	Rademacher & Mengedoht, 2011
4.4 yr	*Dilocarcinus pagei* Stimpson, 1861	Trichodactylidae	GM-RP	Taddei & Herrera, 2010
5 yr	*Eriocheir sinensis* H. Milne Edwards, 1853	Varunidae	C	Herborg et al., 2003
5 yr	*Sesarma jarvisi* Rathbun, 1914	Sesarmidae	GM, LR	Diesel & Horst, 1995
6 yr	*Liberonautes latidactylus* (De Man, 1903)	Potamonautidae	C	Cumberlidge, 1999
10 yr	*Potamonautes lirrangensis* Rathbun, 1904	Potamonautidae	LR, E	Rademacher & Mengedoht, 2011
10 yr	*Parathelphusa pantherina* (Schenkel, 1902)	Gecarcinucidae	LR, E	Rademacher & Mengedoht, 2011
15 yr	*Potamon fluviatile* (Herbst, 1785)	Potamidae	GM-RP	Scalici et al., 2008b

Quoted are the highest values given in the cited references. C, cited data; E, extrapolation; GM, growth model; LR, laboratory rearing; MR, mark-recapture; RP, analysis of reproduction and population data; yr, year.
[a] Cave.
[b] Summer generation.

TABLE II
Life spans of freshwater crayfish and aeglids

Life span	Species	Family	Method	Reference
Astacidea				
1.5 yr	*Cambarellus shufeldtii* (Faxon, 1884)	Cambaridae	C	Walls, 2009
2.5 yr	*Orconectes eupunctus* Williams, 1952	Cambaridae	C	Lukhaup & Pekny, 2008a
3.3 yr	*Parastacus defossus* Faxon, 1898	Parastacidae	GM-RP, LR	Noro & Buckup, 2009
4 yr	*Procambarus clarkii* (Girard, 1852)	Cambaridae	C	Huner, 2002
5 yr	*Cherax quadricarinatus* (Von Martens, 1868)	Parastacidae	C	Sheehy, 1992
6 yr	*Cambarus robustus* Girard, 1852	Cambaridae	LR	Guiaşu & Dunham, 2002
6 yr	*Austropotamobius pallipes* (Lereboullet, 1858)	Astacidae	GM-MR	Neveu, 2000
7 yr	*Paranephrops planifrons* White, 1842	Parastacidae	GM-MR	Parkyn et al., 2002
10 yr[a]	*Orconectes inermis inermis* Cope, 1872	Cambaridae	GM-MR	Weingartner, 1977
11 yr	*Cambaroides japonicus* (De Haan, 1841)	Cambaridae	GM-RP	Kawai et al., 1997
15 yr	*Astacus astacus* (Linnaeus, 1758)	Astacidae	C	Meyer et al., 2007
16 yr	*Pacifastacus leniusculus* (Dana, 1852)	Astacidae	LM	Belchier et al., 1998
16 yr[a]	*Procambarus erythrops* Relyea & Sutton, 1975	Cambaridae	GM-MR	Streever, 1996
20 yr	*Astacoides betsileoensis* Petit, 1923	Parastacidae	GM-MR	Jones et al., 2007
28 yr	*Euastacus armatus* (Von Martens, 1866)	Parastacidae	GM-RP	Gilligan et al., 2007
29 yr	*Paranephrops zealandicus* (White, 1847)	Parastacidae	GM-MR	Whitmore & Huryn, 1999
38 yr[a]	*Orconectes australis australis* (Rhoades, 1941)	Cambaridae	GM-MR	Venarsky et al., 2012
Anomura				
2 yr	*Aegla jarai* Bond-Buckup & Buckup, 1994	Aeglidae	GM-RP	Boos et al., 2006
2.5 yr	*Aegla itacolomiensis* Bond-Buckup & Buckup, 1994	Aeglidae	GM	Silva-Gonçalves et al., 2009
3.3 yr	*Aegla paulensis* Schmitt, 1942	Aeglidae	GM-RP	Cohen et al., 2011

For abbreviations see table I.

individual life span was 1610 days (Vogt, 2010). Therefore, individuals of a species may live much longer in captivity than suggested in tables I and II.

Maximum life span data published for populations of freshwater decapods range from 8 months to 38 years (tables I and II). Longevities of 4-8 months were measured for the summer generation of the crab *Elamenopsis kempi* (Chopra & Das, 1930) in the Shatt-Al-Arab, Iraq (Ali et al., 1995). The longest-lived freshwater decapod reliably aged by a mark-recapture based growth model is the Southern Cave Crayfish *Orconectes australis australis* (Rhoades, 1941) from Alabama, USA, which live for 38 years (Venarsky et al., 2012). The highest value found in the literature is 60 years for the slow-growing Tasmanian crayfish *Astacopsis gouldi* Clark, 1936, which can reach a length of 80 cm inclusive of the chelipeds and a weight of 5 kg (Lukhaup & Pekny, 2008a). Unfortunately, those authors gave no information on the basis of their estimate. Most freshwater decapods seem to live between 1 and 5 years. Approximately 0.9% of the 109 species for which age data are available had a maximum life span of less than 1 year, 63.3% lived between 1 and 5 years, 20.2% lived between 5 and 10 years, and 15.6% lived for more than 10 years.

DEPENDENCY OF LONGEVITY ON TAXONOMY, LATITUDE, ALTITUDE, AND HABITAT

In the marine Decapoda longevity of species and populations depends on taxonomic affiliation, geographic latitude, water depth, and habitat (Vogt, 2012a). The same is true for freshwater decapods but with two differences: altitude is an additional factor that affects life span but water depth is less significant. This section presents examples of life span variation associated with taxonomic affiliation, geographic latitude, altitude, and habitat, and ends with a discussion of possible causes.

Longevity and taxonomic affiliation

The maximum life span in freshwater decapods ranges from 1 to 8 years in palaemonid shrimps, 1 to 15 years in atyid shrimps, 8 months to 15 years in freshwater crabs, 1.5 to 38 years in crayfish, and 2 to 3.3 years in aeglids. These data indicate a relationship between longevity and taxonomic affiliation, particularly with respect to the upper age limits. There are still considerable differences between lineages if life span extending parameters such as geographical latitude, altitude, and habitat (see below) are excluded.

For example, in Italy the atyid shrimp *Atyaephyra desmarestii* (Millet, 1831), the crayfish *Austropotamobius pallipes* (Lereboullet, 1858) and the freshwater crab *Potamon fluviatile* (Herbst, 1785) occur in comparable habitats or even the same habitat but have a maximum life span of 1.6 years, 10 years, and 15 years, respectively (Scalici et al., 2008a, b; Dhaouadi-Hassen & Boumaïza, 2009).

Longevity and latitude, altitude, and water depth

In marine decapods there are many good examples of life span extension in high latitudes (Vogt, 2012a). In freshwater Decapoda such data are scarce although in each of the main groups some species have colonized high latitudes. Freshwater crayfish occur from 67°N to 47°S (Crandall et al., 2000; Jussila & Mannonen, 2004), secondary freshwater crabs from 54°N to 37°S (Chilton, 1914; Normant et al., 2000), freshwater shrimps from 52°N to 47°S (Winterbourn & Mason, 1983; Karge & Klotz, 2008) and aeglids from 20°S to 50°S (Bueno & Shimizu, 2008; Oyanedel et al., 2011).

A life span extending effect of high latitudes can be seen in New Zealand crayfish *Paranephrops zealandicus* (White, 1847), which is one of the southernmost freshwater decapods. This species often exceeds 16 years and has a maximum life span of 29 years (Whitmore & Huryn, 1999), whereas the life span of its tropical and subtropical relatives is significantly shorter: between 1.5 to 4 years (Lukhaup & Pekny, 2008a). Momot (1984) and Walls (2009) presented longevity data for crayfish species from the temperate northern and subtropical southern states of the USA that suggests a positive correlation between geographical latitude and life span as well. Life span extension with latitude in populations of the same species is known for the red swamp crayfish *Procambarus clarkii* (Girard, 1852). In its native range in subtropical Louisiana, USA (~30°N), longevity rarely exceeds 12-18 months (Huner, 2002) whereas this species lives for much longer in temperate Europe ranging from 3.5 to 6.5 years at between 40°N and 49°N (Chucholl, 2011).

The noble crayfish *Astacus astacus* (Linnaeus, 1758) would be a valuable model for a detailed study of the relationship of life span and geographical latitude within a species because it is one of the best investigated and most widespread species occurring from northern Finland (67°N) to Greece and Morroco (33°N), and from lowland waters to high alpine lakes (Skurdal & Taugbøl, 2002; Jussila & Mannonen, 2004). Populations of *A. astacus* in northern Scandinavia (~62°N) show a later onset of reproduction (4-7 years) than do populations in Germany (~50°N; 3-4 years) and also have lower moulting and reproduction frequencies (Hofmann, 1980; Zimmerman, 2009). Such adaptations are usually associated with life span extension but precise longevity data are not yet available.

High altitude also extends the longevity of freshwater decapods. High-altitude habitats have been colonized by species of all higher taxa of freshwater decapods. For instance, the aeglid *Aegla septentrionalis* Bond-Buckup & Buckup, 1994, is found in Argentina at an altitude of 3613 m above sea level (Bond-Buckup et al., 2010), and the freshwater crab *Potamonautes loveni* (Colosi, 1924) from 3800 m on Mt. Elgon in Kenya (Cumberlidge & Clark, 2010). The crayfish *Cambarellus zempoalensis* Villalobos, 1943, occurs in Mexico at 2800 m, which is the highest altitude known for any species of crayfish (Quiroz Castelán et al., 2000). The record holders of freshwater shrimp are *Sinodina* that live in southwestern China at altitudes of 1300-3000 m above sea level. For example, *S. acutipoda* (Liang, 1989), inhabits Lugu Lake at 2685 m (Liang & Cai, 1999).

A strong tendency towards prolongation of longevity with altitude is seen in some species of *Astacoides* from subtropical Madagascar that occur at altitudes between 600 and 2000 m. For example, *Astacoides betsileoensis* Petit, 1923, *A. crosnieri* Hobbs, 1987, and *A. granulimanus* Monod & Petit, 1929 live for more than 20 years (Jones et al., 2007), whereas other species of crayfish that live in subtropical lowlands have a life span between 1.5 and 4 years (Walls, 2009). Similarly, populations of the signal crayfish *Pacifastacus leniusculus* (Dana, 1852) live for up to 6 years in the Sacramento River in lowland California, USA (Shimizu & Goldman, 1983), for 8 years at an altitude of 593 m in Lake Billy Chinook in neighbouring Oregon (Lewis, 1997), and for 12 years at 1900 m in Lake Tahoe, California (Flint, 1975). A good model for the investigation of longevity differences between congeneric species adapted to different altitudes would be the dwarf crayfish pair *Cambarellus zempoalensis*, which occurs in Mexico at 2800 m, and *C. shufeldtii* (Faxon, 1884), which inhabits lowland water bodies in neighbouring Texas and Louisiana, USA (Peterson et al., 1996; Quiroz Castelán et al., 2000). Other promising research subjects are species of *Potamonautes* from East Africa, which display a zonal distribution from lowlands to high mountain areas (Harrison, 1995; Dobson et al., 2008; Cumberlidge & Clark, 2010).

The influence of water depth on longevity is less significant in freshwater Decapoda because they usually live in shallow water habitats. Individual crayfish and aeglids have been found at depths of 200 m and 320 m, respectively (Lewis, 2002; Bond-Buckup et al., 2008) but these individuals are outliers of the main population that live in shallower water. Crayfish are known to migrate temporarily to deep water to forage and to protect themselves from winter storms (Flint, 1977; Coffey & Clayton, 1988). The deepest record for any species of freshwater decapod is the Chilean aeglid *Aegla rostrata* Jara, 1977, from Lake Riñihue, which lives at great depths in the lake as well as in the littoral zone (Jara, 1977).

Real deep water species are perhaps two potamonautid freshwater crabs of the genus *Platythelphusa* from Lake Tanganyika, East Africa. *P. tuberculata* Capart, 1952 occurs on muddy grounds at depth of 50 to 190 m and *P. praelongata* Marijnissen, Schram, Cumberlidge & Michel, 2004, at depths of 40 to 80 m (Marijnissen et al., 2004, 2008). *Plathythelphusa tuberculata* is trophically highly distinct from the shallow water crabs in the lake, and therefore, it may also differ with respect to life history and longevity. Unfortunately, there are no longevity data available for this species and its relatives.

Longevity and habitat

There are some interesting examples published on life span differences among populations of the same species in narrowly limited geographical regions and comparable altitudes, which can be ascribed to habitat differences. For example, populations of the white-clawed crayfish *Austropotamobius pallipes* in a 3.4 km stretch of lowland river in Normandy, France, had life spans of 4-5 years upstream and 5-6 years downstream (Neveu, 2000). Cross et al. (2008) reported on marked differences in longevity of the freshwater shrimp *Xiphocaris elongata* (Guérin-Méneville, 1856) in two neighbouring streams in Puerto Rico, which differed with respect to predator fish and shrimp density. In the stream with predator fish and a low shrimp density *X. elongata* had a life span of 11 years, whereas in the stream without predator fish and a high shrimp density, *X. elongata* had a life span of only 5 years. Different land use can also affect the longevity of freshwater decapods. For example, New Zealand crayfish *Paranephrops planifrons* White, 1842, living in native forest streams had a maximum age of 7 years while other populations of this species living in neighbouring pasture streams had a maximum age of only 4 years (Parkyn et al., 2002).

Long periods of spatial isolation are also factors associated with longevity differences between populations of the same species found in a narrowly limited geographical region. For example females of a landlocked population of *Caridina gracilipes* De Man, 1892, from southern Taiwan have a maximum life span of 14 months, whereas females from a neighbouring amphidromous population live for up to 22 months (Han et al., 2011). Life span prolongation was observed in an isolated population of the river crab *Potamon fluviatile* in the historical centre of Rome, Italy where this species lives for up to 15 years compared to a life-span of 9-10 years seen in this species in rural populations elsewhere in Italy (Scalici et al., 2008b). The broadest differences in longevity within a small geographical region were found among populations of the southern cave crayfish *Orconectes australis australis*, which occur in caves of northeast Alabama, USA. Mark-recapture based growth models revealed minimum life spans for males of 11+ years in Tony Sinks Cave and 22+ years in Hering Cave (Venarsky et al., 2012).

Interestingly, all of the cave-dwelling decapods investigated so far have much higher longevities than their epigean relatives. For example, the Kentucky cave shrimp *Palaemonias ganteri* Hay, 1901, has a maximum life span of 10-15 years (U.S. Fish and Wildlife Service, 2010), whereas its epigean relative, the California freshwater shrimp *Syncaris pacifica* (Holmes, 1895), is estimated to live for only 3 years (U.S. Fish and Wildlife Service, 1998). Females of the Southern Cave Crayfish *Orconectes a. australis* have minimum life spans of 22+ years, depending on cave conditions (Venarsky et al., 2012), whereas the congeneric species *O. placidus* (Hagen, 1870) from surface waters in Alabama live for only 2-3 years (Taylor, 2003). A similar relationship is found between the maximum life spans of the cave crayfish *Procambarus erythrops* (16 years) and the epigean *P. clarkii* (1-2 years) in southern USA (Streever, 1996; Huner, 2002).

Possible causes of life span variation among higher taxa and within species

Differences in longevity of species and populations from different environments can be explained by life history theory that deals with natural selection, fitness, adaptation, and constraints to explain the evolution of life span and the trade-offs that bind the various life history traits together (Stearns, 1992, 2000). Significant evolutionary changes of life history traits such as age at maturity, brood size, interbrood interval, longevity are genetically fixed in less than 30 generations in natural populations of guppies, *Poecilia reticulata* Peters, 1859, a freshwater fish (Reznick et al., 1990).

Differences in longevity between and within the major groups of freshwater Decapoda are probably the result of adaptation to freshwater habitats and evolutionary constraint. Detailed comparisons between freshwater decapod lineages their closest marine relatives, as well as within freshwater decapod lineages could shed more light on this issue but presently there are not enough data available. Moreover, the closest marine relatives of some of the freshwater decapods have not yet been identified. An argument for differential selection in fresh water may be inferred from the fact that shrimps, crayfish, crabs, and aeglids have all independently colonized freshwater habitats in different geographical areas and at different times, starting some 250 million years ago (Pérez-Losada et al., 2002; Porter et al., 2005; Klaus et al., 2011). Support for the influence of evolutionary constraints comes from the Astacidea, which comprises the crayfish families Astacidae, Cambaridae and Parastacidae, and the marine lobsters Nephropidae. Similar to freshwater crayfish, the Nephropidae can reach ages of decades as shown for the European lobster, which lives for over 70 years (Vogt, 2012a).

Freshwater habitats found at high latitudes and high altitudes tend to have cooler water temperatures that are typically associated with slower growth rates, later onset of maturation, reduced moulting, and a reduced reproductive rate, leading to higher life spans in decapods and other poikilothermic animals (Zera & Harshman, 2001; Gilligan et al., 2007; Strahl et al., 2007; Vogt, 2012a). Although the observed effects in the freshwater decapods are triggered by low environmental temperatures they cannot be explained by a temperature-dependent reduction of the metabolic rate alone. They are rather the effect of life history changes that include metabolism, activity, and reproduction, as convincingly demonstrated for the red swamp crayfish *Procambarus clarkii* (see Chucholl, 2011). In temperate regions this subtropical species adopted some K-selected life-history traits such as slow growth, high longevity and large size while retaining some of its typical r-selected characteristics such as early maturation and high fecundity (Chucholl, 2011).

If a freshwater decapod is transferred to cooler water, its breeding period is unavoidably prolonged because a certain amount of degree days (sum of heat) are required to complete embryonic development (Skurdal et al., 2011). Freshwater decapods such as the noble crayfish *Astacus astacus* that reproduce every year at low altitudes reproduce only every second or third year at high latitudes (Hofmann, 1980; Zimmerman, 2009). The reduced allocation of resources for growth and reproduction at lower temperatures enables a higher investment in maintenance and repair that could result in the prolongation of longevity. The longest-lived bilaterian animal known is the ocean quahog, *Arctica islandica* Linnaeus, 1767, which has a maximum life span of 375 years. In this species slow growth, late onset of reproduction, and long life span are correlated with a very efficient maintenance of body proteins when compared to shorter-lived bivalve species that are faster growing and faster reproducing (Strahl et al., 2007).

There is a difference between latitudinal and altitudinal effects, at least if latitude is compared with altitude in the tropics. Freshwater ecosystems at high latitudes have warm temperatures and long day lengths in summer and low temperatures and short day lengths in winter, whereas temperature and day length are similar throughout the year in tropical latitudes. Freshwater crayfish sense day length and radiation intensity by UV-receptors in their compound eyes (Vogt, 2002) and adjust their activity accordingly. In high latitudes, they are active during the summer months and dormant during winter (Pruitt, 1988).

Differences in longevity between populations of the same species in different habitats can be attributed to differences in habitat complexity, food resource availability, predator pressure, and disease. Extrinsic mortality by predation, cannibalism, and disease are of particular relevance (Ricklefs, 1998; Kirkwood & Austad, 2000; Stearns et al., 2000; Abrams, 2004). Theory suggests that higher extrinsic mortality will produce evolutionary conditions that can either shorten or extend life span: life span is prolonged if juveniles are preferably predated but shortened if adults are preferably predated. Experiments with guppies in the wild and in the laboratory corroborated that a higher risk of mortality caused by predators can select for longer-lived organisms (Reznick et al., 1990, 2004). This feature may explain, why freshwater shrimp *Xiphocaris elongata* living in streams with predator fish lives to a greater age than *X. elongata* living in streams that lack predator fish (Cross et al., 2008).

The greatest differences in longevity are seen between cave-dwelling decapods and their epigean relatives. Molecular clock dating suggests that hypogean species diverged from their epigean ancestors one to several million years ago (Trontelj, 2007; Zakšek et al., 2007; Page et al., 2008). Cave animals typically have a scarce and/or erratic food supply, a constantly low temperature and water with low oxygen levels. These factors result in a reduction of metabolism, motility and growth rate, a later onset of maturity, and irregular reproduction in subterranean decapods compared to their epigean relatives (Hüppop, 1985; Streever, 1996; Simčič et al., 2005; Huryn et al., 2008). These energy-saving adaptations have apparently been favoured in the cave environment by long-lasting directional selection, and a positive trade-off between these traits and longevity is probably causative for the particularly long life spans of the subterranean decapods.

Life history traits of decapods found in isolated populations can be influenced by inbreeding, which often is accompanied by inbreeding depression and a shortening of longevity (Casellas et al., 2008). Inbreeding depression may explain the shorter longevity of populations of the landlocked freshwater shrimp *Caridina gracilipes* from Taiwan when compared to neighbouring amphidromous populations. On the other hand, if longevity-promoting genes or mutations happen to be frequent in the founder population then longevity can be prolonged in comparison to the stem population (Willcox et al., 2006). This could have happened in the long-lived isolated population of *Potamon fluviatile* in Rome.

PROTECTION OF EARLY LIFE STAGES AND CONSEQUENCES FOR AGE-SPECIFIC MORTALITY AND LONGEVITY

Freshwater decapods have either an amphidromous life history characterized by extended planktonic development in salt or brackish water, or a completely freshwater life cycle characterized by either abbreviated or direct development and postembryonic brood care (Bauer, 2011). The latter are apparently adaptations to the particularly strong variation of hydrodynamic parameters, physico-chemical factors and phytoplankton availability in freshwater habitats (Anger, 1995; Scholtz & Kawai, 2002; Vogt & Tolley, 2004). Such pronounced life history shifts of freshwater decapods compared to their marine ancestors have a significant impact on their mortality, ageing and longevity.

Ocean-type planktonic larval development

Freshwater species with planktonic development in salt or brackish water are found in a number of atyid and palaemonid shrimps and secondary freshwater crabs. These species usually produce large numbers of small eggs and exhibit a number of planktonic larval stages like their marine relatives. For example, the giant freshwater prawn *Macrobrachium rosenbergii* (De Man, 1879), produces up to 100 000 eggs and has 11 larval stages (Brown et al., 2010) and the Chinese mitten crab *Eriocheir sinensis* H. Milne Edwards, 1853, produces up to 1 million eggs and has 6 larval stages (Herborg et al., 2003).

Abbreviated and direct development

Decapods that spend their entire life cycle in freshwater typically exhibit abbreviated or direct development and have egg numbers in the tens or hundreds and, less frequently, in the lower thousands (Beck & Cowell, 1976; Diesel & Horst, 1995; Liu & Li, 2000; Bond-Buckup et al., 2008; Lukhaup & Pekny, 2008a; Dhaouadi-Hassen & Boumaïza, 2009). Their large eggs have substantial yolk reserves that enable larval development to take place

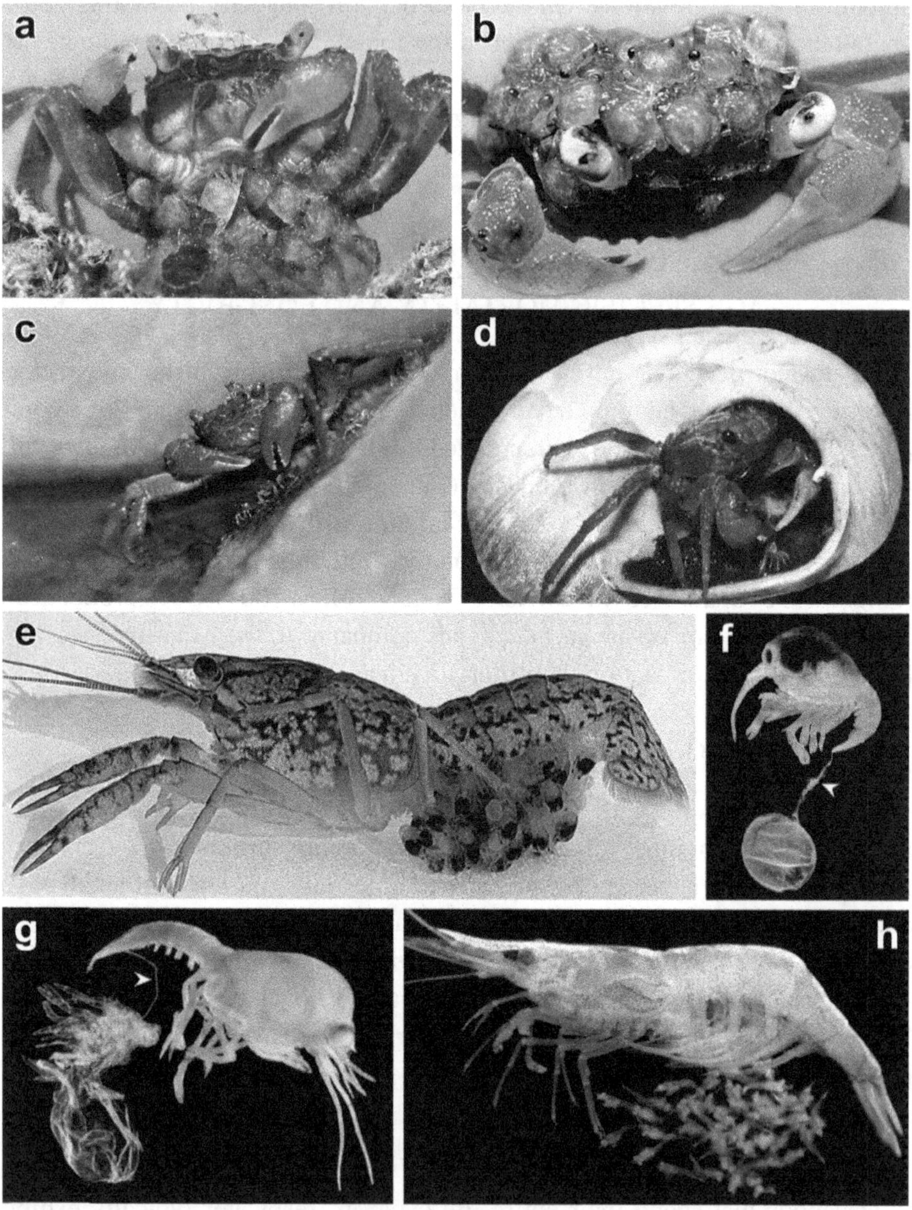

Fig. 1. Brood care in freshwater Decapoda. a, semiterrestrial crab *Geosesarma* sp. (trade name "vampyr") with juveniles under the pleon. In captivity the juveniles are released into freshwater (from Rademacher & Mengedoht, 2011). b, *Geosesarma notophorum* carrying juveniles on top of the carapace. The brood is covered by a film of water and released into a moist substratum (photo: Oliver Mengedoht). c, mother and young of the bromeliad crab *Metopaulias depressus* in a water-filled leaf axil of a bromeliad (from Diesel & Schubart, 2000). d, Jamaican snail crab

completely within the egg and to ensure nutritional independence of the first postembryonic life stages (Anger, 1995, 2001; Vogt, 2008a).

Abbreviated development refers to the production of significantly fewer larval stages that are larger and more advanced than that of their marine relatives with normal planktonic development. Larger larvae reduce the risk of predation of the first free-living stages. Abbreviated development is seen in species of atyid and palaemonid shrimps and in some secondary freshwater crabs (Lucas, 1980; Jalihal et al., 1993; Walsh, 1993; Guinot & Richer de Forges, 1997). The palaemonid genus *Macrobrachium* Bate, 1868, which inhabits the entire range of aquatic habitats from brackish water to fresh water, is a good example of the tendency to abbreviated development in freshwater species. *Macrobrachium rosenbergii* lives in estuarine areas and has 11 larval stages while the hillstream species *M. dayanum* (Henderson, 1893) has only one larval stage (Jalihal et al., 1993).

Some atyid and palaemonid shrimps, and all crayfish, primary freshwater crabs, and aeglids lack planktonic larval stages. For example, the palaemonid *Macrobrachium niphanae* Shokita & Takeda, 1989, produces two benthic zoeae and one megalopa before moulting to the first juvenile (Shokita et al., 1991). The atyid shrimp *Dugastella valentina* (Ferrer Galdiano, 1924) also produces two zoeae and a megalopa, but these develop in the pleonal brood chamber of the female (fig. 1h) (Rodríguez & Cuesta, 2011). In the crayfish, aeglids, primary freshwater crabs, and some species of secondary freshwater crabs, the embryonic and larval development is entirely completed within the egg and the hatchlings have the appearance of juveniles (fig. 1a, f) (Ng & Tan, 1995; Scholtz & Kawai, 2002; López Greco et al., 2004; Vogt & Tolley, 2004; Wehrtmann et al., 2010; Wu et al., 2010; Rademacher & Mengedoht, 2011). This type of development is called direct development.

Posthatching brood care

Posthatching or extended brood care in decapods is an exclusive adaptation to life in fresh water because it is very rare in marine environments (Thiel,

Sesarma jarvisi breeding in a water-filled snail shell (from Diesel & Schubart, 2000). e, female marbled crayfish *Procambarus fallax f. virginalis* carrying juveniles on its pleopods (from Vogt, 2008b). f, safeguarding of hatching in marbled crayfish. The telson thread (arrowhead) keeps the hatchling linked to its egg case, which in situ is attached to a maternal pleopod (from Vogt & Tolley, 2004). g, safeguarding of first moulting in the marbled crayfish. The anal thread (arrowhead) keeps the emerged stage-2 juvenile linked to its exuvia, which in situ is firmly hooked into the pleopodal structures of the mother (from Vogt, 2008a). h, kangaroo shrimp *Dugastella valentina* and its brood. The zoea 2 were removed from the brooding chamber underneath the pleon (from Cuesta et al., 2006).

2000). Extended brood care includes not only the carrying and cleaning of the eggs (a characteristic of all Pleocyemata), but also the care for the early posthatching life stages. In freshwater shrimps, brood care was detected only recently in the atyids *Dugastella valentina* and *D. marocana* Bouvier, 1912, whose larval stages are retained in a brood pouch under the pleon for 5-9 days after which the first juvenile stage is released (fig. 1h) (Cuesta et al., 2006; Huguet et al., 2011; Rodríguez & Cuesta, 2011).

In all aeglids, the young are generally carried on the female pleopods (Bond-Buckup, pers. comm.). Examples are *Aegla uruguayana* Schmitt, 1942, *A. perobae* Hebling & Rodrigues, 1977, and *A. castro* Schmitt, 1942, which carry their young for 3-4, 8-12, and 15 days, respectively (Rodrigues & Hebling, 1978; Swiech-Ayoub & Masunari, 2001; López Greco et al., 2004).

In freshwater crabs, posthatching brood care is widespread but detailed studies are scarce. For instance, in the potamonautid *Liberonautes latidactylus* (De Man, 1903), the pseudothelphusid *Kingsleya ytupora* Magalhaes, 1986, and the potamid *Candidiopotamon rathbunae* (De Man, 1914) the hatchlings and the following juvenile stage are carried on the maternal pleopods for up to 16 days (Cumberlidge, 1999; Liu & Li, 2000; Wehrtmann et al., 2010). Posthatching brood care in the Indian field crab *Spiralothelphusa hydrodroma* (Herbst, 1794) can last for several weeks because the release of the juveniles is adjusted to the onset of monsoon rainfall (Pillai & Subramoniam, 1984). The juveniles of primary freshwater crabs and many secondary freshwater crabs are brooded under the pleon (fig. 1a) but in *Geosesarma notophorum* Ng & Tan, 1995, the hatchlings are carried for approximately 4 days on the dorsal surface of the maternal carapace (fig. 1b) covered by a film of water (Ng & Tan, 1995; Rademacher & Mengedoht, 2011).

Completely different types of extended maternal care are shown by the Jamaican crabs *Metopaulias depressus* Rathbun, 1896, and *Sesarma jarvisi* Rathbun, 1914, which depend on freshwater for both larval (2 zoeae and 1 megalopa) and juvenile development. The bromeliad crab *M. depressus* releases the larvae into water-filled leaf axils of bromeliads and defends them against predators (fig. 1c). Moreover, it supplies food for the young and optimizes water quality in the brooding pools by removing detritus, oxygenation of the water, and buffering of the pH by adding snail shells (Diesel & Schuh, 1993). Maternal care reduces predator-related mortality of the offspring by 60% (Diesel, 1992). Bromeliad crabs can form colonies comprising the mother and three successive batches (Diesel & Schubart, 2000). The Jamaican snail crab *S. jarvisi* releases its larvae into the water filled shells of land snails (fig. 1d) (Diesel & Horst, 1995). Maternal care in *S. jarvisi* usually lasts for 2-3 months and includes protection of the young, refilling the shells with water and providing food (Diesel & Horst, 1995).

Postembryonic brood care in crayfish is universal (fig. 1e). Care of the young in crayfish differs fundamentally from that seen in the other freshwater groups because the first postembryonic stages of crayfish are equipped with special structures such as the telson thread and anal thread that firmly attach them to the maternal pleopods in critical phases of postembryonic life. The telson thread is composed the detached inner layer of the egg shell and a secretion that links the hatching juvenile to the mother via the egg case (fig. 1f) (Scholtz, 1995; Vogt & Tolley, 2004; Vogt, 2008a). When stage-1 juvenile crayfish moult they are secured by the anal thread, which consists of the cuticular lining of the hindgut that is moulted with delay. This safety line keeps the moulting juvenile linked to the mother via the stage-1 exuvia (fig. 1g) (Vogt, 2008a).

In cambarid crayfish, the young are carried at least until the end of juvenile stage 3 when juveniles become free-living and start feeding. The recently released juveniles return regularly to the maternal pleopods for rest and shelter. When shelters are scarce or when conditions are unfavourable brood care is usually extended (Vogt, 2008a). For example, in the red swamp crayfish *Procambarus clarkii* maternal care can extend to 7 juvenile stages and last for 3 months (Gherardi, 2002). The New Zealand crayfish *Paranephrops zealandicus* carrying its eggs for almost a year and its young for another 4-5 months (Whitmore & Huryn, 1999). In cultured marbled crayfish *Procambarus fallax f. virginalis* prehatching brood care lasts for 18-26 days and posthatching brood care for another 15-38 days. The duration of all brooding cycles in the marbled crayfish together correspond to 25% of the adult life span.

The semi-terrestrial crayfish *Engaeus cymus* Clark, 1936, and *E. orramakunna* Horwitz, 1990, protect their offspring for an extremely long time, with a mother and three generations cohabiting in the same burrow (Lukhaup & Pekny, 2008a; Duffy, 2010). Although not yet investigated in detail, this condition resembles the situation in the bromeliad crab (Diesel & Schubart, 2000), and eusocial marine snapping shrimps (Duffy, 2010).

Mortality in brooded life stages, juveniles, and adults

Estimation of the mortality of a population is relatively easy in captivity (except for the embryonic and brooded stages) but is difficult in the wild. In freshwater crayfish the average survival rate from spawning to the release of

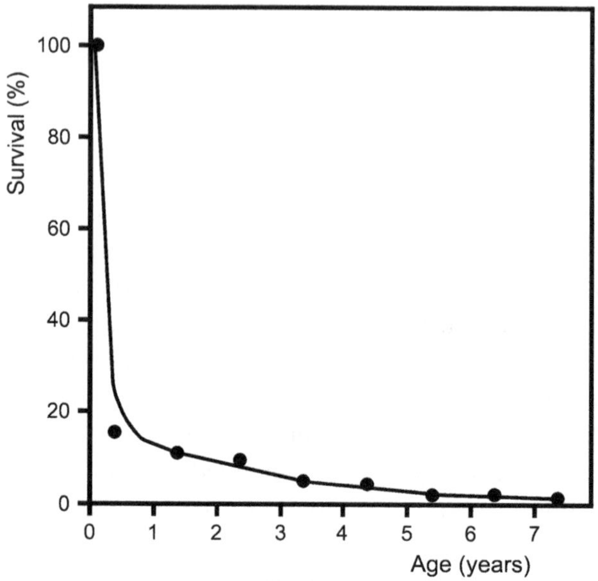

Fig. 2. Estimated survivorship curve of the signal crayfish *Pacifastacus leniusculus* from Lake Billy Chinook, Oregon. Mortality is highest in the first months of free life, i.e. in the postembryonic juvenile stages. The instantaneous mortality rate was Z=0.67, which means that an average of roughly 50% of the population died on an annual basis (from Lewis, 1997; redrawn).

the juveniles is estimated to be 50% (Reynolds, 2002). Thereafter, crayfish populations exhibit a type III survivorship curve where most of the total mortality of a cohort occurs during the first year of life, mainly due to predation and cannibalism (Lewis, 2002), after which the mortality rate is relatively low (fig. 2). Similar survivorship curves were obtained for laboratory populations of the marbled crayfish *Procambarus fallax f. virginalis*. This mortality pattern suggests that in freshwater species with extended brood care individual life expectancy is several weeks or months depending on species and conditions. In marine species and in freshwater species with planktonic larval stages the mean life expectancy is only a few days at best because these planktonic stages suffer heavy predation.

Age specific mortality and the annual mortality rate in a population depend on the longevity of the species, the availability of food, the population density, the predation rate, diseases, and fishing intensity. For instance, a population of the signal crayfish *Pacifastacus leniusculus* in Lake Billy Chinook, Oregon, had a life span of 8 years and an instantaneous mortality rate of Z=0.67, where about 50% of the population died on an annual basis (Lewis, 1997). This rate was twice the mortality rate (Z=0.32) for *P. leniusculus* in the colder

Lake Tahoe environment (life span: 12 years) (Flint, 1975). In the shorter-lived African freshwater prawn *Macrobrachium macrobrachion* (Herklots, 1851) (life span: 1.5 years) in Luubara creek, Nigeria, the instantaneous mortality rate was Z=2.75, which corresponds to an annual mortality rate of roughly 90% of the population (Deekae & Abowei, 2010). The annual mortality rate of the intensely exploited West African river prawn, *Macrobrachium vollenhovenii* (Herklots, 1857) (life span: 3 years) from Dawhenya impoundment, Ghana, was close to 100% (Z=5.36) (Alhassan & Armah, 2011).

Consequences of abbreviated development and brood care for life expectancy and longevity

Abbreviated and/or direct development are characteristic of about 90% of freshwater decapods, and postembryonic brood care has evolved in about 70% of them. The shift from large numbers of offspring with a low survival rate (r-strategy) to a small number of offspring with a high survival rate (K-strategy) in most freshwater decapods has enhanced individual life expectancy by reducing the mortality of the early life stages. But this evolutionary shift did not increase the longevity of freshwater species when compared with marine species. In the 109 freshwater species with a known life span, 0.9% lived for less than 1 year, 63.3% for between 1 and 5 years, 20.2% for between 5 and 10 years, and 15.6% lived for more than 10 years. These figures for freshwater decapods are not fundamentally different from the respective values for 98 marine species with a known age, which are 9.2%, 56.1%, 17.4% and 17.3% (Vogt, 2012a), particularly if the short-lived planktonic penaeids are removed from the data set.

The independent evolution of abbreviated development, direct development, and extended brood care several times in different fresh water decapod groups indicates that these traits are selected for in freshwater environments. The production of small numbers of well-developed offspring and the protection of the vulnerable early life stages is a fitness advantage under the harsh conditions of freshwater habitats, particularly in fast flowing streams. However, the costs associated with these life history traits such as the production of special yolk molecules and food deprivation of females during development of the eggs and young are traded off against fecundity, leaving longevity largely unchanged. This assumption is supported by studies on parental care trade-offs and life-history relationships in insects and birds (Badyaev & Ghalambor, 2001; Gilbert & Manica, 2010). The evolution of long-term associations between mother and one or more generations of juveniles in the bromeliad crab and semiterrestrial crayfish may be explained by the scarcity of suitable habitat and the high risk of mortality for the offspring when searching for a new leaf axil or burrow (Diesel & Schubart, 2000; Duffy, 2010).

SENESCENCE AND ANTI-AGEING MECHANISMS

Senescence refers to age-related changes in structure, physiology and behaviour that have adverse effects on an organism and that increase the likelihood of its death. Senescence is a measure of the speed and pattern of degenerative changes after the first reproduction.

Types of senescence

Senescence is broadly categorized into three types, rapid, gradual, and negligible (Finch, 1990). Negligible senescence means that structural and functional decay is either undetectable or restricted to only a short final life period. The classification of senescence mostly applies to longer-lived and iteroparous species, and makes little sense in annual and semelparous decapods. The mitten crabs *Eriocheir sinensis* and *E. japonica* (De Haan, 1835) show rapid senescence because they die soon after their first breeding season without a further moult (Kobayashi, 2001; Veilleux & de Lafontaine, 2007). The anomuran freshwater crab *Aegla strinatii* Türkay, 1972, which has a life span of 34 months and two reproduction periods, shows a significant fall in reproductive output in the second year, indicating gradual senescence (Rocha et al., 2010).

All of the longer-lived freshwater Decapoda exhibit negligible senescence because they grow indeterminately, that is they continue to moult and grow throughout their lives with no fixed limits (Vogt, 2012a). Moreover, their clutch size is usually positively correlated with body size, i.e. with age (Vogt, 2010; Skurdal et al., 2011). Negligible senescence is also typical for most of the longer-lived marine decapods. Exceptions to this are those species that have a terminal moult like the snow crab *Chionoecetes opilio* (Fabricius, 1788). These species stop growth and moulting at the onset of maturation or around mid-age and suffer particularly from mechanical senescence at older ages (Vogt, 2012a).

Anti-ageing mechanisms

Freshwater decapods have developed several effective anti-ageing mechanisms including moulting (Guiaşu & Dunham, 2002), regeneration of damaged and lost appendages (Buřič et al., 2009), detoxification of free radicals by antioxidant enzymes, glutathione and astaxanthin (Elia et al., 2006; Angeles et al., 2009), lysosomal removal of cellular waste (Vogt, 1994) and the renewal of tissues by life-long stem cell activity (Vogt, 2012b). Moreover, they can detoxify environmental pollutants by metallothioneins and cytochrome P450 (Del Ramo et al., 1995; Ashley et al., 1996) and combat pathogens and neoplastic cells by various immune defence mechanisms (Cerenius et al., 2008; Vogt, 2008c; Gherardi et al., 2010). All of these measures contribute to the improvement of health and an increase in life expectancy.

The most obvious anti-ageing mechanism is moulting, which serves not only for regular renewal of the cuticle but also for replacement of the mouthparts (fig. 3a), the internal grinding and filtering structures of the

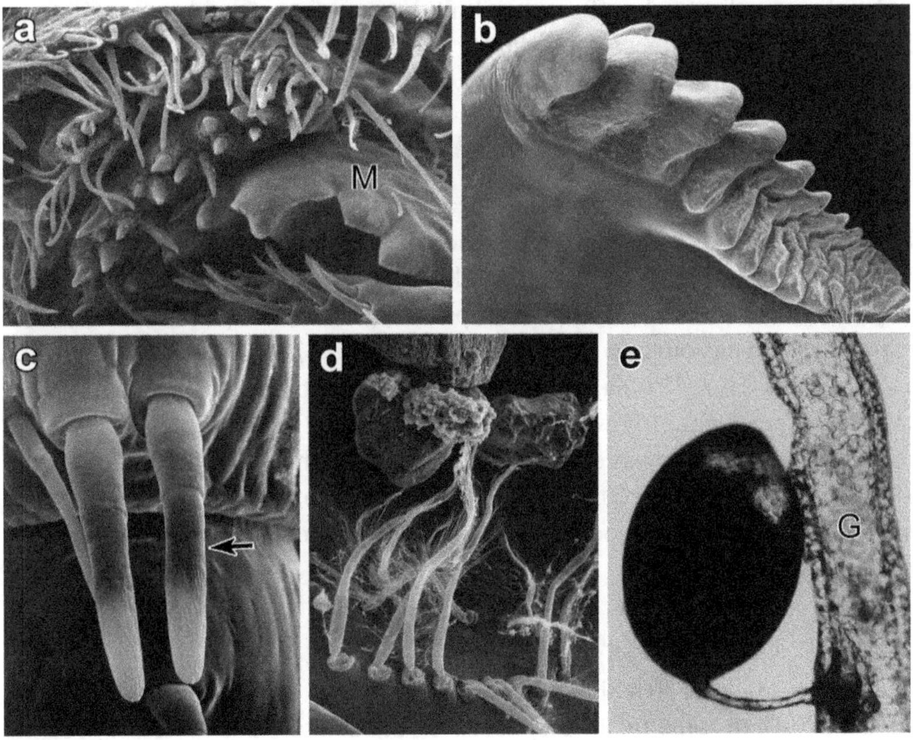

Fig. 3. Examples of structures that are regularly moulted to avoid mechanical senescence. a, mouthparts of juvenile marbled crayfish *Procambarus fallax f. virginalis* (viewed on cutting edges); M, mandible. b, lateral tooth of gastric mill of the noble crayfish *Astacus astacus* (from Vogt, 2002). c, olfactory aesthetascs (arrow) on the 1st antenna of the marbled crayfish. d, detail of a statocyst of the marbled crayfish showing sensory setae glued to sand grains of the statolith. e, gill filament (G) of *Astacus astacus* with an attached cocoon of a branchiobdellid worm (from Vogt, 1999).

stomach (fig. 3b), the olfactory (fig. 3c), gustatory, tactile and hydrodynamic sense organs, the statocysts (fig. 3d), and the external reproductive structures. Moreover, fouling organisms that may impair motility and gas exchange across the gills (fig. 3e) are removed by moulting. Moulting occurs quite often in the lifetime of a freshwater decapod, for instance some 25 times in the marbled crayfish *Procambarus fallax f. virginalis* (maximum age: 4.4 years) (Vogt, 2010) and up to 80 times in the Murray crayfish *Euastacus armatus* (Von Martens, 1866) (maximum age: 28 years) (Gilligan et al., 2007).

ADVANTAGES OF FRESHWATER DECAPODA FOR AGEING RESEARCH

Freshwater decapods offer special advantages for research on ageing and longevity in the Crustacea and even for general biogerontology. Some of these advantages are highlighted in the following.

Resident populations and age determination

One advantage of freshwater Decapoda is the frequent occurrence of resident populations in rivers, lakes and caves. Most of these populations are relatively easily accessible, enable mark and recapture experiments and offer a high chance of the recovery of older specimens. For example, Streever (1996) captured a 16 year-old crayfish *Procambarus erythrops* Relyea & Sutton, 1975, in Sim's Sink cave, Florida, which had been marked by a colleague in an earlier tagging experiment. Likewise, Cross et al. (2008) reported on occasional recaptures of earlier marked specimens of the freshwater shrimp *Xiphocaris elongata* in El Yunque National Forest, Puerto Rico, which had ages up to 18 years, whereas growth models for this species estimate a maximum life span of only 11 years.

Rearing in captivity and maximum life span

Animals often reach a higher life span in captivity than in the wild (Carey & Judge, 2000). In some freshwater decapods aquaculture and laboratory rearing have revised the earlier published maximum life spans upward, for instance from 18 months to 4 years in *Procambarus clarkii* (Huner, 2002). Freshwater decapods enjoy increasing popularity as pets, and therefore, the expertise of pet owners should be exploited for ageing research as well. In Germany dozens of species of freshwater shrimp, crayfish, crabs, and aeglids are available in the aquarium trade and there are several good-quality books on their biology and rearing (Logemann & Logemann, 2007; Karge & Klotz, 2008; Lukhaup & Pekny, 2008a, b; Rademacher & Mengedoht, 2011). Longevity data from reliable aquarist websites sometimes indicate longer life spans than published values. For instance, the life expectancy of the Amano shrimp *Caridina multidentata* Stimpson, 1860, is usually thought to be 2-3 years, but some aquarists have raised individuals for as long as 5 years.

Cave-dwelling species as models for life span extension by low caloric intake

The relationship between life span extension and low caloric intake is a hot topic in biogerontology (Masoro, 2005). Cave dwelling decapods appear particularly suitable to investigate this issue in depth because they have a much higher life span than their epigean relatives. They are adapted to a low and erratic food supply and a constantly low temperature. Cave-dwelling species are found in all freshwater Decapoda, the atyid and palaemonid shrimps, crayfish, brachyuran crabs, and aeglids. Holthuis (1986) listed 121 species of troglobitic freshwater decapods from all over the world (74 shrimps, 30 crayfish, 16 crabs and 1 aeglid) and more species of cavernicolous decapods have been described since then (Hobbs, 1994; Page et al., 2008; Husana et al., 2009; Sket & Zakšek, 2009). Comparison between epigean and hypogean species and genus pairs is a promising approach for the investigation of life span extending physiological, molecular and genetic mechanisms that result from caloric restriction and low temperatures.

Testing for epigenetic and genetic causes of longevity alterations

There is increasing evidence that variation of life history traits can be caused not only by random genetic drift and natural selection but also by epigenetic changes including the methylation of cytosines, modification of histones, and alterations of DNA-binding proteins. Unlike the genetic code, the epigenetic code can be erased and rewritten when environmental conditions change and can be transgenerationally inherited and genetically assimilated under certain conditions (Jaenisch & Bird, 2003; Gilbert & Epel, 2009; Jablonka & Raz, 2009). Epigenetic mechanisms of trait alteration may be particularly relevant for rapid adaptation to changing environments and are thought to also affect longevity.

Good choices for testing the influence of epigenetics on longevity are invasive or transplanted species such as the Chinese mitten crab *Eriocheir sinensis* that has a wide distribution throughout the world. Interestingly, populations of *E. sinensis* from different countries have different longevities. For instance, the life span of *E. sinensis* is around 1 year in the coastal rivers of China and Korea (native range), 2-3 years in the major Chinese rivers, and 4 years in the cooler highlands of China. In Europe, where mitten crabs have spread dramatically after their unintentional introduction in 1912, longevity is 4-5 years. And in San Francisco Bay where the crab was first recorded in 1992 its life span is 3 years (Herborg et al., 2003; Veilleux & de Lafontaine, 2007). If these longevity differences were epigenetically determined they would disappear in a standardized laboratory set-up within about two generations (Reznick et al., 1990) but if they were genetically fixed they would persist.

Allocation of metabolic resources and longevity

The optimal allocation of metabolic resources between competing activities such as growth, reproduction, and maintenance and repair is central to life history theory (Stearns, 1992), and the trade-off between these fitness traits is the mainstay of the disposable soma theory of ageing (Kirkwood & Austad, 2000; Kirkwood, 2008). The allocation pattern can be age-specific and can vary in response to environmental challenges. Cichoń (1997) and Cichoń & Kozłowski (2000) disclosed with their dynamic programming model that the allocation strategy shapes the survivorship curve, age at maturity and maximum life span of a population. The optimal allocation, as determined by natural selection, is that which maximizes the production of progeny in the long-term (Kirkwood, 2008). Indeterminately growing decapods reproduce until they reach old ages, and there is a positive correlation between body size (which itself is roughly correlated with age) and clutch size. Therefore, in these decapods perpetuation of investment into repair until high age should be favoured by natural selection. This situation is quite different from that of animals with a long post-reproductive life period like mammals (Kirkwood & Austad, 2000).

The relationship between longevity and the allocation of metabolic resources towards growth, reproduction, and repair has not been investigated in detail for decapod crustaceans. However, in various species limbs were shown to be regenerated at the expense of growth or reproduction (Juanes & Smith, 1995; this paper, fig. 4) or reproduction. Damaged specimens of the Indian freshwater crab *Spiralothelphusa hydrodroma* allocated energy to regeneration in the non-breeding season but not in the breeding season (Maginnis, 2006).

We have chosen the parthenogenetic marbled crayfish *Procambarus fallax f. virginalis* for in-depth investigation of the resource allocation-longevity

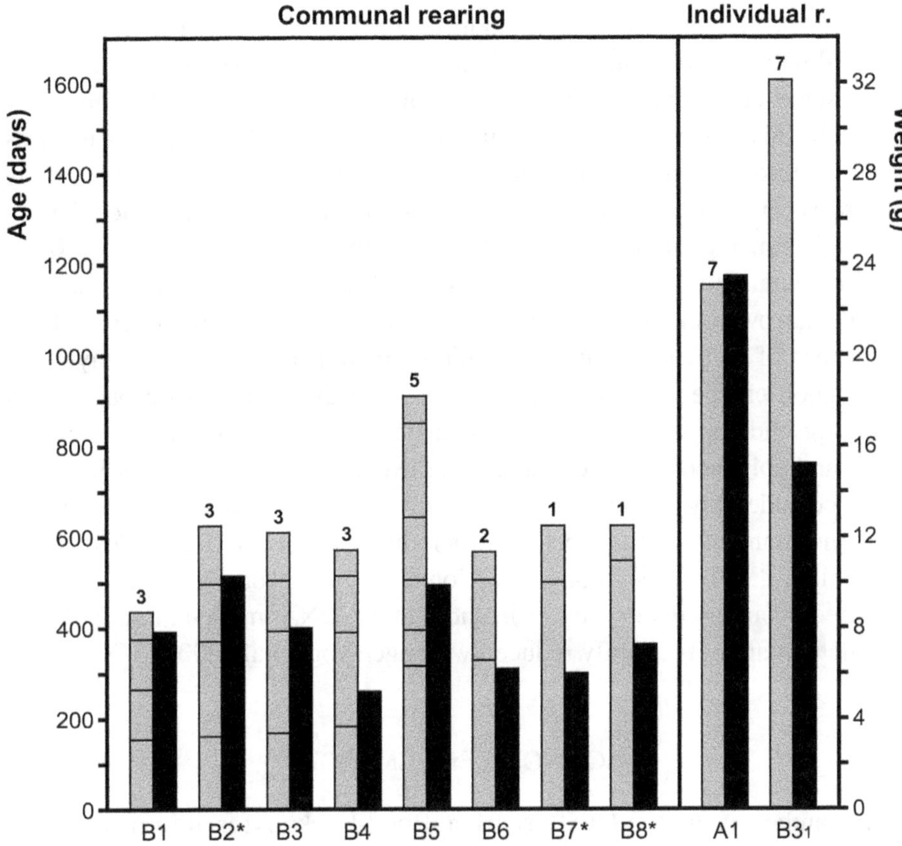

Fig. 4. Variation of longevity, growth, and reproduction among isogenic and identically reared marbled crayfish. Data are shown for eight communally reared batch-mates (B1-B8) and two individually reared specimens (A1 and $B3_1$) fed ad libitum. Life span is indicated by the left column, weight by the right column, the number of reproductive cycles by the numbers on left column, and the time points of oviposition by horizontal bars in the left column. Crayfish $B3_1$, a progeny of female B3, lost several appendages in early adolescence. Specimens indicated by asterisks were sacrificed for biochemical investigation. The other crayfish died a natural death. The graph indicates that life span can vary considerably among isogenic batch-mates (B1-B8), that social stress may reduce longevity (B1-B8 versus A1 and $B3_1$), and that regeneration does not impair longevity ($B3_1$). However, there is a trade-off between repair and growth (compare $B3_1$ and A1). Interestingly, the allocation of resources towards growth, reproduction, and maintenance and repair (reflected by the life span length) varied considerably between communally raised batch-mates (B1-B8), despite their genetic identity (from Vogt, 2009; updated).

issue because this species produces up to 400 genetically identical offspring per batch and can be kept in simple, standardized housing systems under a broad range of environmental conditions (Martin et al., 2007; Vogt, 2008b,

2010, 2011). Our first experiment addressed the question whether genetically identical batch-mates would develop uniform allocation patterns when reared in the same environment with unlimited access to food. The data in fig. 4 show that this is not the case: the allocation of metabolic resources towards growth, reproduction, and maintenance and repair (indirectly reflected by life span) varies considerably among batch-mates despite genetic identity and environmental uniformity. Under conditions of limited food or other environmental challenges this surprisingly broad range of variation may become narrower and the allocation pattern may become more directional.

Because of its genetic uniformity, the parthenogenetic marbled crayfish is also a good choice for laboratory studies of further non-genetic parameters of ageing and longevity such as stochastic developmental variation, social stress, and epigenetic drift with age. First experiments revealed that life span varies considerably among isogenic batch mates even when reared in the same environment and that social stress reduces longevity (fig. 4). Moreover, mechanical damage does not impair longevity if the damaged specimen is protected from predators (fig. 4), and global DNA methylation, a prime epigenetic marker, is slightly reduced with age (Vogt et al., 2008).

ACKNOWLEDGEMENTS

The author is grateful to Georgina Bond-Buckup for information and literature on the Aeglidae, to Jose A. Cuesta for information and literature on brood care in *Dugastella* and for providing fig. 1h, to Monika Rademacher and Oliver Mengedoht for information on minimum age and brood care in aquarium-raised freshwater crabs and for providing figs. 1a and b and to Rudolf Diesel for providing figs. 1c and d. Many thanks also to two anonymous reviewers for valuable comments and suggestions to improve the manuscript and to Neil Cumberlidge for linguistic corrections.

REFERENCES

ABRAMS, P. A., 2004. Mortality and lifespan. Nature, **431**: 1048-1049.

ALHASSAN, E. H. & A. K. ARMAH, 2011. Population dynamics of the African river prawn, *Macrobrachium vollenhovenii*, in Dawhenya impoundment. Turkish Journ. Fish. Aquat. Sci., **11**: 113-119.

ALI, M. H., S. D. SALMAN & A. Y. AL-ADHUB, 1995. Population dynamics of the hymenosomatid crab *Elamenopsis kempi* in a brackish subtidal region of Basrah, Iraq. Sci. Mar., **59**: 1-13.

ANGELES, I. P. JR., Y.-H. CHIEN & A. V. YAMBOT, 2009. Effect of injected astaxanthin on survival, antioxidant capacity, and immune response of the giant freshwater prawn *Macrobrachium rosenbergii* (De Man, 1879) challenged with *Lactococcus garvieae*. Journ. Shellfish Res., **28**: 931-937.

ANGER, K., 1995. The conquest of freshwater and land by marine crabs: adaptations in life-history patterns and larval bioenergetics. Journ. Exp. Mar. Biol. Ecol., **193**: 119-145.

——, 2001. The biology of decapod crustacean larvae. Crustacean Issues, **14**: 1-419. (Balkema, Lisse).

ASHLEY, C. M., M. G. SIMPSON, D. M. HOLDICH & D. R. BELL, 1996. 2,3,7,8-tetrachlorodibenzo-*p*-dioxin is a potent toxin and induces cytochrome P450 in the crayfish, *Pacifastacus leniusculus*. Aquat. Toxicol., **35**: 157-169.

BADYAEV, A. V. & C. K. GHALAMBOR, 2001. Evolution of life histories along elevation gradients: trade-off between parental care and fecundity. Ecology, **82**: 2948-2960.

BAUER, R. T., 2011. Amphidromy and migrations of freshwater shrimps. I. Costs, benefits, evolutionary origins, and an unusual case of amphidromy. In: A. ASAKURA (ed.), New frontiers in crustacean biology. Crustaceana Monographs, **15**: 145-156. (Brill, Leiden).

BECK, J. T. & B. C. COWELL, 1976. Life history and ecology of the freshwater caridean shrimp, *Palaemonetes paludosus* (Gibbes). American Midl. Nat., **96**: 52-65.

BELCHIER, M., L. EDSMAN, M. R. J. SHEEHY & P. M. J. SHELTON, 1998. Estimating age and growth in long-lived temperate freshwater crayfish using lipofuscin. Freshwater Biol., **39**: 439-446.

BOND-BUCKUP, G., C. G. JARA, L. BUCKUP, A. A. P. BUENO, M. PÉREZ-LOSADA & K. A. CRANDALL, 2010. Description of a new species of Aeglidae, and new records of related species from river basins in Argentina (Crustacea, Anomura). Zootaxa, **2343**: 18-30.

BOND-BUCKUP, G., C. G. JARA, M. PÉREZ-LOSADA, L. BUCKUP & K. A. CRANDALL, 2008. Global diversity of crabs (Aeglidae: Anomura: Decapoda) in freshwater. Hydrobiol., **595**: 275-286.

BOOS, H. JR., D. DA SILVA-CASTIGLIONI, K. SCHACHT, L. BUCKUP & G. BOND-BUCKUP, 2006. Crescimento de *Aegla jarai* Bond-Buckup & Buckup (Crustacea, Anomura, Aeglidae). Rev. Brasileira Zool., **23**: 490-496.

BROWN, J. H., M. B. NEW & D. ISMAEL, 2010. Biology. In: M. B. NEW, W. C. VALENTI, J. H. TIDWELL, L. R. D'ABRAMO & M. N. KUTTY (eds.), Freshwater prawns: biology and farming: 18-39. (Blackwell, Oxford).

BUENO, S. L. S. & R. M. SHIMIZU, 2008. Reproductive biology and functional maturity in females of *Aegla franca* (Decapoda: Anomura: Aeglidae). Journ. Crust. Biol., **28**: 652-662.

BURGGREN, W. W. & B. R. MCMAHON (eds.), 1988. Biology of the land crabs: 1-487. (Press Syndicate of the University of Cambridge, Cambridge).

BUŘIČ, M., A. KOUBA & P. KOZÁK, 2009. Chelae regeneration in European alien crayfish *Orconectes limosus* (Rafinesque 1817). Knowl. Managem. Aquat. Ecosyst., **394-395**: 04.

CAREY, J. R., 2003. Longevity. The biology and demography of life span: 1-279. (Princeton University Press, Princeton).

CAREY, J. R. & D. S. JUDGE, 2000. Longevity records: life spans of mammals, birds, amphibians, reptiles, and fish. Odense Monographs on Population Aging, **8**: 1-241. (University Press of Southern Denmark, Odense).

CASELLAS, J., L. VARONA, N. IBÁÑEZ-ESCRICHE, R. QUINTANILLA & J. L. NOGUERA, 2008. Skew distribution of founder-specific inbreeding depression effects on the longevity of Landrace sows. Genet. Res., **90**: 499-508.

CERENIUS, L., B. L. LEE & K. SÖDERHÄLL, 2008. The proPO-system: pros and cons for its role in invertebrate immunity. Trends Immunol., **29**: 263-271.

CHANG, Y.-J., C.-L. SUN, Y. CHEN & S.-Z. YEH, 2012. Modelling the growth of crustacean species. Rev. Fish Biol. Fish., **22**: 157-187.

CHILTON, C., 1914. A fresh-water crab, and its distribution in Australia and New Zealand. Trans. Proc. R. Soc. New Zealand, **47**: 316-320.

CHUCHOLL, C., 2011. Population ecology of an alien "warm water" crayfish (*Procambarus clarkii*) in a new cold habitat. Knowl. Managem. Aquat. Ecosyst., **401**: 29.

CICHOŃ, M., 1997. Evolution of longevity through optimal resource allocation. Proc. R. Soc. London B, **264**: 1383-1388.

CICHOŃ, M. & J. KOZŁOWSKI, 2000. Ageing and typical survivorship curves result from optimal resource allocation. Evol. Ecol. Res., **2**: 857-870.

COFFEY, B. T. & J. S. CLAYTON, 1988. Contrasting deep-water macrophyte communities in two highly transparent New Zealand lakes and their possible association with freshwater crayfish, *Paranephrops* spp. New Zealand Journ. Mar. Freshwater Res., **22**: 225-230.

COHEN, F. P. A., B. F. TAKANO, R. M. SHIMIZU & S. L. S. BUENO, 2011. Life cycle and population structure of *Aegla paulensis* (Decapoda: Anomura: Aeglidae). Journ. Crust. Biol., **31**: 389-395.

CRANDALL, K. A. & J. E. BUHAY, 2008. Global diversity of crayfish (Astacidae, Cambaridae, and Parastacidae-Decapoda) in freshwater. Hydrobiol., **595**: 295-301.

CRANDALL, K. A., J. W. FETZNER JR., C. G. JARA & L. BUCKUP, 2000. On the phylogenetic positioning of the South American freshwater crayfish genera (Decapoda: Parastacidae). Journ. Crust. Biol., **20**: 530-540.

CROSS, W. F., A. P. COVICH, T. A. CROWL, J. P. BENSTEAD & A. RAMÍREZ, 2008. Secondary production, longevity and resource consumption rates of freshwater shrimps in two tropical streams with contrasting geomorphology and food web structure. Freshwater Biol., **53**: 2504-2519.

CUESTA, J. A., E. PALACIOS-THEIL, P. DRAKE & A. RODRÍGUEZ, 2006. A new rare case of parental care in decapods. Crustaceana, **79**: 1401-1405.

CUMBERLIDGE, N., 1999. The freshwater crabs of West Africa: family Potamonautidae: 1-385. (IRD Éditions, Paris).

CUMBERLIDGE, N. & P. F. CLARK, 2010. A redescription of *Potamonautes loveni* (Colosi, 1924), a medically important freshwater crab from western Kenya and eastern Uganda, East Africa (Brachyura, Potamoidea, Potamonautidae). In: P. K. L. NG, P. CASTRO, P. DAVIE & B. RICHER DE FORGES (eds.), Studies on Brachyura: a homage to Danièle Guinot. Crustaceana Monographs, **11**: 61-74. (Brill, Leiden).

DE GRAVE, S., Y. CAI & A. ANKER, 2008. Global diversity of shrimps (Crustacea: Decapoda: Caridea) in freshwater. Hydrobiol., **595**: 287-293.

DE GRAVE, S., N. D. PENTCHEFF, S. T. AHYONG, T.-Y. CHAN, K. A. CRANDALL, P. C. DWORSCHAK, D. L. FELDER, R. M. FELDMANN, C. H. J. M. FRANSEN, L. Y. D. GOULDING, R. LEMAITRE, M. E. Y. LOW, J. W. MARTIN, P. K. L. NG, C. E. SCHWEITZER, S. H. TAN, D. TSHUDY & R. WETZER, 2009. A classification of living and fossil genera of decapod crustaceans. Raffles Bull. Zool., (Suppl.) **21**: 1-109.

DE SILVA, K. H. G. M., 1988. Studies on Atyidae (Decapoda, Caridea) of Sri Lanka III. Aspects of the population ecology of *Caridina simoni* Bouvier, 1904. Crustaceana, **54**: 85-103.

DEL RAMO, J., A. TORREBLANCA, M. MARTÍNEZ, A. PASTOR & J. DÍAZ-MAYANS, 1995. Quantification of cadmium-induced metallothionein in crustaceans by the silver-saturation method. Mar. Environ. Res., **39**: 121-125.

DEEKAE, S. N. & J. F. N. ABOWEI, 2010. The mortality, exploitation rate and longevity of *Macrobrachium macrobrachion* (Herklots, 1851) in Luubara creek, Ogoni land, Niger delta, Nigeria. Curr. Res. Journ. Biol. Sci., **2**: 323-327.

DHAOUADI-HASSEN, S. & M. BOUMAÏZA, 2009. Reproduction and population dynamics of *Atyaephyra desmarestii* (Decapoda, Caridea) from the Sidi Salem Dam Lake (northern Tunisia). Crustaceana, **82**: 129-139.

DIESEL, R., 1992. Maternal care in the bromeliad crab, *Metopaulias depressus*: protection of larvae from predation by damselfly nymphs. Anim. Behav., **43**: 803-812.

DIESEL, R. & D. HORST, 1995. Breeding in a snail shell: ecology and biology of the Jamaican montane crab *Sesarma jarvisi* (Decapoda: Grapsidae). Journ. Crust. Biol., **15**: 179-195.

DIESEL, R. & C. D. SCHUBART, 2000. Die außergewöhnliche Evolutionsgeschichte jamaikanischer Felsenkrabben. BIUZ, **30**: 136-147.

DIESEL, R. & M. SCHUH, 1993. Maternal care in the bromeliad crab *Metopaulias depressus* (Decapoda): maintaining oxygen, pH and calcium levels optimal for the larvae. Behav. Ecol. Sociobiol., **32**: 11-15.

DOBSON, M., A. M. MAGANA, J. M. MATHOOKO & F. K. NDEGWA, 2008. Distribution and abundance of freshwater crabs (*Potamonautes* spp.) in rivers draining Mt Kenya, East Africa. Fund. Appl. Limnol., **168**: 271-279.

DUFFY, J. E., 2010. Crustacean social evolution. Encycl. Anim. Behav., **1**: 421-429.

ELIA, A. C., A. J. M. DÖRR, C. MASTRANGELO, M. PREARO & M. C. ABETE, 2006. Glutathione and antioxidant enzymes in the hepatopancreas of crayfish *Procambarus clarkii* (Girard, 1852) of Lake Trasimeno (Italy). Bull. Française Pêche Piscicult., **380-381**: 1351-1361.

FINCH, C. E., 1990. Longevity, senescence, and the genome: 1-922. (University of Chicago Press, Chicago).

FLINT, R. W., 1975. Growth in a population of the crayfish *Pacifastacus leniusculus* from a subalpine lacustrine environment. Journ. Fish. Res. Board Canada, **32**: 2433-2440.

— —, 1977. Seasonal activity, migration and distribution of the crayfish, *Pacifastacus leniusculus*, in Lake Tahoe. American Midl. Nat., **97**: 280-292.

GHERARDI, F., 2002. Behaviour. In: D. M. HOLDICH (ed.), Biology of freshwater crayfish: 258-290. (Blackwell, Oxford).

GHERARDI, F., C. SOUTY-GROSSET, G. VOGT, J. DIÉGUEZ-URIBEONDO & K. A. CRANDALL, 2010. Infraorder Astacidea Latreille, 1802 p.p.: the freshwater crayfish. In: F. R. SCHRAM & J. C. VON VAUPEL KLEIN (eds.), Treatise on zoology — anatomy, taxonomy, biology. The Crustacea, vol. 9, part A: Eucarida: Euphausiacea, Amphionidacea, and Decapoda (partim): 269-423. (Brill, Leiden).

GILBERT, S. F. & D. EPEL, 2009. Ecological developmental biology: integrating epigenetics, medicine, and evolution: 1-480. (Sinauer Associates, Sunderland).

GILBERT, J. D. J. & A. MANICA, 2010. Parental care trade-offs and life-history relationships in insects. American Nat., **176**: 212-226.

GILLIGAN, D., R. ROLLS, J. MERRICK, M. LINTERMANS, P. DUNCAN & J. KOHEN, 2007. Scoping the knowledge requirements for Murray crayfish (*Euastacus armatus*). Fisheries Final Report Series, **89**: 1-103. (NSW Department of Primary Industries, Narrandera Fisheries Centre, Narrandera, Australia).

GUIAŞU, R. C. & D. W. DUNHAM, 2002. Observations on the timing of moulting events in the crayfish *Cambarus robustus* Girard, 1852 (Decapoda, Cambaridae). Crustaceana, **74**: 1365-1378.

GUINOT, D. & B. RICHER DE FORGES, 1997. Affinités entre les Hymenosomatidae MacLeay, 1838 et les Inachoididae Dana, 1851 (Crustacea, Decapoda, Brachyura). Zoosystema, **19**: 453-502.

HAN, C.-C., C.-S. CHANG, I.-M. CHENG, L.-S. FANG & K. S. TEW, 2011. Population dynamics of a landlocked and amphidromous freshwater shrimp, *Caridina gracilipes* (Decapoda: Caridea) in subtropical waters. Journ. Crust. Biol., **31**: 278-285.

HARRISON, E.-M., 1995. The zonation of freshwater crabs with altitude in a mountain stream in the East Usambara Mountains, Tanzania. Frontier Tanzania Expedition TZ943. Earth and Life Studies: 1-66. (University of Derby, Derby).

HARTNOLL, R. G., 2001. Growth in Crustacea — twenty years on. Hydrobiol., **449**: 111-122.

HERBORG, L.-M., S. P. RUSHTON, A. S. CLARE & M. G. BENTLEY, 2003. Spread of the Chinese mitten crab (*Eriocheir sinensis* H. Milne Edwards) in Continental Europe: analysis of a historical data set. Hydrobiol., **503**: 21-28.

HOBBS, H. H. III, 1994. Biogeography of subterranean decapods in North and Central America and the Caribbean region (Caridea, Astacidea, Brachyura). Hydrobiol., **287**: 95-104.

HOFMANN, J., 1980. Die Flusskrebse: 1-110. (Verlag Paul Parey, Hamburg).

HOLTHUIS, L. B., 1986. Decapoda. In: L. BOTOSANEANU (ed.), Stygofauna mundi. A faunistic, distributional, and ecological synthesis of the world fauna inhabiting subterranean waters (including the marine interstitial): 1-740. (Brill, Leiden).

HUGUET, D., J. E. GARCÍA MUÑOZ, J. E. GARCÍA RASO & J. A. CUESTA, 2011. Extended parental care in the freshwater shrimp genus *Dugastella* Bouvier, 1912 (Decapoda, Atyidae, Paratyinae). Crustaceana, **84**: 251-255.

HUNER, J. V., 2002. *Procambarus*. In: D. M. HOLDICH (ed.), Biology of freshwater crayfish: 541-584. (Blackwell, Oxford).

HÜPPOP, K., 1985. The role of metabolism in the evolution of cave animals. NSS Bulletin, **47**: 136-146.

HUSANA, D. E. M., T. NARUSE & T. KASE, 2009. Two new cavernicolous species of the genus *Sundathelphusa* from western Samar, Philippines (Decapoda: Brachyura: Parathelphusidae). Journ. Crust. Biol., **29**: 419-427.

JABLONKA, E. & G. RAZ, 2009. Transgenerational epigenetic inheritance: prevalence, mechanisms, and implications for the study of heredity and evolution. Quart. Rev. Biol., **84**: 131-176.

JAENISCH, R. & A. BIRD, 2003. Epigenetic regulation of gene expression: how the genome integrates intrinsic and environmental signals. Nat. Genet. Suppl., **33**: 245-254.

JALIHAL, D. R., K. N. SANKOLLI & S. SHENOY, 1993. Evolution of larval developmental patterns and the process of freshwaterization in the prawn genus *Macrobrachium* Bate, 1868 (Decapoda, Palaemonidae). Crustaceana, **65**: 365-376.

JARA, C., 1977. *Aegla rostrata* n. sp., (Decapoda, Aeglidae), nuevo crustáceo dulceacuícola del Sur de Chile. Stud. Neotrop. Fauna Environ., **12**: 165-176.

JONES, J. P. G., F. B. ANDRIAHAJAINA, N. J. HOCKLEY, K. A. CRANDALL & O. R. RAVOAHANGIMALALA, 2007. The ecology and conservation status of Madagascar's endemic freshwater crayfish (Parastacidae; Astacoides). Freshwater Biol., **52**: 1820-1833.

JUANES, F. & L. D. SMITH, 1995. The ecological consequences of limb damage and loss in decapod crustaceans: a review and prospectus. Journ. Exp. Mar. Biol. Ecol., **193**: 197-223.

JUSSILA, J. & A. MANNONEN, 2004. Crayfisheries in Finland, a short overview. Bull. Française Pêche Piscicult., **372-373**: 263-273.

KARGE, A. & W. KLOTZ, 2008. Süßwassergarnelen aus aller Welt (2. Aufl.): 1-216. (Dähne Verlag, Ettlingen).

KAWAI, T., T. HAMANO & S. MATSUURA, 1997. Survival and growth of the Japanese crayfish *Cambaroides japonicus* in a small stream in Hokkaido. Bull. Mar. Sci., **61**: 147-157.

KIRKWOOD, T. B. L., 2008. Understanding ageing from an evolutionary perspective. Journ. Intern. Med., **263**: 117-127.

KIRKWOOD, T. B. L. & S. N. AUSTAD, 2000. Why do we age? Nature, **408**: 233-238.

KLAUS, S., D. C. J. YEO & S. T. AHYONG, 2011. Freshwater crab origins — laying Gondwana to rest. Zool. Anz., **250**: 449-456.

KOBAYASHI, S., 2001. Fecundity of the Japanese mitten crab *Eriocheir japonica* (de Haan). Benthos Res., **56**: 1-7.

LEWIS, S. D., 1997. Life history, population dynamics, and management of signal crayfish in Lake Billy Chinook, Oregon: 1-98. (Master thesis, Oregon State University, Corvallis).

— —, 2002. *Pacifastacus*. In: D. M. HOLDICH (ed.), Biology of freshwater crayfish: 511-540. (Blackwell, Oxford).

LIANG, X. & Y. CAI, 1999. *Sinodina*, a new genus of freshwater shrimps (Crustacea, Decapoda: Atyidae) from southern China, with descriptions of three new species. Raffles Bull. Zool., **47**: 577-590.

LIU, H.-C. & C.-W. LI, 2000. Reproduction in the fresh-water crab *Candidiopotamon rathbunae* (Brachyura: Potamidae) in Taiwan. Journ. Crust. Biol., **20**: 89-99.

LOGEMANN, C. & F. LOGEMANN, 2007. Garnelenfibel. Süßwassergarnelen für Anfänger und Fortgeschrittene (3. Aufl.): 1-96. (Dähne Verlag, Ettlingen).

LÓPEZ GRECO, L. S., V. VIAU, M. LAVOLPE, G. BOND-BUCKUP & E. M. RODRIGUEZ, 2004. Juvenile hatching and maternal care in *Aegla uruguayana* (Anomura, Aeglidae). Journ. Crust. Biol., **24**: 309-313.

LUCAS, J. S., 1980. Spider crabs of the family Hymenosomatidae (Crustacea; Brachyura) with particular reference to Australian species: systematics and biology. Rec. Australian Mus., **33**: 148-247.

LUKHAUP, C. & R. PEKNY, 2008a. Süßwasserkrebse aus aller Welt (2. Aufl.): 1-256. (Dähne Verlag, Ettlingen).

— — & — —, 2008b. Krebse im Aquarium. Haltung und Pflege im Süßwasser (2. Aufl.): 1-155. (Dähne Verlag, Ettlingen).

MAGINNIS, T. L., 2006. The costs of autotomy and regeneration in animals: a review and framework for future research. Behav. Ecol., **17**: 857-872.

MARIJNISSEN, S. A. E., E. MICHEL, M. KAMERMANS, K. OLAYA-BOSCH, M. KARS, D. F. R. CLEARY, E. E. VAN LOON, P. G. RACHELLO DOLMEN & S. B. J. MENKEN, 2008. Ecological correlates of species differences in the Lake Tanganyika crab radiation. Hydrobiol., **615**: 81-94.

MARIJNISSEN, S. A. E., F. R. SCHRAM, N. CUMBERLIDGE & E. MICHEL, 2004. Two new species of *Platythelphusa* A. Milne-Edwards, 1887 (Decapoda, Potamoidea, Platythelphusidae) and comments on the taxonomic position of *P. denticulata* Capart, 1952 from Lake Tanganyika, East Africa. Crustaceana, **77**: 513-532.

MARTIN, P., K. KOHLMANN & G. SCHOLTZ, 2007. The parthenogenetic Marmorkrebs (marbled crayfish) produces genetically uniform offspring. Naturwiss., **94**: 843-846.

MASORO, E. J., 2005. Overview of caloric restriction and ageing. Mech. Ageing Dev., **126**: 913-922.

MCLAUGHLIN, P. A. & T. MURRAY, 1990. *Clibanarius fonticola*, new species (Anomura: Paguridea: Diogenidae) from a fresh-water pool on Espiritu Santo, Vanuatu. Journ. Crust. Biol., **10**: 695-702.

MEYER, K. M., K. GIMPEL & R. BRANDL, 2007. Viability analysis of endangered crayfish populations. Journ. Zool., **273**: 364-371.

MOMOT, W. T., 1984. Crayfish production: a reflection of community energetics. Journ. Crust. Biol., **4**: 35-54.

NEVEU, A., 2000. Étude des populations d'*Austropotamobius pallipes* (Crustacea, Astacidae) dans un ruisseau forestier de Normandie. I. Structures démographiques et croissance: stabilité et variabilité au cours de six années. Bull. Française Pêche Piscicult., **356**: 71-98.

NG, P. K. L. & C. G. S. TAN, 1995. *Geosesarma notophorum* sp. nov. (Decapoda, Brachyura, Grapsidae, Sesarminae), a terrestrial crab from Sumatra, with novel brooding behaviour. Crustaceana, **68**: 390-395.

NORMANT, M., A. WISZNIEWSKA & A. SZANIAWSKA, 2000. The Chinese mitten crab *Eriocheir sinensis* (Decapoda: Grapsidae) from Polish waters. Oceanologia, **42**: 375-383.

NORO, C. K. & L. BUCKUP, 2009. The growth of *Parastacus defossus* (Crustacea: Decapoda: Parastacidae). Zoologia, **26**: 54-60.

OH, C.-W., H.-L. SUH, K.-Y. PARK, C.-W. MA & H.-S. LIM, 2002. Growth and reproductive biology of the freshwater shrimp *Exopalaemon modestus* (Decapoda: Palaemonidae) in a lake of Korea. Journ. Crust. Biol., **22**: 357-366.

OYANEDEL, A., C. VALDOVINOS, N. SANDOVAL, C. MOYA, G. KIESSLING, J. SALVO & V. OLMOS, 2011. The southernmost freshwater anomurans of the world: geographic distribution and new records of patagonian aeglids (Decapoda, Aeglidae). Journ. Crust. Biol., **31**: 396-400.

PAGE, T. J., W. F. HUMPHREYS & J. M. HUGHES, 2008. Shrimps down under: evolutionary relationships of subterranean crustaceans from Western Australia (Decapoda: Atyidae: *Stygiocaris*). PLoS ONE, **3**: e1618.

PARKYN, S. M., K. J. COLLIER & B. J. HICKS, 2002. Growth and population dynamics of crayfish *Paranephrops planifrons* in streams within native forest and pastoral land uses. New Zealand Journ. Mar. Freshwater Res., **36**: 847-861.

PÉREZ-LOSADA, M., G. G. JARA, G. BOND-BUCKUP, M. L. PORTER & K. A. CRANDALL, 2002. Phylogenetic position of the freshwater anomuran family Aeglidae. Journ. Crust. Biol., **22**: 670-676.

PETERSON, M. S., J. F. FITZPATRICK JR. & S. J. VANDERKOOY, 1996. Distribution and habitat use by dwarf crayfishes (Decapoda: Cambaridae: *Cambarellus*). Wetlands, **16**: 594-598.

PILLAI, C. K. & T. SUBRAMONIAM, 1984. Monsoon-dependent breeding in the field crab *Paratelphusa hydrodromous* (Herbst). Hydrobiol., **119**: 7-14.

PORTER, M. L., M. PÉREZ-LOSADA & K. A. CRANDALL, 2005. Model-based multi-locus estimation of decapod phylogeny and divergence times. Mol. Phyl. Evol., **37**: 355-369.

PRUITT, N. L., 1988. Membrane lipid composition and overwintering strategy in thermally acclimated crayfish. American Journ. Physiol., Regul. Physiol., **254**: R870-R876.

QUIROZ CASTELÁN, H., M. DIAS VARGAS, R. TREJO ALBARRÁN & E. E. ELIZALDE ARRIAGA, 2000. Aspectos sobre la abundancia y distribución de los principales grupos de la fauna bentónica en el lago "Zempoala", Morelos, México. Ciencia y Mar, **10**: 39-50.

RADEMACHER, M. & O. MENGEDOHT, 2011. Krabben-Fibel. Die schönsten Krabben für das Aquaterrarium: 1-96. (Dähne Verlag, Ettlingen).

REYNOLDS, J. D., 2002. Growth and reproduction. In: D. M. HOLDICH (ed.), Biology of freshwater crayfish: 152-191. (Blackwell, Oxford).

REZNICK, D. A., H. BRYGA & J. A. ENDLER, 1990. Experimentally induced life-history evolution in a natural population. Nature, **346**: 357-359.

REZNICK, D. N., M. J. BRYANT, D. ROFF, C. K. GHALAMBOR & D. E. GHALAMBOR, 2004. Effect of extrinsic mortality on the evolution of senescence in guppies. Nature, **431**: 1095-1099.

RICKLEFS, R. E., 1998. Evolutionary theories of ageing: confirmation of a fundamental prediction, with implications for the genetic basis and evolution of life span. American Nat., **152**: 24-44.

ROCHA, S. S., R. M. SHIMIZU & S. L. S. BUENO, 2010. Reproductive biology in females of *Aegla strinatii* (Decapoda: Anomura: Aeglidae). Journ. Crust. Biol., **30**: 589-596.

RODRIGUES, W. & N. J. HEBLING, 1978. Estudos biológicos em *Aegla perobae* Hebling & Rodrigues, 1977 (Decapoda, Anomura). Rev. Brasileira Biol., **38**: 383-390.

RODRÍGUEZ, A. & J. A. CUESTA, 2011. Morphology of larval and first juvenile stages of the kangaroo shrimp *Dugastella valentina* (Crustacea, Decapoda, Caridea), a freshwater atyid with abbreviated development and parental care. Zootaxa, **2867**: 43-58.

SCALICI, M., A. BELLUSCIO & G. GIBERTINI, 2008a. Understanding population structure and dynamics in threatened crayfish. Journ. Zool., **275**: 160-171.

SCALICI, M., D. MACALE, F. SCHIAVONE, F. GHERARDI & G. GIBERTINI, 2008b. Effect of urban isolation on the dynamics of river crabs. Fundam. Appl. Limnol., **172**: 167-174.

SCHOLTZ, G., 1995. The attachment of the young in the New Zealand freshwater crayfish *Paranephrops zealandicus* (White, 1847) (Decapoda, Astacida, Parastacidae). New Zealand Nat. Sci., **22**: 81-89.

SCHOLTZ, G. & T. KAWAI, 2002. Aspects of embryonic and postembryonic development of the Japanese freshwater crayfish *Cambaroides japonicus* (Crustacea, Decapoda) including a hypothesis on the evolution of maternal care in the Astacida. Acta Zool., **83**: 203-212.

SHEEHY, M. R. J., 1992. Lipofuscin age-pigment accumulation in the brains of ageing field- and laboratory-reared crayfish *Cherax quadricarinatus* (von Martens) (Decapoda: Parastacidae). Journ. Exp. Mar. Biol. Ecol., **161**: 79-89.

SHIMIZU, S. J. & C. R. GOLDMAN, 1983. *Pacifastacus leniusculus* (Dana) production in the Sacramento River. Freshwater Crayfish, **5**: 210-228.

SHOKITA, S., M. TAKEDA, S. SITTILERT & T. POLPAKDEE, 1991. Abbreviated larval development of a fresh-water prawn, *Macrobrachium niphanae* Shokita and Takeda (Decapoda: Palaemonidae), from Thailand. Journ. Crust. Biol., **11**: 90-102.

SILVA-GONÇALVES, R. DA, G. BOND-BUCKUP & L. BUCKUP, 2009. Crescimento de *Aegla itacolomiensis* (Crustacea, Decapoda) em um arroio da Mata Atlântica no sul do Brasil. Iheringia Sér. Zool., **99**: 397-402.

SIMČIČ, T., S. LUKANČIČ & A. BRANCELJ, 2005. Comparative study of electron transport system activity and oxygen consumption of amphipods from caves and surface habitats. Freshwater Biol., **50**: 494-501.

SKET, B. & V. ZAKŠEK, 2009. European cave shrimp species (Decapoda: Caridea: Atyidae), redefined after a phylogenetic study; redefinition of some taxa, a new genus and four new *Troglocaris* species. Zool. Journ. Linnean Soc., **155**: 786-818.

SKURDAL, J., D. O. HESSEN, E. GARNÅS & L. A. VØLLESTAD, 2011. Fluctuating fecundity parameters and reproductive investment in crayfish: driven by climate or chaos? Freshwater Biol., **56**: 335-341.

SKURDAL, J. & T. TAUGBØL, 2002. *Astacus*. In: D. M. HOLDICH (ed.), Biology of freshwater crayfish: 467-510. (Blackwell, Oxford).

STEARNS, S. C., 1992. The evolution of life histories: 1-249. (Oxford University Press, Oxford).

——, 2000. Life history evolution: successes, limitations, and prospects. Naturwiss., **87**: 476-486.

STEARNS, S. C., M. ACKERMANN, M. DOEBELI & M. KAISER, 2000. Experimental evolution of aging, growth, and reproduction in fruitflies. PNAS, **97**: 3309-3313.

STRAHL, J., E. PHILIPP, T. BREY, K. BROEG & D. ABELE, 2007. Physiological aging in the Icelandic population of the ocean quahog *Arctica islandica*. Aquat. Biol., **1**: 77-83.

STREEVER, W. J., 1996. Energy economy hypothesis and the troglobitic crayfish *Procambarus erythrops* in Sim's Sink Cave, Florida. American Midl. Nat., **135**: 357-366.

SWIECH-AYOUB, B. DE P. & S. MASUNARI, 2001. Biologia reprodutiva de *Aegla castro* Schmitt (Crustacea, Anomura, Aeglidae) no Buraco do Padre, Ponta Grossa, Paraná, Brasil. Rev. Brasileira Zool., **18**: 1019-1030.

TADDEI, F. G. & D. R. HERRERA, 2010. Crescimento do caranguejo *Dilocarcinus pagei* Stimpson, 1861 (Crustacea, Brachyura, Trichodactylidae) na Represa Barra Mansa, Mendonça, SP. Bol. Inst. Pesca São Paulo, **36**: 99-110.

TAYLOR, C. A., 2003. Conservation assessment for a crayfish (*Orconectes placidus*): 1-11. (USDA Forest Service, Eastern Region, Milwaukee).

THIEL, M., 2000. Extended parental care behavior in crustaceans — a comparative overview. In: J. C. VON VAUPEL KLEIN & F. R. SCHRAM (eds.), The biodiversity crisis and Crustacea. Crustacean Issues, **12**: 211-226. (Balkema, Rotterdam).

TRONTELJ, P., 2007. The age of subterranean crayfish species. A comment on Buhay & Crandall (2005): Subterranean phylogeography of freshwater crayfishes shows extensive gene flow and surprisingly large population sizes. Mol. Ecol., **16**: 2841-2843.

U.S. FISH AND WILDLIFE SERVICE, 1998. Recovery plan for the California freshwater shrimp (*Syncaris pacifica* Holmes 1895): 1-94. (U.S. Fish and Wildlife Service, Portland).

— —, 2010. Kentucky cave shrimp (*Palaemonias ganteri*). 5-year review: summary and evaluation: 1-16. (U.S. Fish and Wildlife Service, Frankfort).

VEILLEUX, É. & Y. DE LAFONTAINE, 2007. Biological synopsis of the Chinese mitten crab (*Eriocheir sinensis*). Canadian Manuscr. Rep. Fish. Aquat. Sci., **2812**: 1-45.

VENARSKY, M. P., A. D. HURYN & J. P. BENSTEAD, 2012. Re-examining extreme longevity of the cave crayfish *Orconectes australis* using new mark-recapture data: a lesson on the limitations of iterative size-at-age models. Freshwater Biol., **57**: 1471-1481.

VOGT, G., 1994. Life-cycle and functional cytology of the hepatopancreatic cells of *Astacus astacus* (Crustacea, Decapoda). Zoomorphol., **114**: 83-101.

— —, 1999. Diseases of European freshwater crayfish, with particular emphasis on interspecific transmission of pathogens. In: F. GHERARDI & D. M. HOLDICH (eds.), Crayfish in Europe as aliens species. Crustacean Issues, **11**: 87-103. (Balkema, Rotterdam).

— —, 2002. Functional anatomy. In: D. M. HOLDICH (ed.), Biology of freshwater crayfish: 53-151. (Blackwell, Oxford).

— —, 2008a. Investigation of hatching and early post-embryonic life of freshwater crayfish by in vitro culture, behavioral analysis, and light and electron microscopy. Journ. Morphol., **269**: 790-811.

— —, 2008b. The marbled crayfish: a new model organism for research on development, epigenetics and evolutionary biology. Journ. Zool., **276**: 1-13.

— —, 2008c. How to minimize formation and growth of tumours: potential benefits of decapod crustaceans for cancer research. Int. Journ. Cancer, **123**: 2727-2734.

— —, 2009. Research on aging and longevity in the parthenogenetic marbled crayfish, with special emphasis on stochastic developmental variation, allocation of metabolic resources, regeneration, and social stress. In: J. V. BENTELY & M. A. KELLER (eds.), Handbook on longevity: genetics, diet and disease: 353-372. (Nova Science Publishers, Hauppauge).

— —, 2010. Suitability of the clonal marbled crayfish for biogerontological research: a review and perspective, with remarks on some further crustaceans. Biogerontol., **11**: 643-669.

— —, 2011. Marmorkrebs: natural crayfish clone as emerging model for various biological disciplines. Journ. Biosci., **36**: 377-382.

— —, 2012a. Ageing and longevity in the Decapoda (Crustacea): a review. Zool. Anz., **251**: 1-25.

— —, 2012b. Hidden treasures in stem cells of indeterminately growing bilaterian invertebrates. Stem Cell Rev. Rep., **8**: 305-317.

VOGT, G., M. HUBER, M. THIEMANN, G. VAN DEN BOOGAART, O. J. SCHMITZ & C. D. SCHUBART, 2008. Production of different phenotypes from the same genotype in the same environment by developmental variation. Journ. Exp. Biol., **211**: 510-523.

VOGT, G. & L. TOLLEY, 2004. Brood care in freshwater crayfish and relationship with the offspring's sensory deficiencies. Journ. Morphol., **262**: 566-582.

WALLS, J. G., 2009. Crawfishes of Louisiana: 1-241. (Louisiana State University Press, Baton Rouge).

WALSH, C. J., 1993. Larval development of *Paratya australiensis* Kemp, 1917 (Decapoda: Caridea: Atyidae), reared in the laboratory, with comparisons of fecundity and egg and larval size between estuarine and riverine environments. Journ. Crust. Biol., **13**: 456-480.

WEHRTMANN, I. S., C. MAGALHÃES, P. HERNÁEZ & F. L. MANTELATTO, 2010. Offspring production in three freshwater crab species (Brachyura: Pseudothelphusidae) from the Amazon region and Central America. Zoologia, **27**: 965-972.

WEINGARTNER, D., 1977. Production and trophic ecology of two crayfish species cohabiting an Indiana cave: 1-323. (Ph.D. thesis, Michigan State University, East Lansing).

WHITMORE, N. & A. D. HURYN, 1999. Life history and production of *Paranephrops zealandicus* in a forest stream, with comments about the sustainable harvest of a freshwater crayfish. Freshwater Biol., **42**: 467-478.

WILLCOX, D. C., B. J. WILLCOX, W.-C. HSUEH & M. SUZUKI, 2006. Genetic determinants of exceptional human longevity: insights from the Okinawa Centenarian Study. AGE, **28**: 313-332.

WINTERBOURN, M. J. & K. MASON, 1983. Freshwater life: streams, ponds, swamps, lakes and rivers. Mobil New Zealand Nature Series: 1-76. (Reed, Wellington).

WU, H., J. XUE & N. CUMBERLIDGE, 2010. An extra embryonic phase in the true freshwater crab *Sinopotamon yangtsekiense* Bott, 1967 (Decapoda, Potamidae). Chinese Journ. Oceanol. Limnol., **28**: 725-730.

YAM, R. S. W. & D. DUDGEON, 2005. Inter- and intraspecific differences in the life history and growth of *Caridina* spp. (Decapoda: Atyidae) in Hong Kong streams. Freshwater Biol., **50**: 2114-2128.

YEO, D. C. J., P. K. L. NG, N. CUMBERLIDGE, C. MAGALHÃES, S. R. DANIELS & M. R. CAMPOS, 2008. Global diversity of crabs (Crustacea: Decapoda: Brachyura) in freshwater. Hydrobiol., **595**: 275-286.

ZAKŠEK, V., B. SKET & P. TRONTELJ, 2007. Phylogeny of the cave shrimp *Troglocaris*: evidence of a young connection between Balkans and Caucasus. Mol. Phylogenet. Evol., **42**: 223-235.

ZERA, A. J. & L. G. HARSHMAN, 2001. The physiology of life history trade-offs in animals. Ann. Rev. Ecol. Syst., **32**: 95-126.

ZIMMERMAN, J. K. M., 2009. Conditions for sustainable crayfish populations in northern Sweden: 1-30. (Rapport, Mid-Sweden University, Sundsvall).

First received 22 June 2011.
Final version accepted 29 July 2013.

FRESHWATER DECAPOD CONSERVATION: RECENT PROGRESS AND FUTURE CHALLENGES

BY

NEIL CUMBERLIDGE[1])
Department of Biology, Northern Michigan University, Marquette, MI 49855, U.S.A.

ABSTRACT

Freshwater ecosystems around the world support a highly diverse fauna that includes significant numbers of decapod crustaceans (freshwater crabs, crayfish, and shrimps) many of which are of economic importance. However, freshwater habitats and animals that depend on them are now under increasing threat. Recent International Union for the Conservation of Nature (IUCN) Red List assessments of the world's freshwater crabs and crayfish revealed significant numbers of species threatened with extinction. The long-term survival of many freshwater species is becoming more precarious as wetland habitats are increasingly degraded and threats to freshwater biodiversity intensify. The majority of threatened decapods are restricted-range endemics living in habitats threatened by unprecedented human demands for water and food resulting in alteration of drainage patterns, pollution, and over-harvesting. Current strategies for slowing the decline of the world's threatened freshwater decapod species include the sustainable management of their freshwater habitats and the collection of more baseline data on their diversity, population and distribution patterns, and conservation status.

INTRODUCTION

It is estimated that more than 126 000 species worldwide rely on freshwater ecosystems for their survival (Balian et al., 2008, 2010), but the rapid deterioration of freshwater habitats driven by human population growth, agriculture, and deforestation is having a negative impact on aquatic biodiversity (e.g., see Dudgeon et al., 2006; Strayer, 2006; Wong et al., 2007). Moreover, alteration of water availability and shifts in water temperatures associated with global climate change are likely to increase the number of threatened species (Carpenter et al., 1992; Thomas et al., 2004; Foden et al., 2008). In addition, the

[1]) e-mail: ncumberl@nmu.edu

rate of species loss from inland waters is believed to be greater than that for terrestrial or marine ecosystems, with freshwater species (both vertebrates and invertebrates) among the most threatened of all (Stiassny, 1999; Gibbons et al., 2000; Revenga et al., 2005; Dudgeon et al., 2006; Strayer, 2006; Strayer & Dudgeon, 2010). Freshwater ecosystems not only provide habitats for a rich diversity of species, they also filter and store clean drinking water, control floods and erosion, and provide food and building materials. The goods and services provided by the world's tropical rivers and wetlands have been valued at $5.58 and $70 billion per year, respectively (Schuyt & Brander, 2004; Neiland & Béné, 2008).

Reliable global species inventories and distributional information for nearly all freshwater groups are now available from the Freshwater Animal Diversity Assessment (Balian et al., 2008). Fresh waters support more than 126 000 animal species, representing almost 7% of all life on Earth. Invertebrates dominate freshwater animal communities (107 295 species, 84.5%), and these far outnumber vertebrates (18 235 species, 14.5%) (Balian et al., 2008; Hoffman et al., 2010). The majority of freshwater invertebrates are insects (60.4%), followed by crustaceans (10%), arachnids (5%), and molluscs (4%). In general, species diversity is highest in warm tropical waters, and lowest in cooler high latitude and recently-glaciated systems (Collen et al., 2008b; Strayer & Dudgeon, 2010).

Among the reasons that fresh waters are so rich in biodiversity is the isolating nature of the freshwater habitats themselves. Relatively few freshwater species have a wide geographic range, and most tend to be local endemics with a limited ability to disperse through freshwater systems, often being restricted to a single lake or drainage basin (Strayer, 2006; Balian et al., 2008; Strayer & Dudgeon, 2010). Isolation and adaptation to specific local conditions drive speciation and have led to the rich freshwater biodiversity seen today, but this has also contributed to a general vulnerability to increasing threats from human activities. While the human-caused extinction rate for the better-studied marine and terrestrial vertebrates now exceeds the background extinction rate (Ehrlich & Pringle, 2008; Rockstroom et al., 2009; Butchart et al., 2010; Mace et al., 2010) only a few species of freshwater invertebrates are known to be either Extinct (EX) or Extinct In The Wild (EW). It is unlikely that this low number reflects the true rate of extinction, and it is more probable that it is an artifact resulting from the lack of data and documentation.

Among the objectives of the Convention on Biological Diversity (CBD) was a commitment to achieve by 2010 a significant reduction of the rate of biodiversity loss at the global, regional, and national levels, which would in turn contribute to poverty alleviation and benefit all life on Earth (UNEP, 2002). The CBD inspired the United Nations General Assembly in 2006 to designate 2010 as the International Year of Biodiversity, a year long celebration of biological diversity and its value for life on Earth. However, by the end of 2010, the CBD objectives had not been met and the rate of biodiversity loss had merely slowed and had not been not halted or reversed. As a result there remain huge gaps in basic biodiversity information for many groups, especially the invertebrates (Hoffman et al., 2010; Stuart et al., 2010).

DISTRIBUTION OF BIODIVERSITY IN FRESH WATER

Balian et al. (2008) demonstrated that the species-rich freshwater animal fauna is unevenly distributed throughout the world's eight major zoogeographical regions (table I). The majority of freshwater species are found in the Palaearctic (24%), Neotropical (21%), and Oriental regions (18%), fewer in the Nearctic (14%), Afrotropical (13%), and Australasian realms (9%), and fewer still in the Pacific (0.9%) and Antarctic (0.1%) regions. This pattern changes when the focus is shifted to consider the global distribution of the 2832 species of freshwater decapods (table II): over three quarters of species occur in the Oriental (41%), Neotropical (20%), and Nearctic (15%) regions; there are less in Australasia (12%), the Afrotropical (8%) and Palaearctic (6%) regions, fewer in the Pacific (2%), and none in Antarctica (Balian et al., 2008). While the numbers of species in the biodiversity-rich Oriental, Neotropical, and Afrotropical regions will undoubtedly increase when exploration in fresh water is intensified, it is unlikely that the proportions of species found in each

TABLE I

Percentage of all freshwater animals, and all freshwater decapods found in the eight zoogeographical regions recognized by Balian et al. (2008)

Region	% of global freshwater fauna	% of global freshwater decapods
Palaearctic	24%	6%
Neotropics	21%	20%
Oriental	18%	41%
Nearctic	14%	15%
Afrotropics	13%	8%
Australasia	9%	12%
Pacific	0.9%	2%
Antarctica	0.1%	0%

TABLE II

The numbers of species of freshwater decapod crustaceans by zoogeographic region, with the world species totals for each taxon, and totals for the freshwater decapods as a group as understood in 2008. Data sources: Aeglidae (Bond-Buckup, 2008; Perez-Losada et al., 2009), Astacidea (Crandall & Buhay, 2008), Brachyura (Yeo et al., 2008) and Caridea (De Grave et al., 2008). PA, Palaearctic; NA, Nearctic; AT, Afrotropical; NT, Neotropical; OL, Oriental; AU, Australasian; PAC, Pacific; and ANT, Antarctic

Decapod group	PA	NA	AT	NT	OL	AU	PAC	ANT	World
Aeglidae	0	0	0	68	0	0	0	0	68
Astacidea	38	382	9	64	0	151	0	0	638
Brachyura	97	19	149	340	818	89	24	0	1476
Caridea	47	17	92	109	349	87	25	0	655
Total	182	481	250	518	1167	327	49	0	2837

of these regions will change much. The almost three thousand species of freshwater decapods comprise only a small proportion of the 11 900 species of crustaceans found in fresh waters (Balian et al., 2008). The 1476 species of brachyurans constitute the largest group of freshwater decapods, comprising the primary freshwater crabs (Potamoidea sensu Cumberlidge & Ng, 2009) plus other crabs that live in fresh water. The next most species rich groups are the freshwater shrimps (Caridea, 655 species), the crayfish (Astacoidea, Parastacoidea, 638 species), and the anomuran freshwater crabs (Aeglidae, 68 species) (table II).

Other trends revealed by the study of Balian et al. (2008) highlighted imbalances in biodiversity knowledge between zoogeographical regions, types of ecosystems, and taxonomic groups. In general, there is more information available for temperate than tropical species, more for terrestrial than aquatic species, and more for vertebrates than invertebrates (Collen et al., 2008a). Tropical freshwater invertebrates are therefore among the least studied groups, although some initial progress has been made in some groups such as the molluscs (Lydeard et al., 2004; Strong et al., 2008), dragonflies (Kalkman et al., 2008; Clausnitzer et al., 2009), brachyuran freshwater crabs (Yeo et al., 2008; Cumberlidge et al., 2009), crayfish (Taylor et al., 2007; Crandall & Buhay, 2008), and anomuran freshwater crabs (Pérez-Losada et al., 2002, 2009).

IUCN RED LIST ASSESSMENTS

Although there are still large gaps in taxon coverage, the International Union for the Conservation of Nature (IUCN) Red List of Threatened Species (www.redlist.org) is the internationally recognized authority for making conservation assessments of animals, plants, and fungi, with the website receiving more than a million visits per month. Red List conservation assessments use a formalized and standardized methodology that follows uniform protocols (Vié et al., 2008; IUCN, 2010). These assessments are carried out under the auspices of the IUCN Species Survival Commission (SSC), which is one of the six Commissions of the IUCN. The SSC advises the IUCN on species conservation and mobilizes action for species threatened with extinction. Red Listing began in the 1960s when the focus was solely on threatened species, with assessments depending heavily on expert opinion. The SSC developed networks of experts who used formal protocols for Red Listing and concentrated on conservation assessments of specific taxa. There are now more than 100 SSC specialist groups that bring together more than 7500 taxonomic specialists, researchers, government officials, wildlife veterinarians, zoological gardens and botanical institute employees, marine biologists, and protected area managers. Specialist groups focus on plant, bird, mammal, fish, amphibian, reptile, and invertebrate taxa, with decapods represented by the Freshwater Crab and Crayfish Specialist Group, formed only in 2009.

Birds were the first group to be assessed globally using the Red List protocols (in 1988) and there have been a number of follow-up assessments since then in 1994, 2000, 2004, and 2008 (Butchart et al., 2004; BirdLife International, 2008). The mammals were the next group to be assessed globally (in 1996 and 2008) (Schipper et al., 2008), followed by the amphibians in 2004 (Stuart et al., 2004). The Red List Index, based on the IUCN Red List of Threatened Species, is an indicator of the changing state of global biodiversity that monitors trends in extinction risk over time by conducting conservation assessments at regular intervals (Butchart et al., 2005). Most of the initial Red List studies were carried out on charismatic vertebrate groups and at first the invertebrates were largely overlooked (Collen et al., 2008a). Global assessments for entire invertebrate groups are now becoming available with the reef-building corals (Carpenter et al., 2008), the freshwater crabs (Cumberlidge et al., 2009), and the crayfish (IUCN, 2010). Global conservation assessments for other invertebrate groups such as the cephalopods and saltwater lobsters are currently underway (Collen et al., 2008a). A similar study of freshwater shrimp was completed in 2013. However, global conservation assessments for freshwater groups that have an overwhelming number of species (e.g., fish, dragonflies, and molluscs) are logistically problematic. In response to this the Red List Index (sampled approach) (SRLI) was developed whereby a representative sample (about 1500 species) of the global fauna is assessed, rather than every single species (Baillie et al., 2008). SRLI methods have now been used to assess groups of vertebrates, invertebrates, plants, and fungi. Whichever method is used, global or SRLI conservation assessments will continue to require attention because they should ideally be repeated at regular intervals to capture a more broadly representative picture of trends in extinction risk over time.

Despite these efforts, only a small proportion of life on earth has been subjected to Red List protocols, and the task ahead is daunting if we are to reach the goal of a conservation assessment for each of the planet's 1.9 million described species. Realistically, this will probably not be achieved this century, especially with recent estimates that there may be 8.7 million species of eukaryotes on the planet (Mora et al., 2011). By the end of 2010 we only had Red List conservation assessments for 47 666 species of vertebrates and invertebrates, which represents only a tiny fraction (2.5%) of the 1.9 million species described (Stuart et al., 2010). Furthermore, most of the species that have been assessed so far are vertebrates (27 882 out of the 64 788 known species), while only a small fraction (7615 species or 0.4%) of the 1.3 million species of invertebrates have been assessed. Stuart et al. (2010) estimated that conservation assessments for 45 344 species of invertebrates would be needed for these to function as a representative sample of the group that would be useful for monitoring trends in extinction risk over time. However, at the end of 2010 we are still far behind this target because we still need assessments for another 37 729 species of invertebrates. More than half of the 7615 invertebrate species that have been assessed so far belong to freshwater groups such as dragonflies (Kalkman et al., 2008; Clausnitzer et al., 2009), molluscs (Strong et al., 2008), brachyuran freshwater crabs (Yeo et al., 2008; Cumberlidge et al., 2009), and crayfish (Crandall & Buhay, 2008; IUCN, 2010). It is estimated that it would cost about US$20 million to fund the Red Listing effort to expand coverage to reach the targeted representative sample for the invertebrates (Stuart et al., 2010).

PRODUCTS OF RED LIST ASSESSMENTS

The products of Red List assessments include species lists and inventories, worldwide species distribution maps, endemic species distribution maps, as well as information on population trends, habitat preferences, threats, and literature references (Cumberlidge, 2005, 2009; Cumberlidge & Daniels, 2007; Cumberlidge et al., 2010; Darwall et al., 2011). Red List assessments also help pinpoint areas of species richness and identify the locations of threatened species thereby spotlighting countries that need increased biodiversity research and conservation interventions. Combined Red List data from different groups can reveal areas of overlapping concentrations of species richness and threatened species (Collen et al., 2008a). Red List assessments repeated at intervals can detect trends in threat levels over time that may be linked to the effects of global climate change on their habitat. This biodiversity information also drives further research including new exploration, species discovery, updating of biotic inventories, studying museum collections, assessing species abundance, measuring trends in population levels, and carrying out DNA analyses. Red Lists and biodiversity information can be used by conservation biologists for freshwater ecosystem management and protection, for developing conservation measures for threatened species, and for making species recovery plans. Red List assessments have also created a demand for taxonomic and ecological research that has in turn exposed the long-term decline in the number of invertebrate taxonomists, and a general decline in the prestige and support for taxonomy (Wheeler, 2008).

FRESHWATER DECAPOD CONSERVATION ASSESSMENTS

Freshwater crabs. — The most recent data available indicate that the brachyuran freshwater crabs comprise over 1360 species worldwide (Peter K. L. Ng, pers. comm.). Most of these are the primary freshwater crabs (in five families, Pseudothelphusidae, Trichodactylidae, Potamonautidae, Potamidae, and Gecarcinucidae). There are also a smaller number of brachyurans (about 200 species) (Peter K. L. Ng, pers. comm.) that live in freshwater habitats that belong to five families that include marine, as well as freshwater, species (Yeo et al., 2008). Using the data provided by Cumberlidge et al. (2009) most species of primary freshwater crabs are found in the Palaearctic/Oriental/Australasian realms (930 species), followed by the Neotropics (298 species), and the Afrotropics (132 species). All Neotropical, Palaearctic, Oriental, Afrotropical and Australasian freshwater crabs are endemic to the zoogeographical region where they occur. The rate of endemism at the country level is also high: 84% of the Palaearctic, Oriental, and Australasian species, 76% of the Neotropical species, and 54% of the Afrotropical species are country endemics (Cumberlidge et al., 2009).

The global true freshwater crab conservation assessment by Cumberlidge et al. (2009) dealt with 1280 species in 220 genera and five families (table IIIA) that occur in 122 countries and six zoogeographic regions (Ng et al., 2008;

TABLE III

Summary of the IUCN Red List assessments for (A) brachyuran freshwater crabs (global study, based on Cumberlidge et al., 2009), (B) crayfish (global study, based on the IUCN Red List), and (C) anomuran freshwater crabs (based on Pérez-Losada et al., 2009). The table also shows the number of species per family, and the total species in each group. Red List criteria: LC = Least Concern; NT = Near Threatened; VU = Vulnerable; EN = Endangered; CR = Critically Endangered; EW = Extinct In The Wild; EX = Extinct; DD = Data Deficient

Family	Total species	LC	NT	VU	EN	CR	EW	DD
A. Brachyuran freshwater crabs (2009 data)								
Gecarcinucidae	344	107	7	33	26	27	0	144
Potamidae	505	139	8	38	12	3	0	305
Potamonautidae	133	72	2	16	10	2	0	31
Pseudothelphusidae	251	72	1	33	3	2	0	140
Trichodactylidae	47	35	0	3	1	0	0	8
Total	1280	425	18	123	52	34	0	628
B. Crayfish								
Astacidae	10	3	0	1	1	1	1	3
Cambaridae	411	219	26	19	34	19	3	91
Parastacidae	148	49	8	12	27	22	0	30
Total	569	271	34	32	62	42	4	124
C. Anomuran freshwater crabs								
Aeglidae	68	41	2	11	1	4	2	7
Total species	3766	1433	106	321	229	156	10	1511

Cumberlidge & Ng, 2009; Cumberlidge et al., 2009: Appendix A). That study revealed unexpectedly high numbers of threatened species of freshwater crabs in the Neotropics (Pseudothelphusidae and Trichodactylidae), the Afrotropics (Potamonautidae), and the Palaearctic, Oriental, and Australasian realms (Potamidae, and Gecarcinucidae). Only one third of the described species (33%, 425 species) are not threatened, i.e. Least Concern (LC), whereas 209 species (16.3%) have an elevated risk of extinction in one of the three threat categories: Critically Endangered (CR, 32 species), Endangered (EN, 52 species), and Vulnerable (VU, 123 species). Although none could be confirmed to be Extinct (EX) or Extinct In The Wild (EW), almost half of all described species (49%, 628 species) are Data Deficient (DD) and too poorly known to assess. Out of the 122 countries where freshwater crabs occur, 43 have threatened species in need of protection. The majority of threatened species are restricted-range endemics living in habitats subjected to deforestation, alteration of drainage patterns, and pollution, and many live either in highland areas or on islands, and are either burrow-living terrestrial or semi-terrestrial air-breathing species, or species that live in specialized habitats such as caves, karsts, and phytotelms (Cumberlidge et al., 2009).

Crayfish. — Crandall & Buhay (2008) listed 638 species of crayfish worldwide (table II), whereas the global crayfish conservation assessment (table IIIB) included 569 species in three families (Cambaridae, Astacidae, and Parastacidae) found in 49 countries and five zoogeographic regions (IUCN, 2010). The vast majority of the 411 species of the family Cambaridae are found the Nearctic region (USA, Canada, and northern Mexico) east of the Rocky Mountains, except for the four species of *Cambaroides* found in eastern Russia, China, Korea and Japan. The crayfish family Astacidae has only 10 species, with five Palaearctic species (*Astacus* and *Autropotamobius*) from western Europe, and five Nearctic species (*Pacifastacus*) from the USA and Canada west of the Rocky Mountains. The family Parastacidae is distributed over three southern continents from South America (10 species), Madagascar (9 species), and Australasia (151 species), and none are found in continental Africa (Taylor et al., 2007; Crandall & Buhay, 2008; table II).

The following is based on the IUCN Red List of Threatened Species that provides conservation assessments for 569 species of crayfish worldwide (table IIIB). The family Parasticidae from Australia, New Guinea and New Zealand includes high numbers of threatened species (61 species, 41%), but none of the species from South America and Madagascar are threatened. Three out of the ten species of Palaearctic Astacidae are threatened (30%), but none of the five species of Nearctic astacid crayfish are threatened. Almost one fifth (72/411 species, 18%) of the species-rich Nearctic cambarid crayfish are threatened, while the four Palaearctic cambarid species from East Asia are all DD. Overall, only one third of the assessed species of crayfish (61%, 271/445 species) are LC, whereas 136/445 species (31%) have an elevated risk of extinction and were assessed in one of the three threat categories: CR (42 species), EN (62 species), and VU (32 species). Although no species of crayfish is Extinct (EX), four species are Extinct In The Wild (EW), and about a fifth of all described species (22%, 124/569 species) are too poorly known to assess (DD). Some 39 out of the 47 countries where crayfish occur have at least one threatened species of crayfish in need of protection.

Aeglidae. — The Aeglidae are a group of 85 species of anomuran decapods (Keith Crandall, pers. comm.) that are found in five Neotropical countries in temperate South America (Chile, Brazil, Argentina, Uruguay, and Bolivia) (table II). The aeglids all belong to a single genus, *Aegla*, and are known in the vernacular as ‘freshwater crabs’ (Bond-Buckup et al., 2008). These decapods are referred to in this work as 'anomuran freshwater crabs' to distinguish them from the brachyuran freshwater crabs. Conservation assessments have been done for 68 species of aeglids (Pérez-Losada et al., 2009), but none are presently on the IUCN Red List. Despite the relatively small fraction of the fauna assessed, half of the species have an elevated risk of extinction (16/68 species, 24%), with four species assessed as CR, one as EN, and eleven as VU (table IIIC). Almost two thirds of the assessed species (41/68 species, 61%) are not threatened (LC), and although none were found to be Extinct, two species are Extinct In The Wild (EW), and seven species are too poorly known to assess (DD) (Bond-Buckup et al., 2008; Pérez-Losada et al., 2009).

Caridea. — The 655 species of freshwater caridean shrimps belong to eight families/subfamilies, with the vast majority belonging to the Atyidae (359 species, most of which are in the genus *Caridina*) and Palaemonidae (276 species, most of which are in the genus *Macrobrachium*) (De Grave et al., 2008). Freshwater shrimp are found in seven zoogeographical regions, but are absent from Antarctica, with the majority in the Oriental region, and the fewest in the Nearctic and Palearctic realms (table II). A global conservation assessment of 760 species and 38 subspecies of freshwater shrimps was completed in 2013 and is available on the IUCN Red List (www.iucnredlist.org).

COMPARISONS OF THREATENED DECAPOD GROUPS

About one third of the assessed species of brachyuran freshwater crabs and crayfish (31% and 32% respectively), 27% of freshwater carideans, and 24% of anomuran freshwater crabs are threatened with extinction, in one of the three threatened categories (CR, EN, VU) (table IIIA-C). The one-third proportion of threatened brachyuran freshwater crabs and crayfish is similar to that found for the 845 species of reef-building corals (33%, Kent et al., 2008), but is a much higher than for the 1498 species of dragonflies (13%, Clausnitzer et al., 2009). To put this in context of all of the invertebrate groups that have been assessed globally (freshwater crabs, freshwater shrimps, crayfish, and reef-building corals) one-third are threatened with extinction, which exceeds that for all other groups that have been assessed globally except for amphibians (Stuart et al., 2004). With conservation assessments of all freshwater crabs, freshwater shrimps, and crayfish species now on-line (www.redlist.org), there is a need to complete similar global assessments for the freshwater aeglids and carideans. The number of threatened species of decapods may be even greater when Red List assessments are available for the estimated thousands of species of freshwater invertebrates (including decapod crustaceans) that are still awaiting description (Balian et al., 2008).

The above figures for the numbers of threatened species of decapods do not tell the whole story because of the high numbers of poorly known (DD) species in this, and in almost every vertebrate and invertebrate group that has been studied (except for the birds). We simply do not know the extinction threat levels of DD species because only a few specimens are known and the necessary biological information is missing. However, we can estimate extinction threats by making future projections under three different assumptions of how DD species would convert when enough data have been collected to assess them (Collen et al., 2008b). These three assumptions are: if all remain DD, if all prove to be LC, or if all prove to be threatened (VU, EN, or CR). Using the 2009 study of the brachyuran freshwater crabs as an example (Collen et al., 2008b), if all DD species remain as they are then 209 out of 651 assessed species (32%) would be threatened, if all DD species are upgraded to LC, then 209 out of 1280 species (16%) would be threatened, and if all DD species are threatened then an alarming 938 out of 1280 species (65%) would be imperiled (Hoffman et al., 2010). In the case of the crayfish, if all DD species remain as they are then 140 out of 445 assessed species (32%) would be threatened, if all DD species are upgraded to LC then 140 out of 569 species (25%) would be threatened, and if all DD species are threatened then almost half (264 out of 569 species, 46%) would be threatened with extinction. Within the vertebrates, amphibians have the highest numbers of DD species (1533 out of 6260 species), whereby 56% of amphibians are potentially highly threatened (Stuart et al., 2004; Hoffman et al., 2010). In contrast, the birds are very well known and only a single species is data deficient (BirdLife International, 2008; Collen et al., 2008a).

THREATS TO FRESHWATER DECAPODS

The main threats to freshwater decapods include habitat destruction, flow modification (from dams and alteration of drainage patterns), sedimentation (from soil erosion), pollution (from mining, pesticides, and fertilizers that cause eutrophication and toxic algal blooms), over-harvesting of species, and competition from introduced species (Dudgeon et al., 2006; Cumberlidge et al., 2009). Rising temperatures and shifts in runoff and precipitation patterns associated with global climate change further compound these threats and make it difficult to predict extinction risks. The displacement of native species by introduced alien species (Strayer, 2010) is illustrated by the dominance of US crayfish (*Procambarus clarkii*) in East African rivers over native species of freshwater crabs (*Potamonautes loveni*), where introduced crayfish not only caused crab populations to crash, they also caused a decline in the population of clawless otters that fed on the crabs (Ogada, 2006; Foster & Harper, 2007). Even LC species of decapod crustaceans that are abundant enough to support local commercial fisheries, such as species of freshwater prawns (e.g., *Macrobrachium* sp.), crayfish (e.g., *Procambarus clarkii* in the US), and freshwater crabs (e.g., *Platythelphusa armata* and *Potamonautes lirrangensis* in Africa, and *Sayamia germaini* and *Somanniathelphusa pax* in Indochina) could become threatened should populations decline from either over-harvesting or habitat alteration (Allan et al., 2005; Cumberlidge, 2011). Similarly threatened are those species of colorful and attractive brachyuran freshwater crabs that are collected for the aquarium trade such as *P. armata* and *P. lirrangensis* from two African Rift Valley lakes, and *Heterothelphusa fatum* and *Demanietta khirikhan* from Thailand.

FUTURE PROSPECTS

In the past there has generally been a mismatch between areas that are biodiversity rich (mostly tropical countries), and the location of specialists, museums, and libraries needed to study biodiversity (mostly located in developed countries). This generalization is no longer true for the freshwater crabs because a great deal of taxonomic expertise and resources are now also found in North, Central, and South America, as well as in Africa, Europe, Asia, and Australia. Nevertheless, taxonomists, wherever they are based, still need to refer to type specimens and other material drawn from the more than three billion specimens held in museums throughout the world. One solution to this problem is to increase the on-line availability of the rich specimen-level datasets of museum collections around the world. It is encouraging that on-line databases of hundreds of thousands of museum specimens are indeed becoming available on museum websites, as are electronic copies of historically important and current taxonomic publications (e.g., Assembling the Tree of Life, ATOL, Decapoda, http://decapoda.nhm.org; Biodiversity Heritage Library, http://www.biodiversitylibrary.org/). Equally important are electronically accessible digital photographs of type specimens (e-types), but these are now available for just 2-3% of museum types (http://insects.oeb.harvard.edu/etypes/index.htm). Electronic biodiversity information of this nature is the foundation for the field of biodiversity informatics, and is now freely available on a host of web-based biodiversity sites (Encyclopedia of Life, Global Biodiversity Information Facility, IUCN Red List, FishBase 2000, Discover Life, Zoobank, Animal Diversity Web, Species 2000, ArKive, Catalogue of Life). All of this biodiversity information can be captured either in spreadsheets or specimen-level relational databases (such as Mantis, the Manager of Taxonomic Information and Specimens (http://insects.oeb.harvard.edu/mantis/)) and is increasingly used to make meta-analyses of a wide range of biodiversity-related topics (Wheeler, 2008).

ACKNOWLEDGEMENTS

I am extremely grateful to both Michael Türkay (Senckenberg Museum, Frankfurt, SMF) and Peter K. L. Ng (National University of Singapore), who led the organization of the Senckenberg Freshwater Decapod Conference, held in Frankfurt, Germany in December 2010. Thanks are also extended to the numerous colleagues at the SMF who all made invaluable contributions to the success of the meeting. The Peter White Scholar Award (NMU) is thanked for its support.

REFERENCES

ALLAN, J. D., R. ABELL, Z. HOGAN, C. REVENGA, B. W. TAYLOR, R. WELCOMME & K. WINEMILLER, 2005. Overfishing of inland waters. BioScience, **55**: 1041-1051.

BAILLIE, J. E. M., B. COLLEN, R. AMIN, H. R. AKÇAKAYA, S. H. M. BUTCHART, N. BRUMMITT, T. R. MEAGHER, M. RAM, C. HILTON-TAYLOR & G. M. MACE, 2008. Towards monitoring global biodiversity. Conservation Letters, **1**: 18-26.

BALIAN, E., I. HARRISON, S. BUTCHART, P. CHAMBERS, J. CORDEIRO, N. CUMBERLIDGE, F. DE MOOR, C. GASCON, H. JAMES, V. KALKMAN, P. P. VAN DIJK & D. C. J. YEO, 2010. Chapter 2. Diversity of species in freshwater systems. In: R. A. MITTERMEIER, T. A. FARRELL, I. J. HARRISON, A. J. UPGREN & T. BROOKS (eds.), Freshwater — the essence of life: 50-89. (Conservation International and Cemex Conservation Book Series, Earth in Focus Editions, China).

BALIAN, E. V., H. SEGERS, C. LÉVÈQUE & K. MARTENS, 2008. The freshwater animal diversity assessment: an overview of the results. Hydrobiologia, **595**: 627-637.

BIRDLIFE INTERNATIONAL, 2008. State of the world's birds: indicators for our changing world. (BirdLife International, Cambridge, UK).

BOND-BUCKUP, G., C. G. JARA, M. PÉREZ-LOSADA, L. BUCKUP & K. A. CRANDALL, 2008. Global diversity of crabs (Aeglidae: Anomura: Decapoda) in freshwater. Hydrobiologia, **595**(1): 267-273.

BUTCHART, S. H. M., A. J. STATTERSFIELD, J. E. M. BAILLIE, L. A. BENNUN, S. N. STUART, H. R. AKÇAKAYA, C. HILTON-TAYLOR & G. M. MACE, 2005. Using Red List indices to measure progress towards the 2010 target and beyond. Philosophical Transactions of the Royal Society of London, **360**: 255-268.

BUTCHART, S. H. M., A. J. STATTERSFIELD, L. A. BENNUN, S. M. SHUTES, H. R. AKÇAKAYA, J. E. M. BAILLIE, S. N. STUART, C. HILTON-TAYLOR & G. M. MACE, 2004. Measuring global trends in the status of biodiversity: Red List indices for birds. PLoS Biology, **2**: e383.

BUTCHART, S. H. M., M. WALPOLE, B. COLLEN, A. VAN STRIEN, J. P. W. SCHARLEMANN, R. A. E. ALMOND, J. E. M. BAILLIE, B. BOMHARD, B. BROWN, J. BRUNO, et al., 2010. Global biodiversity: indicators of recent declines. Science, **328**: 1164-1168.

CARPENTER, K. E., M. ABRAR, G. AEBY, R. B. ARONSON, S. BANKS, A. BRUCKNER, A. CHIRIBOGA, J. CORTÉS, J. C. DELBEEK, L. DEVANTIER, G. J. EDGAR, A. J. EDWARDS, D. FENNER, H. M. GUZMÁN, B. W. HOEKSEMA, G. HODGSON, O. JOHAN, W. Y. LICUANAN, S. R. LIVINGSTONE, E. R. LOVELL, J. A. MOORE, D. O. OBURA, D. OCHAVILLO, B. A. POLIDORO, W. F. PRECHT, M. C. QUIBILAN, C. REBOTON, Z. T. RICHARDS, A. D. ROGERS, J. SANCIANGCO, A. SHEPPARD, C. SHEPPARD, J. SMITH, S. STUART, E. TURAK, J. E. N. VERON, C. WALLACE, E. WEIL & E. WOOD, 2008. One-third of reef-building corals face elevated extinction risk from climate change and local impacts. Science, **321**: 560-563.

CARPENTER, S. R., S. G. FISHER, N. B. GRIMM & J. F. KITCHELL, 1992. Global change and freshwater ecosystems. Annual Review of Ecology and Systematics, **23**: 119-139.

CLAUSNITZER, V., V. J. KALKMAN, M. RAM, B. COLLEN, J. E. M. BAILLIE, M. BEDJANI, W. R. T. DARWALL, K.-D. B. DIJKSTRA, R. DOW, J. HAWKING, H. KARUBE, E. MALIKOVA, D. PAULSON, K. SCHUÅNTTE, F. SUHLING, R. J. VILLANUEVA, N. VON ELLENRIEDER & K. WILSON, 2009. Odonata enter the biodiversity crisis debate: the first global assessment of an insect group. Biological Conservation, **142**: 1864-1869.

COLLEN, B., M. RAM, N. DEWHURST, V. CLAUSNITZER, V. KALKMAN, N. CUMBERLIDGE & J. E. M. BAILLIE, 2008a. Broadening the coverage of biodiversity assessments. In: J.-C. VIÉ, C. HILTON-TAYLOR & S. N. STUART (eds.), The 2008 review of the IUCN Red List of Threatened Species. (IUCN Gland, Switzerland).

COLLEN, B., M. RAM, T. ZAMIN & L. MCRAE, 2008b. The tropical biodiversity data gap: addressing disparity in global monitoring. Tropical Conservation Science, **1**: 75-88.

CRANDALL, K. A. & J. E. BUHAY, 2008. Global diversity of crayfish (Astacidae, Cambaridae, and Parastacidae—Decapoda) in freshwater. Hydrobiologia, **595**: 295-301.

CUMBERLIDGE, N., 2005. The status and distribution of freshwater crabs. In: W. R. T. DARWALL, K. G. SMITH, T. LOWE & J.-C. VIÉ (compilers), The status and distribution of freshwater biodiversity in Eastern Africa. (IUCN, Gland, Switzerland and Cambridge, UK).

— —, 2009. The status and distribution of freshwater crabs. In: K. G. SMITH, M. D. DIOP, M. NIANE & W. R. T. DARWALL (compilers), The status and distribution of freshwater biodiversity in Western Africa: 56-70. (IUCN, Gland, Switzerland and Cambridge, UK).

— —, 2010. The status and distribution of freshwater crabs. In: N. GARCÍA, A. CUTTELOD & D. ABDUL MALAK (eds.), The status and distribution of freshwater biodiversity in Northern Africa: 71-78. (IUCN Gland, Switzerland, Cambridge, UK, and Malaga, Spain).

— —, 2011. Freshwater crabs of Africa: diversity, distribution, and conservation. In: W. R. T. DARWALL, K. G. SMITH, D. J. ALLEN, R. A. HOLLAND, I. J. HARRISON & E. G. E. BROOKS (eds.), The diversity of life in African freshwaters: under water, under threat. An analysis of the status and distribution of freshwater species throughout mainland Africa: 178-199. (IUCN, Cambridge, UK and Gland, Switzerland).

CUMBERLIDGE, N. & S. R. DANIELS, 2007. A conservation assessment of the freshwater crabs of southern Africa (Brachyura: Potamonautidae). African Journal of Ecology, **46**: 74-79.

— — & — —, 2009. The status and distribution of freshwater crabs. In: D. TWEDDLE (ed.), Freshwater biodiversity of Southern Africa: 66-79. (IUCN, Gland, Switzerland and Cambridge, UK).

CUMBERLIDGE, N. & P. K. L. NG, 2009. Systematics, evolution, and biogeography of the freshwater crabs. In: J. W. MARTIN, K. A. CRANDALL & D. L. FELDER (eds.), Decapod crustacean phylogenetics. Crustacean Issues, **18**: 491-504. (CRC Press, Boca Raton, Florida).

CUMBERLIDGE, N., P. K. L. NG, D. C. J. YEO, C. MAGALHAES, M. R. CAMPOS, F. ALVAREZ, T. NARUSE, S. R. DANIELS, L. J. ESSER, F. Y. K. ATTIPOE, F.-L. CLOTILDE-BA, W. R. T. DARWALL, A. MCIVOR, M. RAM & B. COLLEN, 2009. Freshwater crabs and the biodiversity crisis: importance, threats, status, and conservation challenges. Biological Conservation, **142**: 1665-1673.

CUMBERLIDGE, N., P. K. L. NG, D. C. J. YEO, T. NARUSE, K. S. MEYER & L. J. ESSER, 2010. Diversity, endemism, and conservation of the freshwater crabs of China (Brachyura: Potamidae and Gecarcinucidae). Integrative Zoology, **6**: 45-55.

DARWALL, W. R. T., K. SMITH, T. LOWE & J.-C. VIÉ, 2005. The status and distribution of freshwater biodiversity in Eastern Africa. IUCN SSC Freshwater Biodiversity Assessment Programme: i-viii, 1-36. (IUCN, Gland, Switzerland and Cambridge, UK).

DARWALL, W. R. T., K. G. SMITH, D. J. ALLEN, R. A. HOLLAND, I. J. HARRISON & E. G. E. BROOKS, 2011. The diversity of life in African freshwaters: under water, under threat. An analysis of the status and distribution of freshwater species throughout mainland Africa: i-xiii, 1-347. (IUCN, Cambridge, UK and Gland, Switzerland).

DE GRAVE, S., Y. CAI & A. ANKER, 2008. Global diversity of shrimps (Caridea) in freshwater. Hydrobiologia, **595**: 287-293.

DUDGEON, D., A. H. ARTHINGTON, M. O. GESSNER, Z.-I. KAWABATA, D. KNOWLER, C. LÉVÊQUE, R. J. NAIMAN, A.-H. PRIEUR-RICHARD, D. SOTO, M. L. J. STIASSNY & C. A. SULLIVAN, 2006. Freshwater biodiversity: importance, threats, status and conservation challenges. Biological Reviews, **81**: 163-182.

EHRLICH, P. R. & R. M. PRINGLE, 2008. Where does biodiversity go from here? A grim business-as-usual forecast and a hopeful portfolio of partial solutions. Proceedings of the National Academy of Sciences of the United States of America, **105**(Suppl. 1): 11579-11586.

FODEN, W. B., G. M. MACE, J.-C. VIÉ, A. ANGULO, S. H. M. BUTCHART, L. DE VANTIER, H. T. DUBLIN, A. GUTSCHE, S. N. STUART & E. TURAK, 2008. Species susceptibility to climate change impacts. In: J.-C. VIÉ, C. HILTON-TAYLOR & S. N. STUART (eds.), The 2008 review of the IUCN Red List of Threatened Species. (IUCN, Gland, Switzerland).

FOSTER, J. & D. HARPER, 2007. Status and ecosystem interactions of the invasive Louisianan red swamp crayfish *Procambarus clarkii* in East Africa. In: F. GHERARDI (ed.), Biological invaders in inland waters: profiles, distribution, and threats: 91-101. (Springer, Dordrecht).

GIBBONS, J. W., D. E. SCOTT, T. RYAN, K. BUHLMANN, T. TUBERVILLE, J. GREENE, T. MILLS, Y. LEIDEN, S. POPPY, C. WINNE & B. METTS, 2000. The global decline of reptiles, déja vu amphibians. BioScience, **50**: 653-666.

HOFFMANN, M., C. HILTON-TAYLOR, A. ANGULO, M. BÖHM, T. M. BROOKS, S. H. M. BUTCHART, K. E. CARPENTER, J. CHANSON, B. COLLEN, N. A. COX, et al., 2010. The impact and shortfall of conservation on the status of the world's vertebrates. Science, **330**: 1503-1509.

IUCN, 2010. IUCN Red List of Threatened Species. Version 2010.4. www.iucnredlist.org.

KALKMAN, V. J., V. CLAUSNITZER, K.-D. B. DIJKSTRA, A. G. ORR, D. R. PAULSON & J. VAN TOL, 2008. Global diversity of dragonflies (Odonata) in freshwater. Hydrobiologica, **595**(1): 351-363.

LYDEARD, C., R. H. COWIE, W. F. PONDER, A. E. BOGAN, P. BOUCHET, S. A. CLARK, K. S. CUMMINGS, T. J. FREST, O. GARGOMINY, D. G. HERBERT, R. HERSHLER, K. E. PEREZ, B. ROTH, M. SEDDON, E. E. STRONG & F. G. THOMPSON, 2004. The global decline of nonmarine mollusks. BioScience, **54**(4): 321-330.

MACE, G. M., W. CRAMER, S. DIAZ, D. P. FAITH, A. LARIGAUDERIE, P. LE PRESTRE, M. PALMER, C. PERRINGS, R. J. SCHOLES, M. WALPOLE, B. A. WALTHER, J. E. M. WATSON & H. A. MOONEY, 2010. Biodiversity targets after 2010. Current Opinion in Environmental Sustainability, **2**: 3-8.

MORA, C., D. P. TITTENSOR, S. ADL, A. G. B. SIMPSON & B. WORM, 2011. How many species are there on Earth and in the ocean? PLoS Biology, **9**(8): e1001127.

NEILAND, A. E. & C. BÉNÉ, 2008. Tropical river fisheries valuation: background papers to a global synthesis. The WorldFish Center Studies and Reviews, **1836**: 290 pp. (The WorldFish Center, Penang, Malaysia).

NG, P. K. L., D. GUINOT & P. DAVIE, 2008. Systema Brachyuorum: Part I. An annotated checklist of extant brachyuran crabs of the world. Raffles Bulletin of Zoology, (Supplement) **17**: 1-286.

OGADA, M. O., 2006. Effects of the Louisiana crayfish invasion and other human impacts on the African clawless otter. (Ph.D. thesis, Kenyatta University, Kenya).

PÉREZ-LOSADA, M., G. BOND-BUCKUP, C. G. JARA & K. A. CRANDALL, 2009. Conservation assessment of southern South America freshwater ecoregions based on endemic crabs of the genus *Aegla*. Conservation Biology, **23**(3): 692-702.

PÉREZ-LOSADA, M., C. G. JARA, G. BOND-BUCKUP & K. A. CRANDALL, 2002. Conservation phylogenetics of Chilean freshwater crabs *Aegla* (Anomura, Aeglidae): assigning priorities for aquatic habitat protection. Biological Conservation, **105**: 345-353.

REVENGA, C., I. CAMPBELL, R. ABELL, P. DE VILLIERS & M. BRYER, 2005. Prospects for monitoring freshwater ecosystems towards the 2010 targets. Philosophical Transactions of the Royal Society, (B) Biological Sciences, **360**: 397-413.

ROCKSTROOM, J., W. STEFFEN, K. NOONE, A. PERSSON, F. S. CHAPIN III, E. LAMBIN, T. M. LENTON, M. SCHEFFER, C. FOLKE, H. SCHELLNHUBER, B. NYKVIST, C. A. DE WIT, T. HUGHES, S. VAN DER LEEUW, H. RODHE, S. SOÅNRLIN, P. K. SNYDER, R. COSTANZA, U. SVEDIN, M. FALKENMARK, L. KARLBERG, R. W. CORELL, V. J. FABRY, J. HANSEN, B. H. WALKER, D. LIVERMAN, K. RICHARDSON, C. CRUTZEN & J. FOLEY, 2009. A safe operating space for humanity. Nature, **461**: 472-475.

SCHIPPER, J., J. CHANSON, F. CHIOZZA, N. COX, M. HOFFMANN, V. KATARIYA, J. LAMOREUX, A. RODRIGUES, S. N. STUART, H. J. TEMPLE, J. E. M. BAILLIE, L. BOITANI, T. E. LACHER, R. A. MITTERMEIER, A. T. SMITH, et al., 2008. The status of the world's land and marine mammals: diversity, threat and knowledge. Science, **322**: 225-230.

SCHUYT, K. & L. BRANDER, 2004. The economic value of the world's wetlands. WWF living waters: conserving the source of life: 1-31. (Gland, Switzerland).

STIASSNY, M. L. J., 1999. The medium is the message: freshwater biodiversity in peril. In: J. CRACRAFT & F. T. GRIFO (eds.), The living planet in crisis: biodiversity science and policy: 53-71. (Columbia University Press, New York).

STRAYER, D. L., 2006. Challenges for freshwater invertebrate conservation. Journal of the North American Benthological Society, **25**: 271-287.

— —, 2010. Alien species in fresh waters: ecological effects, interactions with other stressors, and prospects for the future. Freshwater Biology, **55**(Suppl. 1): 152-174.

STRAYER, D. L. & D. DUDGEON, 2010. Freshwater biodiversity conservation: recent progress and future challenges. Journal of the North American Benthological Society, **29**: 344-358.

STRONG, E. E., O. GARGOMINY, W. F. PONDER & P. BOUCHET, 2008. Global diversity of gastropods (Gastropoda; Mollusca) in freshwater. Hydrobiologia, **595**: 149-166.

STUART, S. N., J. S. CHANSON, N. A. COX, B. E. YOUNG, A. S. L. RODRIGUES, D. L. FISCHMAN & R. W. WALLER, 2004. Status and trends of amphibian declines and extinctions worldwide. Science, **306**: 1783-1786.

STUART, S. N., E. O. WILSON, J. A. MCNEELY, R. A. MITTEMEIER & J. P. RODRIGUEZ, 2010. The barometer of life. Science, **328**: 177.

TAYLOR, C. A., G. A. SCHUSTER, J. E. COOPER, R. J. DISTEFANO, A. G. EVERSOLE, P. HAMR, H. H. HOBBS III, H. W. ROBISON, C. E. SKELTON & R. F. THOMA, 2007. A reassessment of the conservation status of crayfishes of the United States and Canada after 10+ years of increased awareness. Fisheries, **32**: 372-389.

THOMAS, C. D., A. CAMERON, R. E. GREEN, M. BAKKENES, L. J. BEAUMONT, Y. C. COLLINGHAM, B. F. N. ERASMUS, M. F. D. SIQUEIRA, A. GRAINGER, L. HANNAH, L. HUGHES, B. HUNTLEY, A. S. V. JAARSVELD, G. F. MIDGLEY, L. MILES, et al., 2004. Extinction risk from climate change. Nature, **427**: 145-148.

UNEP, 2002. Report on the sixth meeting of the Conference of the Parties to the Convention on Biological Diversity (UNEP/CBD/COP/20/Part 2) Strategic Plan Decision VI/26.

VIÉ, J.-C., C. HILTON-TAYLOR, C. POLLOCK, J. RAGLE, J. SMART, S. N. STUART & R. TONG, 2008. The IUCN Red List: a key conservation tool. In: J.-C. VIÉ, C. HILTON-TAYLOR & S. N. STUART (eds.), The 2008 review of the IUCN Red List of Threatened Species: 1-13. (IUCN, Gland, Switzerland).

WHEELER, Q. (ed.), 2008. The new taxonomy. Systematic Association Special Publications, **76**: i-xi, 1-237. (CRC Press, Boca Raton, Florida).

WONG, C. M., C. E. WILLIAMS, J. PITTOCK, U. COLLIER & P. SCHELLE, 2007. World's top 10 rivers at risk. (WWF International, Gland, Switzerland).

YEO, D. C. J., P. K. L. NG, N. CUMBERLIDGE, C. MAGALHÃES, S. R. DANIELS & M. R. CAMPOS, 2008. Global diversity of crabs (Crustacea: Decapoda: Brachyura) in freshwater. Hydrobiologia, **595**: 275-286.

First received 1 July 2011.
Final version accepted 29 July 2013.

AN OVERVIEW OF THE AFROTROPICAL FRESHWATER CRAB FAUNA: DIVERSITY, BIOGEOGRAPHY, AND CONSERVATION (BRACHYURA, POTAMOIDEA, POTAMONAUTIDAE AND POTAMIDAE)

BY

NEIL CUMBERLIDGE[1])
Department of Biology, Northern Michigan University, Marquette, MI 49855, U.S.A.

ABSTRACT

The Afrotropical region has a rich, highly diverse, and distinctly recognisable freshwater crab fauna comprising 145 species in 20 genera and two families. The entire Afrotropical freshwater crab fauna is endemic to Africa and its islands, and is distributed over 47 countries in eight distinct regions. Major centers of species richness include the Rift Valley of eastern Africa, the Upper Guinea forests of western Africa, and the Lower Guinea forests of central Africa. The IUCN Red List indicates that the majority (71%) of the 103 species of potamonautid and potamid Afrotropical freshwater crabs are of Least Concern and just two species are Near Threatened. Some twenty-eight species (27%) are listed in one of three threatened categories, either as Vulnerable (16 species), Endangered (10 species), or Critically Endangered (2 species). Excluded are the 33 species found to be Data Deficient because of a lack of information on distribution and population levels, and the 18 species discovered since the last assessment in 2009.

INTRODUCTION

This work focuses on the freshwater crabs found in the Afrotropical zoogeographical zone, a vast area of 24.5 million km^2 that comprises the 44 African countries south of the Sahara desert plus three islands/island groups in the western Indian Ocean (Socotra (Yemen), the Seychelles, and Madagascar) (fig. 1). Not included here are the freshwater crabs of the North African countries of Algeria, Libya, Egypt, Morocco, Western Sahara, and Tunisia, because these lie in the Palaearctic zoogeographical zone (Yeo et al., 2008;

[1]) e-mail: ncumberl@nmu.edu

Fig. 1. The eight regions in the Afrotropical zoogeograpical zone.

Cumberlidge, 2009b; Cumberlidge et al., 2009). Freshwater crabs are common throughout the Afrotropical region in almost all freshwater habitats such as large rivers, mountain streams, lakes, and wetlands, and wherever there is permanent surface water (Bott, 1955; Cumberlidge, 1999; Yeo et al., 2008). Those species that live in seasonally arid areas or are semi-terrestrial in nature spend the day in burrows and move about on land at night (Cumberlidge, 1986, 1991a, 1999), but these essentially aquatic crustaceans are notably absent from the more arid parts of the region where there are prolonged dry periods and little permanent surface water (e.g., the Sahara and Kalahari deserts, and the Horn of Africa).

Despite the fact that freshwater crabs are among the most important invertebrates inhabiting Afrotropical fresh waters, their study has been largely overlooked until relatively recently. Rathbun (1904, 1905, 1906) and Chace (1942) were the first authors to describe the freshwater crab fauna of Africa and its islands as part of global treatments of the group, while Bott (1955, 1965) produced the first monographic works that focused on continental Africa and Madagascar, respectively. Unfortunately, these contributions are now out of date and are no longer reliable to use to assess the Afrotropical freshwater crab faunal composition (Cumberlidge, 1999). This situation is now changing and there is currently an upsurge of interest in the biology of the freshwater crabs that has resulted in a steep increase in the known biodiversity of the Afrotropical zone.

A number of works are now available on the taxonomy, identification, phylogeny, diversity, distribution patterns, and conservation status of Afrotropical freshwater crabs. For example, there are modern studies that summarize the freshwater crab faunas and provide identification keys to the species for west Africa (Cumberlidge, 1999), Tanzania (Reed & Cumberlidge, 2006), Lake Tanganyika (Cumberlidge et al., 1999; Marijnissen et al., 2004; Reed & Cumberlidge, 2006), Angola (Cumberlidge & Tavares, 2006), southern Africa (Cumberlidge & Tavares, 2006; Cumberlidge & Daniels, 2007; Daniels & Bayliss, 2012; Daniels et al., 2014; Phiri & Daniels, 2014), central Africa (Cumberlidge & Boyko, 2000; Cumberlidge et al., 2002; Cumberlidge & Reed, 2004; Cumberlidge & Meyer, 2011; Meyer & Cumberlidge, 2011), east Africa (Cumberlidge & Vannini, 2004; Dobson, 2004; Marijnissen et al., 2004, 2005; Reed & Cumberlidge, 2004, 2006; Cumberlidge & Dobson, 2008; Cumberlidge, 2009; Cumberlidge & Clark, 2010a, b), and northeastern Africa (Cumberlidge, 2009b; Cumberlidge & Meyer, 2010; Cumberlidge & Clark, 2012). Taxonomic works are also available for the freshwater crab faunas of Madagascar (Cumberlidge & Sternberg, 2002; Cumberlidge et al., 2007; Cumberlidge, 2008; Cumberlidge & Meyer, 2009), the Seychelles (Ng et al., 1995; Cumberlidge, 2008; Daniels, 2011; Cumberlidge & Daniels, 2014), and Socotra (Apel & Brandis, 2000; Cumberlidge & Wranik, 2002; Cumberlidge, 2008). Although a majority of the Afrotropical species are quite well studied, there are still some that are known from either the type locality only, or from just a few localities, and in these cases further collections are necessary to ascertain their actual distributions. Consequently, there are still gaps in our knowledge and workers in many parts of the Afrotropical zone needing to identify material and compile distribution records must still either refer to a patchwork of different articles, or in some cases examine the original type specimens. These are difficult tasks for the non-specialist, and a single comprehensive monograph of the Afrotropical freshwater crabs is still a few years away.

Other important milestones in the modern understanding of the Afrotropical freshwater crab fauna include the production of phylogenies that support both the monophyly of the 142 species in the Potamonautidae Bott, 1970a (Sternberg et al., 1999; Cumberlidge & Sternberg, 2002; Daniels et al., 2006; Klaus et al., 2006; Cumberlidge et al., 2008; Cumberlidge & Ng, 2009), and the emerging idea that this family is endemic to both continental sub-Saharan

Africa (except for the Nile valley) and the Seychelles and Madagascar (Cumberlidge, 2008; Cumberlidge & Ng, 2009). Previous opinions treated the freshwater crabs of the Seychelles (Ng et al., 1995) and some species from Madagascar (Bott, 1965) as western outliers of the Asian family Gecarcinucidae. However, the present consensus is that all of the members of the two subfamilies of the Potamonautidae are entirely Afrotropical. The Potamonautinae Bott, 1970a includes seven genera from continental Africa, and the Deckeniinae Ortmann, 1897, includes 11 genera from continental Africa, the Seychelles, and Madagascar. The Deckeniinae comprises two tribes, the Hydrothelphusini Colosi, 1920 (with seven genera from Madagascar), and the Deckeniini Ortmann, 1897 (with four genera, three from continental Africa and one from the Seychelles). The Deckeniini in turn has two subtribes, the Deckeniina Ortmann, 1897 (*Deckenia* Hilgendorf, 1869, and *Seychellum* Ng, Števčić & Pretzmann, 1995) and the Globonautina Bott, 1969 (*Globonautes* Bott, 1959, and *Afrithelphusa* Bott, 1969) (Cumberlidge et al., 2008). The three species of Afrotropical Potamidae Ortmann, 1896 (*Socotra* and *Socotrapotamon* from Socotra) belong to the subfamily Potamiscinae Bott, 1970b (Cumberlidge, 2008; Cumberlidge et al., 2008).

Another important milestone was the production by Yeo et al. (2008) and Ng et al. (2008) of the first reliable modern species lists for this region that superseded those of Bott (1955, 1965). These comprehensive species lists in turn were updated by Cumberlidge et al. (2009) and are regularly revised as new species are described. It was this improved taxonomic and phylogenetic groundwork (Cumberlidge, 2005, 2009c, 2010, 2011a, b; Cumberlidge & Daniels, 2007, 2009) that enabled the first comprehensive assessment of the conservation status of the entire Afrotropical freshwater crab fauna (Cumberlidge et al., 2009).

DIVERSITY

The updated figures for the Afrotropical freshwater crab fauna indicate that it currently comprises 145 species in 20 genera and two families (table I) (Cumberlidge, 1999, 2009a, 2011a; Cumberlidge & Sternberg, 2002; Daniels et al., 2006; Cumberlidge et al., 2008, 2009; Ng et al., 2008; Yeo et al., 2008; present work) and there is every prospect that the species-count will continue to increase as taxonomic discrimination improves and exploration continues. The Afrotropical freshwater crab fauna is highly endemic at the family, genus, and species levels with 100% endemicity for the 142 species in

TABLE I

The most recent numbers of species in each of the 20 genera of freshwater crabs found in the Afrotropical zone. Also shown are the number of species and genera found in each of the eight Afrotropical regions recognised in the present work. N = northeastern Africa; W = west Africa; C = central Africa; E = east Africa; S = southern Africa; M = Madagascar; SE = Seychelles; SO = Socotra. All genera are regionally endemic except for those marked by *

Genus	No. of species	N	W	C	E	S	M	SE	SO
Afrithelphusa	4	–	4	–	–	–	–	–	–
Boreathelphusa	1	–	–	–	–	–	1	–	–
*Deckenia**	2	1	–	–	2	–	–	–	–
Erimetopus	2	–	–	2	–	–	–	–	–
Foza	3	–	–	–	–	–	3	–	–
Globonautes	1	–	1	–	–	–	–	–	–
Hydrothelphusa	4	–	–	–	–	–	4	–	–
Liberonautes	8	–	8	–	–	–	–	–	–
Louisea	2	–	–	2	–	–	–	–	–
Madagapotamon	1	–	–	–	–	–	1	–	–
Malagasya	2	–	–	–	–	–	2	–	–
Marojejy	1	–	–	–	–	–	1	–	–
*Platythelphusa**	9	–	–	4	8	4	–	–	–
*Potamonautes**	84	6	5	27	33	25	–	–	–
*Potamonemus**	3	–	1	3	–	–	–	–	–
*Sudanonautes**	11	1	7	9	1	–	–	–	–
Seychellum	1	–	–	–	–	–	–	1	–
Skelosophusa	3	–	–	–	–	–	3	–	–
Socotra	1	–	–	–	–	–	–	–	1
Socotrapotamon	2	–	–	–	–	–	–	–	2
Totals	145	8	26	48	44	29	15	1	3

the predominant family, the Potamonautidae (Bott, 1955, 1959, 1964, 1969, 1970a, b; Cumberlidge, 1999; Cumberlidge et al., 2009). The three potamid species from Socotra are also endemic to the region at the genus and species levels (but not at the family level).

The Potamonautidae includes two subfamilies, the Potamonautinae (with 119 species and 7 genera from continental sub-Saharan Africa), and the Deckeniinae (with 23 species and 11 genera from sub-Saharan Africa as well as Madagascar and the Seychelles), that represent two separate evolutionary lineages (Cumberlidge, 1999; Daniels et al., 2006; Cumberlidge et al., 2008; Yeo et al., 2008). The most species-rich genera of the Potamonautinae are *Potamonautes* (84 species), *Sudanonautes* (11 species), *Platythelphusa* (9 species), and *Liberonautes* (8 species) (Cumberlidge et al., 2008, 2009; Yeo et al., 2008). The most species-rich genera of the Deckeniinae are *Afrithelphusa* (4 species),

Hydrothelphusa (4 species), *Skelosophusa* (3 species), and *Foza* (3 species) (Cumberlidge et al., 2008, 2009; Yeo et al., 2008). Four of the eleven deckeniine genera are found in continental sub-Saharan Africa and have a markedly disjunct distribution, with the west African *Globonautes* and *Afrithelphusa* being separated from *Deckenia* of northeast and east Africa by over 3000 km of land. The three mainland African deckeniine genera are in turn separated by substantial stretches of ocean from *Seychellum* on the Seychelles, whose closest relative is *Deckenia* from the northeast and east African mainland. These latter two genera form a clade that is sister to the west African *Afrithelphusa* and *Globonautes* (Daniels et al., 2006; Cumberlidge, 2008; Cumberlidge & Ng, 2009). Finally, these four deckeniine genera are sister to the monophyletic group formed by the Malagasy genera (Daniels et al., 2006; Cumberlidge, 2008; Cumberlidge & Ng, 2009).

The Afrotropical Potamidae comprises one subfamily, the predominantly Asian subfamily Potaminscinae, with three species and two genera (*Socotra* and *Socotrapotamon*) that are all endemic to Socotra Island, Yemen (Apel & Brandis, 2000; Cumberlidge & Wranik, 2002; Yeo & Ng, 2003; Daniels et al., 2006).

REGIONAL PATTERNS OF DISTRIBUTION

The composition of the Afrotropical freshwater crab fauna is not uniform; it is different in different parts of the continent, and varies with ecosystem, aquatic drainage basins, and vegetation cover. This means that the Afrotropical freshwater crabs are distinctly regional in their distribution, and fall into eight major regions: five on continental Africa (west, central, north-eastern, east, and southern), plus Madagascar, the Seychelles, and Socotra island (fig. 1). Each of these eight Afrotropical sub-regions has a distinct freshwater crab fauna, with different species richness and diversity (table I), and there are relatively few species (25/145, 17%) that have a widespread distribution ranging across two or more regions (table II). The two most species-rich regions are central Africa (47 species) and east Africa (44 species) followed by west Africa (26 species), southern Africa (28 species) and Madagascar (15 species). The regions with the fewest species are northeast Africa (8 species), Socotra (3 species), and the Seychelles (1 species). This pattern of diversity is different at the genus level, with the most diverse regions being Madagascar (7 genera), central Africa and west Africa (6 genera each) and east Africa (4 genera), followed by northeast Africa (3 genera), Socotra (2

TABLE II
Species of Afrotropical freshwater crabs (five genera and 25 species) with a widespread distribution that encompasses more than one of the eight regions used in the present study. N = northeastern Africa; W = west Africa; C = central Africa; E = east Africa; S = southern Africa

	Taxon	Regions
1	*Deckenia imitatrix*	E/N
2	*Platythelphusa armata*	E/S/C
3	*Platythelphusa conculcata*	E/S/C
4	*Platythelphusa maculata*	E/S
5	*Platythelphusa polita*	E/C
6	*Platythelphusa tuberculata*	E/C
7	*Potamonautes adeleae*	C/S
8	*Potamonautes anchietae*	S/C
9	*Potamonautes bayonianus*	S/C
10	*Potamonautes berardi*	E/N
11	*Potamonautes lirrangensis*	E/C/S
12	*Potamonautes loveridgei*	E/S
13	*Potamonautes montivagus*	E/S
14	*Potamonautes mutandensis*	C/E
15	*Potamonautes niloticus*	E/N
16	*Potamonautes obesus*	E/S
17	*Potamonautes perparvus*	C/E
18	*Potamonautes platynotus*	E/C/S
19	*Potamonautes suprasulcatus*	E/S
20	*Potamonemus sachsi*	W/C
21	*Sudanonautes africanus*	W/C
22	*Sudanonautes aubryi*	W/C
23	*Sudanonautes floweri*	W/C/E/N
24	*Sudanonautes granulatus*	W/C
25	*Sudanonautes monodi*	W/C

genera), southern Africa (1 genus), and the Seychelles (1 genus) (tables III-IX). Although central Africa (47 species) and east Africa (44 species) are the most species-rich regions, Madagascar (15 species, 7 genera, one subfamily) and west Africa (26 species, 6 genera, two subfamilies) stand out as being exceptionally diverse (table I). The vast majority of Afrotropical freshwater crab genera (75%) have a relatively restricted range and are endemic to a particular region. For example, all seven Malagasy genera, plus *Seychellum* (the Seychelles), and *Socotra* and *Socotrapotamon* (Socotra) are all endemic to their respective Indian Ocean islands, while *Afrithelphusa*, *Globonautes*, and *Liberonautes* are endemic to west Africa and *Louisea* and *Erimetopus* are both found only in central Africa. Only five Afrotropical genera have a wide distribution, occurring in more than one region of continental Africa (table II). For example, *Potamonautes* is found in all five regions of continental Africa, *Sudanonautes* is found in four regions, while *Platythelphusa* occurs in three regions, and *Deckenia* and *Potamonemus* both occur in two regions.

THE AFROTROPICAL REGIONAL FAUNAS

1. West Africa. — The west African region as defined here comprises 15 countries: Benin, Burkina Faso, Cote d'Ivoire, the Gambia, Ghana, Guinea, Guinea Bissau, Liberia, Mali, Mauritania, Niger, Nigeria Senegal, Sierra Leone, and Togo (fig. 1). These countries have a combined area of 6 139 131 km^2, making west Africa the largest of the Afrotropical regions (table III). This treatment of west Africa is slightly different from that of Cumberlidge (1999) who included Cameroon in the region, whereas the latter country is dealt with here as part of central Africa, in order to conform to the regional divisions of Africa used by the IUCN (Cumberlidge, 2009c, 2011a). West Africa is home to 26 species of freshwater crabs assigned to six genera and two subfamilies, the Potamonautinae (*Liberonautes* (8 species), *Potamonautes* (5 species), *Potamonemus* (1 species), and *Sudanonautes* (7 species)) and the Deckeniinae (*Afrithelphusa* (4 species) and *Globonautes* (1 species)) (Bott, 1955, 1959, 1960, 1964, 1969, 1970a, b; Monod, 1977, 1980; Cumberlidge & Clark, 1992; Cumberlidge, 1993, 1999; Ďuriš & Koch, 2010). Most species of west Africa's freshwater crabs (20/26, 77%) are endemic, as are three of

TABLE III

The most recent numbers of species and genera of freshwater crabs found in the eight Afrotropical regions, and the total area of each region (km^2)

Region	No. of species	No. of genera	Area (km^2)
West Africa	26	6	6 139 131
Southern Africa	29	2	5 984 739
Central Africa	48	6	5 346 335
Northeastern Africa	8	3	4 674 010
East Africa	44	4	1 813 075
Madagascar	15	7	587 041
Socotra, Yemen	3	2	3625
Seychelles	1	1	445
Afrotropical total			24 548 401

TABLE IV

Distribution of the 26 species and six genera of freshwater crabs among the fifteen countries of west Africa, with the number of regional endemics (in parentheses), and the area of each country (km^2)

Country	No. of species	No. of genera	Area (km^2)
Benin	2 (1)	1 (0)	112 622
Burkina Faso	1 (1)	1 (0)	274 200
Cote d'Ivoire	6 (6)	2 (1)	322 462
The Gambia	2 (1)	2 (1)	11 295
Ghana	6 (5)	3 (1)	238 533
Nigeria	10 (3)	3 (0)	923 768
Guinea	7 (7)	3 (3)	245 857
Guinea Bissau	1 (1)	1 (1)	36 125
Liberia	9 (9)	2 (2)	111 369
Mali	2 (2)	2 (1)	1 240 000
Mauritania	2 (2)	2 (1)	1 030 000
Niger	1 (1)	1 (0)	1 267 000
Senegal	4 (4)	2 (1)	197 160
Sierra Leone	3 (3)	2 (2)	71 740
Togo	4 (5)	2 (0)	57 000
West Africa total			6 139 131

its six genera (*Liberonautes*, *Globonautes*, and *Afrithelphusa*) (Cumberlidge, 1999) (table IV). In fact, only *Potamonemus sachsi* and five species of *Sudanonautes* (*S. africanus*, *S. aubryi*, *S. floweri*, *S. granulatus*, *S. monodi*) are also found in central Africa (Bott, 1955; Cumberlidge, 1999). Nigeria is only the fourth largest country in west Africa but it is the most species-rich (10/26 species, 3 genera) although only three of its species are regional endemics (3/26, 12%). On the other hand, west Africa's biodiversity hotspot includes three relatively small countries in the Upper Guinea forest that all host regional endemics: Liberia (9/26 species, 2 genera), Guinea (7 species, 3 genera), and Cote d'Ivoire (6 species, 2 genera) (table IV). The most species-poor countries in this region include the large and predominantly desert countries of Mauritania (2 species, 2 genera), Mali (2 species, 2 genera), and Niger (1 species) each with over a million square kilometres of territory, and the three smallest countries in the region: Gambia (2 species, 2 genera), Burkina Faso (1 species), and Guinea-Bissau (1 species) (table IV).

Of the three endemic genera of west African freshwater crabs, *Liberonautes* has a wide range from Senegal to Cote d'Ivoire, while the four species of *Afrithelphusa* are found only in Guinea and Sierra Leone, and the single species of *Globonautes* is found only in Liberia and Guinea (Cumberlidge,

1999). All five of the west African species of *Potamonautes* (*P. ecorssei*, *P. lipkei*, *P. reidi*, *P. senegalensis*, and *P. triangulus*) are endemic to this region, and most of these have a relatively restricted distributional range, except for *P. ecorssei* which is found from Gambia and Senegal to northern Nigeria (Cumberlidge, 1999). Two of the west African species of *Sudanonautes* (*S. kagoroensis* and *S. nigeria*) are endemic to Nigeria, while *S. africanus*, *S. aubryi*, *S. floweri*, and *S. granulatus* each has a wider distribution extending well outside of the region (table IV).

2. Central Africa. — The central African region as defined here includes nine countries: Angola (Cabinda), Cameroon, Central African Republic, Chad, Congo, D. R. Congo, Equatorial Guinea (including Bioko), Gabon, and the island nations of São Tomé and Príncipe (fig. 1). These countries have a combined area of more than 5.3 million km^2, making central Africa the third largest of the Afrotropical regions (table III). The distinctly recognizable freshwater crab fauna of central Africa is the most species-rich in the entire Afrotropical zone (table V), comprising 48 species of freshwater crabs assigned to six genera and one subfamily, the Potamonautinae (Rathbun, 1921; Balss, 1936; Bott, 1955; Cumberlidge, 1999; Cumberlidge et al., 1999; Cumberlidge & Boyko, 2000; Cumberlidge & Reed, 2004; Marijnissen et al., 2004). The faunal composition is as follows: *Erimetopus* (2 species), *Louisea* (2 species), *Sudanonautes* (9 species), *Potamonemus* (3 species), *Potamonautes* (28 species), and *Platythelphusa* (4 species). The D. R. Congo (35/47 species, 4 genera) is by far the largest and the most species-rich country in central Africa and has

TABLE V

Distribution of the 48 species and six genera of freshwater crabs among the nine countries of central Africa, with the number of regional endemics (in parentheses), and the area of each country (km^2)

Country	No. of species	No. of genera	Area (km^2)
Angola (Cabinda)	1 (0)	1 (0)	7823
Cameroon	13 (7)	4 (1)	469 440
Central African Republic	4 (0)	2 (0)	622 984
Chad	2 (0)	1 (0)	1 259 200
Congo	11 (1)	3 (1)	342 000
D. R. Congo	35 (22)	4 (1)	2 345 410
Equatorial Guinea + Bioko	4 (0)	2 (0)	30 017
Gabon	8 (0)	2 (0)	267 665
Sao Tome, Principe	2 (2)	1 (0)	1796
Central Africa total			5 346 335

the most regionally-endemic species (22/47, 45%), while Cameroon (fourth largest), is the second most species-rich (13/42 species, 4 genera). Next are Congo (11 species, 3 genera) and Gabon (8 species, 2 genera), the fifth and sixth largest countries, respectively. The predominantly desert country of Chad (2 species, 1 genus), is among the most species-poor countries in this region despite the fact that it is the second largest, while Sao Tome and Principe (2 species, 1 genus) is the smallest country in the region (1796 km^2). The Angolan Province of Cabinda (7823 km^2) situated near southern Congo has only a single species, *S. floweri*. Central Africa's freshwater crab fauna is highly endemic at the species level (31/47, 74%) with two regionally endemic genera, *Louisea* (2 species) and *Erimetopus* (2 species) (Cumberlidge 1999; Cumberlidge & Reed, 2004) (table V). Twenty-one out of the 28 central African species of *Potamonautes* are endemic to the region, and most of these endemics have a relatively restricted distributional range. In contrast, 16 species in four genera (*Platythelphusa*, *Potamonemus*, *Potamonautes*, and *Sudanonautes*) are also found outside of central Africa (Bott, 1955; Cumberlidge, 1999; Cumberlidge et al., 2008, 2009) (table II).

3. Northeastern Africa. — This region as defined here includes six countries: Djibouti, Ethiopia, Eritrea, Somalia, Sudan, and South Sudan (fig. 1). These countries have a combined area of more than 4.6 million km^2, making this the fourth largest of the Afrotropical regions (table III). Only eight species of freshwater crabs are found in this vast and largely arid region, and two countries (Djibouti and Eritrea) have no reported species (table VI). The eight species are assigned to three genera and two subfamilies, the Potamonautinae (*Potamonautes* (6 species) and *Sudanonautes* (1 species)) and the Deckeniinae (*Deckenia*, 1 species) (Bott, 1955; Cumberlidge, 1999; Cumberlidge et al.,

TABLE VI

Distribution of the eight species and three genera of freshwater crabs among the six countries of northeastern Africa, with the number of regional endemics (in parentheses), and the area of each country (km^2)

Country	No. of species	No. of genera	Area (km^2)
Djibouti	0	0	23 200
Ethiopia	5 (3)	1 (0)	1 100 000
Eritrea	0	0	125 000
Somalia	2 (0)	2 (0)	920 000
Sudan	2 (0)	1 (0)	1 886 065
South Sudan	3 (0)	2 (0)	619 745
Northeastern Africa total			4 674 010

2008, 2009). The northeastern African freshwater crabs live mainly in the Nile River basin and the Ethiopian Highlands (Flower, 1931; Chace, 1942; Bott, 1955; Cumberlidge, 2009b; Cumberlidge et al., 2009; Cumberlidge & Meyer, 2010). Some species (*P. niloticus* and *P. berardi*) are widespread and common throughout the Nile river basin from the headwaters south of the equator north through Uganda, Ethiopia, South Sudan and Sudan, as far north as the delta in Egypt where the Nile flows into the Mediterranean Sea (Cumberlidge, 2009b). Another species whose range extends into South Sudan is the widespread central African species *S. floweri*, while two other widespread species (*P. obesus* and *D. imitatrix*) are found in coastal Somalia and Kenya. Four species (*P. antheus*, *P. ignestii*, *P. holthuisi*, and *P. kundudo*) have a limited distribution and all are endemic to Ethiopia (the second largest country) (Cumberlidge & Clark, 2012). The vast desert countries of Sudan and Somalia are relatively species-poor and have no endemic species of freshwater crabs (table IV).

4. East Africa. — This region as defined here includes five countries: Burundi, Kenya, Rwanda, Tanzania, and Uganda (fig. 1). These countries have a combined area of more than 1.8 million km^2, making east Africa the fifth largest of the Afrotropical regions (table III). Despite its relatively small area East Africa has a rich and distinctly recognizable freshwater crab fauna comprising 44 species of freshwater crabs assigned to four genera and two subfamilies: the Potamonautinae (*Platythelphusa* (8 species), *Potamonautes* (33 species), and *Sudanonautes* (1 species)) and the Deckeniinae (*Deckenia* (2 species)) (Rathbun, 1935; Bott, 1955; Cumberlidge, 1997, 1998, 2009a, b; Cumberlidge et al., 1999, 2008, 2009; Corace et al., 2001; Dobson, 2004; Cumberlidge & Vannini, 2004; Marijnissen et al., 2004, 2005; Reed & Cumberlidge, 2004, 2006; Darwall et al., 2005; Cumberlidge & Dobson, 2008). The region's freshwater crab fauna is highly endemic at the species level (33/44, 77%), but none of these genera are regional endemics (table VII). Eleven species have a distributional range that extends outside of this region: four species are also found in northeast Africa (*D. imitatrix*, *P. berardi*, *P. niloticus*, and *P. obesus*), seven species are also found in central Africa (*P. idjwiensis*, *P. lirrangensis*, *P. mutandensis*, *Platythelphusa armata*, *P. conculcata*, *P. polita*, and *P. tuberculata*), and seven species are also found in southern Africa (*P. lirrangensis*, *P. montivagus*, *P. obesus*, *P. suprasulcatus*, *Platythelphusa armata*, *P. conculcata*, and *P. maculata*). Tanzania (26/44 species, 3 genera) is by far the largest and the most species-rich country in east Africa and has the most regionally endemic species (18/44 species, 3 genera), while Kenya, the second

TABLE VII
Distribution of the 44 species and four genera of freshwater crabs among the five countries of east Africa, with the number of regional endemics (in parentheses), and the area of each country (km^2)

Country	No. of species	No. of genera	Area (km^2)
Burundi	6 (0)	2 (0)	25 650
Kenya	17	2 (0)	580 000
Rwanda	4 (0)	1 (0)	26 338
Tanzania	31 (23)	3 (0)	945 087
Uganda	13	2 (0)	236 000
East Africa total			1 813 075

largest country in this region, is the second most species-rich (17/44 species, 2 genera). The next most species-rich country is Uganda (13/44 species, 2 genera), the third largest country in the region. Rwanda (4/44 species, 1 genus) and Burundi (1 species) are the two smallest and most species-poor countries in this region.

5. Southern Africa. — This region as defined here includes ten countries: Angola, Botswana, Lesotho, Malawi, Mozambique, Namibia, South Africa, Swaziland, Zambia, and Zimbabwe (fig. 1). These countries have a combined area of more than 5.9 million km^2, making southern Africa the second largest of the Afrotropical regions (table III). Southern Africa has its own distinctly recognizable freshwater crab fauna comprising 29 species in two genera (*Potamonautes* (25 species) and *Platythelphusa* (4 species)) in one subfamily, the Potamonautinae (Stewart et al., 1995; Stewart, 1997a, b; Daniels et al., 1998, 2001, 2002; Stewart & Cook, 1998; Gouws et al., 2000, 2001; Reed & Cumberlidge, 2004, 2006; Cumberlidge & Tavares, 2006; Cumberlidge et al., 2009). Eighteen species are endemic to the region, and eleven species have a distribution outside of the region (tables II, VIII). South Africa is the most species-rich country in southern Africa with 14/29 species, half of which (7/14 species) are country endemics. Four of these South African endemics (*P. brincki*, *P. granulatus*, *P. parvicorpus*, and *P. parvispinus*) are found in the isolated mountain streams and the middle stretches of rivers associated with the fynbos vegetation zone in the Cape Fold Mountains of the Western Cape Province (Cumberlidge & Daniels, 2007). The other four South African endemics are found in KwaZulu-Natal in mountain streams (*P. clarus* and *P. depressus*), the middle stretches of the rivers (*P. dentatus*), and the marshy, low-lying wetlands (*P. lividus*). Zambia (7 species) is the second

TABLE VIII

Distribution of the 29 species and two genera of freshwater crabs among the ten countries of southern Africa, with the number of regional endemics (in parentheses), and the area of each country (km^2)

Country	No. of species	No. of genera	Area (km^2)
Angola	5 (4)	1 (0)	1 246 700
Botswana	2 (1)	1 (0)	582 000
Lesotho	1 (1)	1 (0)	30 355
Malawi	5 (3)	1 (0)	118 484
Mozambique	3 (2)	1 (0)	801 537
Namibia	3 (2)	1 (0)	824 292
South Africa	14 (14)	1 (0)	1 220 813
Swaziland	1 (1)	1 (0)	17 364
Zambia	7 (1)	2 (0)	752 614
Zimbabwe	3 (2)	1 (0)	390 580
Southern Africa total			5 984 739

most species-rich country in southern Africa with one regionally-endemic species, *Platythelphusa praelongata* from Lake Tanganyika. With five species, Angola is the third most species-rich country and two of its species are country endemics, followed by Malawi with four species that include two country endemics. The lowest species-richness (from one to three species) is found in Botswana (2 species), Lesotho (1 species), Mozambique (3 species), Namibia (3 species), Swaziland (1 species), and Zimbabwe (3 species) none of which have endemic species of freshwater crabs. Recent as-yet unpublished discoveries of likely new species of freshwater crabs from Malawi and Mozambique are likely to increase these numbers again (S. R. Daniels, pers. comm.). Freshwater crab diversity is also low in the Namib and Kalahari deserts where crabs are restricted to permanent water sources on the margins of these arid lands, and there are no endemic species. Freshwater crab diversity is unexpectedly low in the major aquatic ecosystems of the region such as the Orange, Limpopo, Cunene, Okavango and Zambezi River basins where there are only common widespread species (*P. bayonianus* and *P. warreni*) and no endemics (Cumberlidge & Daniels, 2007).

6. Madagascar. — This is the largest island in the western Indian Ocean (587 041 km^2, table III) and has the most diverse freshwater crab fauna comprising 15 species of potamonautid freshwater crabs that are all in a single subfamily, the Deckeniinae (Bott, 1965; Ng & Takeda, 1994; Cumberlidge & Sternberg, 2002; Cumberlidge et al., 2004, 2007, 2008; Daniels et al., 2006; Reed & Cumberlidge, 2006, 2008, 2010; Cumberlidge & Meyer,

TABLE IX

The most recent numbers of species and genera of freshwater crabs found in the three Indian Ocean islands of Madagascar, Seychelles, and Socotra, with the number of regional endemics (in parentheses), and the area of each country (km^2)

Country	No. of species	No. of genera	Area (km^2)
Madagascar	15 (15)	7 (7)	587 041
Seychelles	1 (1)	1 (1)	445
Socotra, Yemen	3 (3)	2 (2)	3625

2009). The Malagasy freshwater crabs are assigned to seven genera: *Hydrothelphusa* (4 species), *Skelosophusa* (3 species), *Foza* (3 species), *Malagasya* (2 species), *Boreathelphusa* (1 species), *Madagapotamon* (1 species), and *Marojejy* (1 species). Madagascar's freshwater crab species are 100% endemic at the species and genus levels (tables I, IX). The 15 species of Malagasy freshwater crabs are unevenly distributed on the island and are most diverse in Antsiranana Province in northern Madagascar, an area with both dry and moist forests that includes the Tsaratanana and Marojejy mountains (Cumberlidge & Reed, 2004). The Malagasy freshwater crabs form a monophyletic group within the Deckeniinae, implying a single ancestral colonization of the island. The closest relatives of the Madagascan crabs are *Seychellum*, *Deckenia*, *Globonautes*, and *Afrithelphusa* (Daniels et al., 2006; Cumberlidge et al., 2008).

7. Seychelles. — These are a group of more than a hundred granitic and coralline islands, cays, and atolls in the Indian Ocean that extend between 160 and 1300 km north of Madagascar, 480 and 1600 km east of the east African coast of Somalia, Kenya, and Tanzania, and some 3000 km from India (fig. 1). The Seychelles islands occupy the smallest area of the Afrotropical region (445 km^2, table III) and host a single species of freshwater crab, *Seychellum alluaudi* (Milne-Edwards & Bouvier, 1893) (Potamonautidae: Deckeniinae) (table IX). The recent study by Daniels (2011) of the freshwater crabs of the Seychelles indicates that there may be two other species of *Seychellum* found on these islands (Cumberlidge & Daniels, 2014). At present, *Seychellum* is found only on the five largest and oldest granitic Seychelles islands (Silhouette, Mahé, Praslin, La Digue, and Frégate), and is absent from the more recent but still isolated coralline Seychelles islands (Ng et al., 1995; Cumberlidge, 2008; Daniels, 2011). The closest affinities of *Seychellum* are with the east African genus *Deckenia*, and *Seychellum* is more distant from the Malagasy genera.

Significantly, there are no close evolutionary links either between Seychellum and the continental African Potamonautinae, or between *Seychellum* and the Indian and Southeast Asian Gecarcinucidae (Daniels et al., 2006; Klaus et al., 2006; Cumberlidge et al., 2008; Klaus et al., 2009). The fact that the closely related genera *Seychellum* (Seychelles) and *Deckenia* (east Africa) are now separated by a deep oceanic barrier (the western Indian Ocean), coupled with a relatively young age for these genera that easily postdates Gondwanan fragmentation (Daniels et al., 2006), lends support to hypotheses that propose that the ancestors of *Seychellum* dispersed overseas from Africa to the Seychelles (Cumberlidge & Ng, 2009).

8. Socotra. — This island (3625 km^2, table III) lies just off the coast of Somalia and is politically part of the country of Yemen. Socotra has three endemic species of potamid freshwater crabs, *Socotra pseudocardisoma* Cumberlidge & Wranik, 2002, *Socotrapotamon socotrensis* (Rathbun, 1904), and *S. nojidensis* Apel & Brandis, 2000, all of which are endemic at the genus level (table IX). Interestingly, despite close proximity of Socotra to Somalia on the African mainland the two Socotran potamid genera are not part of the African potamonautid freshwater crab radiation. Morphological studies place *Socotra* and *Socotrapotamon* close to the Palaearctic and Oriental potamid subfamily Potaminae (Apel & Brandis, 2000; Cumberlidge & Wranik, 2002; Yeo & Ng, 2003), but the molecular studies of Daniels et al. (2006) group the two Socotran genera together in a clade that is located within the Oriental potamid subfamily the Potamiscinae close to *Johora tiomanensis* (Ng & Tan, 1984) from Malaysia and *Geothelphusa albogilva* Shy, Ng & Yu, 1994 from Taiwan. Shih et al. (2009) also placed *Socotrapotamon* in the Potamiscinae, as a sister group to an "eastern Asia" subclade, which included species from east Asia, the Philippines, and continental parts of Southeast Asia.

DISTRIBUTION PATTERNS

Potamonautes. — With 84/145 species, *Potamonautes* MacLeay, 1838, is by far the largest and most cosmopolitan genus in the entire Afrotropical zone and it is found in all five regions of continental Africa, distributed throughout sub-Saharan Africa from the Cape to Cairo and from Senegal to the Horn of Africa. *Potamonautes* is abundant in almost all available freshwater bodies, and is clearly well-adapted and successful (Bott, 1951, 1955, 1959, 1960, 1964, 1970a; Stewart et al., 1995; Stewart, 1997a, b; Cumberlidge,

1997, 1998, 1999, 2009a, b; Cumberlidge & Boyko, 2000; Corace et al., 2001; Cumberlidge et al., 2002, 2009; Cumberlidge & Vanini, 2004; Reed & Cumberlidge, 2004, 2006; Cumberlidge & Tavares, 2006; Cumberlidge & Dobson, 2008; Cumberlidge & Clark, 2010a, b, 2012; Cumberlidge & Meyer, 2010). *Potamonautes* is absent from Madagascar (Cumberlidge & Sternberg, 2002), the Seychelles (Cumberlidge & Ng, 2009), and Socotra (Cumberlidge, 2008). Some species of *Potamonautes* (*P. ballayi*, *P. ecorssei*) are widespread and are associated with the major river basins throughout the continent, some (*P. senegalensis*) have adopted a semi-terrestrial air-breathing habit, while others (*P. rukwanzi*) have a narrow distribution and are endemic to a single body of water. The largest number of species of *Potamonautes* occur in east Africa (33 species), in the forested Congo River basin (24 species), and in southern Africa (20 species). There are six species of *Potamonautes* in northeastern Africa, and only five species in west Africa.

Sudanonautes. — The majority of the 11 species of *Sudanonautes* are distributed in central and west Africa, with one species (*S. floweri*) reaching as far east as Uganda in east Africa, and South Sudan in northeast Africa (Cumberlidge, 1999). Species of *Sudanonautes* occur in most of the major ecosystems of west and central Africa (tropical rainforest, Guinea and Sudan savanna), and are found in most aquatic habitats (standing water, streams, and major rivers) and on land. The majority of species of *Sudanonautes* are found in the forested region of southeast Nigeria, southwest Cameroon, Congo, Gabon, the D. R. Congo, and on the island of Bioko. *Sudanonautes africanus*, *S. nigeria*, *S. granulatus*, *S. orthostylis*, *S. chavanesii* and *S. faradjensis* occur exclusively in rainforest habitats, while *S. aubryi* and *S. floweri* are found in both rainforest and woodland savanna (Cumberlidge, 1999). *Sudanonautes kagoroensis* occurs only in Guinea savanna in Nigeria (Cumberlidge, 1991b), while the semi-terrestrial air-breathing *S. monodi* is the only species found in both Guinea and in dry Sudan savanna (Cumberlidge, 1986, 1999). Eight of the 11 species of *Sudanonautes* are widespread throughout west and central Africa, while the other two species, *S. orthostylis* and *S. sangha*, each have a restricted distribution in central Africa and in Nigeria, west Africa (Cumberlidge, 1999; Cumberlidge & Boyko, 2000).

Liberonautes. — The eight species of *Liberonautes* are all restricted to west Africa: the western limit of the genus is Sénégal, the eastern limit is Ghana, and the northern limit is Mali (Cumberlidge, 1999; Cumberlidge & Huguet, 2001). One species (*L. latidactylus*) is widespread over the entire range of the genus, while the remaining seven species each have a limited distribution in

Liberia, Guinea, and Cote d'Ivoire. Species of *Liberonautes* are found mostly in rainforest, although *L. latidactylus* is also found in both Guinea and Sudan savanna zones. Most species are found in aquatic habitats (streams or major rivers) while *L. paludicolis* is more terrestrial, *L. rubigimanus* and *L. nimba* are found at high altitudes, and *L. chaperi* and *L. nanoides* are found exclusively in the major rivers (Cumberlidge, 1999).

Other genera. — The three species of *Potamonemus* are found in west and central Africa in the forested highlands of southeast Nigeria, southwest Cameroon, and western Togo. These species occur in small streams and probably leave the water at night to feed (Cumberlidge & Clark, 1992; Cumberlidge, 1993, 1999). *Globonautes macropus*, the Liberian tree hole crab, occurs only in the western part of the Upper Guinea forest in Liberia and Guinea, and lives in the small water reservoirs in tree holes in closed canopy forest (Cumberlidge, 1991a, 1996a, b, 1999; Cumberlidge & Sachs, 1991). The four species of *Afrithelphusa* occur only in the western part of the Upper Guinea forest of west Africa. *Afrithelphusa gerhildae* and *A. monodosa* are both found only in Guinea, while *A. afzelii* and *A. leonensis* are both known only from Sierra Leone (Cumberlidge, 1987, 1991a, 1996a, b, 1999). The two species of *Louisea* are found only in southwest Cameroon in the forested highlands and the moist lowland rainforests (Cumberlidge, 1994, 1999). One of the two species of *Erimetopus* (*E. brazzae*) is relatively widespread in the lower Congo River, while the other species, *E. vandenbrandeni*, is more restricted in its distribution (Cumberlidge & Reed, 2004; Cumberlidge et al., 2009). All nine species of *Platythelphusa* are endemic to Lake Tanganyika (Cumberlidge et al., 1999; Marijnissen et al., 2004; Reed & Cumberlidge, 2006), while both species of *Deckenia* are endemic to northeast and east Africa (Ng et al., 1995; Marijnissen et al., 2005; Reed & Cumberlidge, 2006). All 15 species of Malagasy freshwater crabs are endemic to Madagascar (Cumberlidge & Sternberg, 2002; Cumberlidge et al., 2007; Cumberlidge & Meyer, 2009), while *Seychellum* (the Seychelles), and *Socotra* and *Socotrapotamon* (Socotra) are endemic to their home islands.

CONSERVATION STATUS

The conservation status of the Afrotropical freshwater crab fauna as it was understood in 2009 (136 species, 20 genera, 2 families) was assessed by Cumberlidge et al. (2009) using the IUCN Red List Categories and

Criteria at the global scale (IUCN, 2003). The conservation status of 103 out of 136 species of potamonautid and potamid Afrotropical freshwater crabs can be found at the IUCN Red List website (www.iucnredlist.org) (table X), but assessments were not possible for the 33 species treated as data deficient (DD) due to a lack specimens, and locality and population data (Cumberlidge et al., 2009). This website does not yet include the 18 species of Afrotropical freshwater crabs that have been recognised since the study of Cumberlidge et al. (2009). The 18 Afrotropical species that have not yet had their conservation status assessed are: *Foza ambohitra* Cumberlidge & Meyer, 2009, *Potamonautes barbarai* Phiri & Daniels, 2014, *P. barnardi* Phiri & Daniels, 2014, *P. biballensis* Bott, 1951, *P. bourgaultae* Cumberlidge & Meyer, 2011, *P. elgonensis* Cumberlidge & Clark, 2010a, *P. holthuisi* Cumberlidge & Meyer, 2010, *P. kundudo* Cumberlidge & Clark, 2012, *P. lipkei* Duris &

TABLE X

Results of the conservation assessments carried out in 2009 for 103 species of Afrotropical freshwater crabs in 20 genera by Red List category. Assessments were not done for the 33 DD species that were excluded from the 2009 study. LC = Least Concern; NT = Near Threatened; VU = Vulnerable; EN = Endangered; CR = Critically Endangered; DD = Data Deficient (based on Cumberlidge et al., 2009: Appendix 1). Genera are regionally endemic are marked by *

Genus	No. of species	LC	NT	VU	EN	CR	DD
*Afrithelphusa**	4	–	0	0	1	–	3
*Boreathelphusa**	1	–	–	1	–	–	–
Deckenia	2	–	2	–	–	–	–
*Erimetopus**	2	1	–	–	–	–	1
*Foza**	2	2	–	–	–	–	–
*Globonautes**	1	–	–	–	1	–	–
*Hydrothelphusa**	4	4	–	–	–	–	–
*Liberonautes**	8	3	–	1	2	2	–
*Louisea**	2	–	–	–	2	–	–
*Madagapotamon**	1	–	–	1	–	–	–
*Malagasya**	2	1	–	–	–	–	1
*Marojejy**	1	–	–	–	–	–	1
Platythelphusa	9	8	–	–	–	–	1
Potamonautes	76	42	–	12	4	–	18
Potamonemus	3	1	–	1	–	–	1
Sudanonautes	11	8	–	–	–	–	3
*Seychellum**	1	1	–	–	–	–	–
*Skelosophusa**	3	–	–	–	–	–	3
*Socotra**	1	1	–	–	–	–	–
*Socotrapotamon**	2	1	–	–	–	–	1
Species per Red List Category	136	73	2	16	10	2	33

Koch, 2010, *P. machedoi* Bott, 1964, *P. mulanjeensis* Daniels & Bayliss, 2012, *P. mutareensis* Phiri & Daniels, 2013, *P. namuliensis* Daniels & Bayliss, 2012, *P. regnieri* Rathbun, 1904, *P. subukia* Cumberlidge & Dobson, 2008, *P. williamsi* Cumberlidge & Clark, 2010a, *Seychellum silhouette* Cumberlidge & Daniels, 2014, and *S. mahefregate* Cumberlidge & Daniels, 2014.

Detailed regional conservation assessments of the continental African freshwater crabs are available for east Africa (Darwall et al., 2005), southern Africa (Cumberlidge & Daniels, 2007), west Africa (Cumberlidge, 2009c), central Africa (Cumberlidge, 2011b), and north Africa (Cumberlidge, 2010). A comprehensive overview of the conservation status of the sub-Saharan freshwater crabs is also available (Cumberlidge, 2011a). The majority (71%) of the 103 assessed species of Afrotropical freshwater crabs belonging to twelve genera were judged to be Least Concern (LC), and most of these live in rivers, marshy lowlands, or mountain streams in the forested parts of the region (Cumberlidge, 1999; Cumberlidge et al., 2009). Forty two out of the 76 species of *Potamonautes* assessed in 2009 are LC, as are one of the three species of *Potamonemus*, eight of the 11 species of *Sudanonautes*, eight of the nine species of *Platythelphusa*, all four species of *Hydrothelphusa*, three of the eight species of *Liberonautes*, two of the three species of *Foza*, one each of the two species of *Erimetopus*, *Malagasya*, and *Socotrapotamon*, and *Seychellum alluaudi* and *Socotra pseudocardisoma* (tables X, XI). Twenty-eight of the 103 species as-

TABLE XI

Conservation assessments for the 136 species of Afrotropical freshwater crabs included in the 2009 study, showing numbers in each Red List (RL) category in each region, with the overall total in each category (based on Cumberlidge et al., 2009: Appendix 1). Not assessed = those species that have been described since the 2009 study. The numbers for the Total RL species per category include species found in more than one region. N = northeastern Africa; W = west Africa; C = central Africa; E = east Africa; S = southern Africa; M = Madagascar; SE = Seychelles; SO = Socotra

Red List Category (2009 data)	N	W	C	E	S	M	SE	SO	Total RL species per category
Critically Endangered	0	2	0	0	0	0	0	0	2
Endangered	0	4	5	2	0	0	0	0	11
Vulnerable	1	4	1	7	3	2	0	0	18
Near Threatened	1	0	0	2	0	0	0	0	3
Least Concern	5	10	26	26	20	7	1	2	97
Data Deficient	0	5	13	4	5	5	0	1	33
Total species	7	25	45	41	28	14	1	3	0
Not Assessed in 2009	1	1	4	3	0	1	0	0	0

sessed in 2009 (27%) were listed in one of three threatened categories, either as Vulnerable (VU) (16 species), Endangered (EN) (10 species), or Critically Endangered (CR) (2 species) (tables X, XI). Just two species assessed in 2009 (2%) were assessed as Near Threatened (NT).

One third of the species examined by Cumberlidge et al. (2009) were judged to be Data Deficient (DD) and were excluded from the conservation assessments of 2009 (tables X, XI) due to a lack of specimens and lack of recent information on their extent of occurrence, habitat, ecological requirements, population size, population trends, and long-term threats (Cumberlidge et al., 2009). Data Deficient species are found in the vast majority of the Afrotropical regions, except for northeast Africa and the Seychelles (table X) and it is of great concern that many of these DD species are known only from a few individuals collected many years ago, and that no new specimens have been found recently. In some cases the DD status may be due to under-sampling and at least some DD species may prove to be restricted range endemics that are vulnerable to habitat loss. As a result, the proportion of species in a threatened category (28%) could well prove to be an underestimate if any of the DD species are subsequently found to be threatened (Collen et al., 2008; Cumberlidge et al., 2009). No species of Afrotropical freshwater crabs from Africa were found to be Extinct (EX) or Extinct In The Wild (EW), but a species cannot be formally assessed as extinct until exhaustive surveys probing its disappearance have been carried out.

CONCLUSIONS

The recent upsurge of interest in freshwater crab conservation means that the biology and distribution patterns of the Afrotropical freshwater crabs are now better known, as are the potential threats to their long-term survival (Collen et al., 2008; Cumberlidge et al., 2009; Balian et al., 2010; Darwall et al., 2011). With 28/103 species of the non-DD species of Afrotropical freshwater crabs assessed in 2009 as being at risk of global extinction, the long-term survival of the region's largely endemic freshwater crab fauna is a concern (Cumberlidge et al., 2009). Nevertheless, it is hoped that conservation recovery plans for threatened species will be developed for those species identified to be in need of conservation action through the Red List assessment process (Cumberlidge et al., 2009; Darwall et al., 2011). Significant areas of this vast continent still remain insufficiently explored and new species of freshwater crabs are sure to be discovered if collection efforts in remote areas are intensified and taxonomic advances become more available in the form of identification keys to freshwater crabs all of the regions. Although taxonomic knowledge has advanced considerably in recent years, and museum collections of freshwater crabs have improved, a great deal of work still needs to be done. There is a great need for further surveys that are likely to discover undescribed species, refine our knowledge of species distributions, define specific habitat requirements, describe population levels and trends, and identify specific threats to Africa's important and unique freshwater crab fauna. This information is required for new and updated conservation assessments, and all newly-collected specimens have the potential to provide fresh tissue for further molecular phylogenetic and systematic studies that is vital if we are to further advance our understanding of the Afrotropical freshwater crabs.

ACKNOWLEDGEMENTS

I am extremely grateful to both Michael Türkay (Senckenberg Museum, Frankfurt, SMF) and Peter K. L. Ng (National University of Singapore), who led the organization of the Senckenberg Freshwater Decapod Conference, held in Frankfurt, Germany in December 2010. Thanks are also extended to the numerous colleagues at the SMF who all made invaluable contributions to the success of the meeting. The Peter White Scholar Award (NMU) is thanked for its support.

REFERENCES

APEL, M. & D. BRANDIS, 2000. A new species of freshwater crab (Crustacea: Brachyura: Potamidae) from Socotra Island and description of *Socotrapotamon* n. gen. Fauna Arabia, **18**: 133-144.

BALIAN, E., I. HARRISON, S. BUTCHART, P. CHAMBERS, J. CORDEIRO, N. CUMBERLIDGE, F. DE MOOR, C. GASCON, H. JAMES, V. KALKMAN, P. P. VAN DIJK & D. C. J. YEO, 2010. Chapter 2. Diversity of species in freshwater systems. In: R. A. MITTERMEIER, T. A. FARRELL, I. J. HARRISON, A. J. UPGREN & T. BROOKS (eds.), Freshwater — the essence of life: 50-89. (Conservation International and Cemex Conservation Book Series, Earth in Focus Editions, China).

BALSS, H., 1936. Beiträge zur Kenntnis der Potamonidae (Süßwasserkrabben) des Kongogebietes. Revue de Zoologie et de Botanique Africaines, **28**(2): 165-204.

BOTT, R., 1951. *Potamonautes biballensis* Rathbun aus einer Hohle bei Thysville (Belgisch Congo). Revue Zoologique et Botanique d'Afrique, **XLIV**(3): 233-235.

— —, 1955. Die Süßwasserkrabben von Afrika (Crust., Decap.) und ihre Stammesgeschichte. Annales du Musée Royal du Congo Belge, (C, Zoologie) **1**(3)(3): 209-352.

— —, 1959. Potamoniden aus West-Afrika (Crust., Dec.). Bulletin de l'Institut français d'Afrique noire, **21**, série A(**3**): 994-1008.

— —, 1960. Crustacea (Decapoda): Potamonidae. In: B. HANSTRÖM et al. (eds.), South African animal life. Results of the Lund University Expedition in 1950-1952, **7**: 13-18.

— —, 1964. Decapoden aus Angola unter besonderer Berücksichtigung der Potamoniden (Crust. Decap.) und einem Anhang: Die Typen von *Thelphusa pelii* Herklots, 1861. Publicaçoes culturais da Companhia de Diamantes de Angola, Lisboa, **69**: 23-34.

— —, 1965. Die Süßwasserkrabben von Madagaskar. Bullétin du Muséum national d'Histoire naturelle, Paris, **37**(2), 335-350.

— —, 1969. Präadaptation, Evolution und Besiedlungsgeschichte der Süßwasserkrabben der Erde. Natur und Museum, **99**: 266-275.

— —, 1970a. Betrachtungen über die Entwicklungsgeschichte und Verbreitung der Süßwasser-Krabben nach der Sammlung des Naturhistorischen Museums in Genf/Schweiz. Revue Suisse de Zoologie, **77**(2): 327-344.

— —, 1970b. Die Süßwasserkrabben von Europa, Asien, Australien und ihre Stammesgeschichte. Eine Revision der Potamoidea und der Parathelphusoidea (Crustacea, Decapoda). Abhandlungen der Senckenbergischen Naturforschenden Gesellschaft, **526**: 1-338.

CHACE, F. A., 1942. III. Decapod Crustacea. In: Scientific results of the fourth expedition to forested areas in eastern Africa. Bulletin of the Museum of Comparative Zoology, Harvard College, **91**: 185-233.

COLLEN, B., M. RAM, N. DEWHURST, V. CLAUSNITZER, V. KALKMAN, N. CUMBERLIDGE & J. E. M. BAILLIE, 2008. Broadening the coverage of biodiversity assessments. In: J.-C. VIÉ, C. HILTON-TAYLOR & S. N. STUART (eds.), The 2008 review of the IUCN Red List of Threatened Species: 1-9. (IUCN Gland, Switzerland).

COLOSI, G., 1920. I Potamonidi conservati del R. Museo Zoologico di Torino. Bolletino dei Musei di Zoologia ed Anatomia comparata della R. Universita di Torino, **35**(734): 1-39.

CORACE, R. G., N. CUMBERLIDGE & R. GARMS, 2001. A new species of freshwater crab from Rukwanzi, East Africa. Proceedings of the Biological Society of Washington, **114**: 178-187.

CUMBERLIDGE, N., 1986. Ventilation of the branchial chambers in the amphibious West African fresh-water crab *Sudanonautes* (*Convexonautes*) *aubryi monodi* (Balss, 1929) (Brachyura, Potamonautidae). Hydrobiologia, **134**: 53-65.

— —, 1987. Notes on the taxonomy of West African gecarcinucids of the genus *Globonautes* Bott, 1959 (Decapoda, Brachyura). Canadian Journal of Zoology, **65**(9): 2210-2215.

— —, 1991a. The respiratory system of *Globonautes macropus* (Rathbun 1898), a terrestrial freshwater crab from Liberia (Parathelphusoidea, Gecarcinucidae). Crustaceana, **61**: 69-80.

— —, 1991b. *Sudanonautes kagoroensis*, a new species of freshwater crab (Decapoda: Potamoidea: Potamonautidae) from Nigeria. Canadian Journal of Zoology, **69**: 1938-1944.

— —, 1993. Two new species of *Potamonemus* Cumberlidge and Clark, 1992, (Brachyura, Potamoidea, Potamonautidae) from the rain forests of West Africa. Journal of Crustacean Biology, **13**(3): 571-584.

— —, 1994. *Louisea*, a new genus of fresh-water crab (Brachyura, Potamoidea, Potamonautidae) *Globonautes macropus edeaensis* Bott, 1969, from Cameroon. Proceedings of the Biological Society of Washington, **107**(1): 122-131.

— —, 1996a. A taxonomic revision of freshwater crabs (Brachyura, Potamoidea, Gecarcinucidae) from the Upper Guinea forest of West Africa. Crustaceana, **69**(6): 681-695.

— —, 1996b. On the Globonautinae Bott, 1969, freshwater crabs from West Africa (Brachyura: Potamoidea: Gecarcinucidae). Crustaceana, **69**(7): 809-820.

— —, 1997. The African and Madagascan freshwater crabs in the Museum of Natural History, Vienna (Crustacea: Decapoda: Brachyura: Potamoidea). Annalen des Naturhistorischen Museums in Wien, **99B**: 571-589.

— —, 1998. The African and Madagascan freshwater crabs in the Zoologische Staatssammlung, Munich (Crustacea: Decapoda: Brachyura: Potamoidea). Spixiana, **21**(3): 193-214.

— —, 1999. The freshwater crabs of West Africa, family Potamonautidae. Faune et flore tropicales, **35**: 1-382. (Éditions de l'Institut de Recherche pour le Développement, Paris).

— —, 2005. The status and distribution of freshwater crabs. In: W. R. T. DARWALL, K. G. SMITH, T. LOWE & J.-C. VIÉ (compilers), The status and distribution of freshwater biodiversity in Eastern Africa. (IUCN, Gland, Switzerland and Cambridge, UK).

— —, 2008. Insular species of Afrotropical freshwater crabs (Crustacea: Decapoda: Brachyura: Potamonautidae and Potamidae) with special reference to Madagascar and the Seychelles. Contributions to Zoology, **77**(2): 71-81.

— —, 2009a. A revision of the freshwater crabs of Mt. Kenya and the Aberdare Mountains, Kenya, East Africa (Brachyura: Potamoidea: Potamonautidae). Zootaxa, **1981**: 29-42.

— —, 2009b. Chapter 27. Freshwater crabs and shrimps (Crustacea: Decapoda) of the Nile Basin. In: H. J. DUMONT (ed.), The Nile. Origin, environments, limnology and human use. Monographiae Biologicae, **89**: 547-561. (Springer, New York).

— —, 2009c. The status and distribution of freshwater crabs. In: K. G. SMITH, M. D. DIOP, M. NIANE & W. R. T. DARWALL (compilers), The status and distribution of freshwater biodiversity in Western Africa: 56-70. (IUCN, Gland, Switzerland and Cambridge, UK).

— —, 2010. The status and distribution of freshwater crabs. In: N. GARCÍA, A. CUTTELOD & D. ABDUL MALAK (eds.), The status and distribution of freshwater biodiversity in Northern Africa: 71-78. (IUCN Gland, Switzerland, Cambridge, UK and Malaga, Spain).

— —, 2011a. Freshwater crabs of Africa: diversity, distribution, and conservation. In: W. R. T. DARWALL, K. G. SMITH, D. J. ALLEN, R. A. HOLLAND, I. J. HARRISON & E. G. E. BROOKS (eds.), The diversity of life in African freshwaters: under water, under threat. An analysis of the status and distribution of freshwater species throughout mainland Africa: 178-199. (IUCN, Cambridge, UK and Gland, Switzerland).

— —, 2011b. The status and distribution of freshwater crabs. In: E. G. E. BROOKS, D. J. ALLEN & W. R. T. DARWALL (compilers), The status and distribution of freshwater biodiversity in Central Africa: 71-78. (IUCN Gland, Switzerland and Cambridge, UK).

CUMBERLIDGE, N. & C. B. BOYKO, 2000. Freshwater crabs (Brachyura: Potamoidea: Potamonautidae) from the rainforests of the Central African Republic. Proceedings of the Biological Society of Washington, **3**(2): 406-419.

CUMBERLIDGE, N. & P. F. CLARK, 1992. A new genus and species of freshwater crab from Cameroon, West Africa (Crustacea, Brachyura, Potamoidea, Potamonautidae). Bulletin of the British Museum of Natural History (Zoology), London, **58**(2): 149-156.

— — & — —, 2010a. Two new species of freshwater crabs from Mt. Elgon, East Africa (Brachyura: Potamoidea: Potamonautidae) with a redescription of *Potamonautes niloticus* (A. Milne-Edwards, 1887) and notes on their natural history. Journal of Natural History, **44**: 1807-1829.

— — & — —, 2010b. A redescription of *Potamonautes loveni* (Colosi, 1924), a medically important freshwater crab from western Kenya and eastern Uganda, East Africa (Brachyura: Potamoidea: Potamonautidae). In: P. CASTRO, P. J. F. DAVIE, P. K. L. NG & B. RICHER

DE FORGES (eds.), Studies on Brachyura: a homage to Danièle Guinot. Crustaceana Monographs, **11**: 61-74.

— — & — —, 2012. The freshwater crabs of Ethiopia, northeastern Africa, with the description of a new *Potamonautes* cave species (Brachyura: Potamonautidae). Contributions to Zoology, **81**: 235-251.

CUMBERLIDGE, N., P. F. CLARK & J. BAILLIE, 2002. A new species of freshwater crab (Brachyura: Potamoidea: Potamonautidae) from Príncipe, Gulf of Guinea, Central Africa. Bulletin of the British Museum of Natural History (Zoology), London, **68**(1): 13-18.

CUMBERLIDGE, N. & S. R. DANIELS, 2007. A conservation assessment of the freshwater crabs of southern Africa (Brachyura: Potamonautidae). African Journal of Ecology, **46**: 74-79.

— — & — —, 2009. The status and distribution of freshwater crabs. In: D. TWEDDLE (ed.), Freshwater biodiversity of Southern Africa: 66-79. (IUCN, Gland, Switzerland and Cambridge, UK).

— — & — —, 2014. Recognition of two new species of freshwater crabs from the Seychelles based on molecular evidence (Potamoidea: Potamonautidae). Invertebrate Systematics, **28**(1): 17-31.

CUMBERLIDGE, N., S. R. DANIELS & R. V. STERNBERG, 2008. A revision of the higher taxonomy of the Afrotropical freshwater crabs (Decapoda: Brachyura) with a discussion of their biogeography. Biological Journal of the Linnean Society, **93**: 399-413.

CUMBERLIDGE, N. & M. DOBSON, 2008. A new species of freshwater crab (Crustacea: Decapoda: Brachyura: Potamonautidae) from western Kenya. Proceedings of the Biological Society of Washington, **121**(4): 468-474.

CUMBERLIDGE, N. & D. HUGUET, 2003. Les Crustacés Décapodes du Nimba et sa région. Memoires du Muséum national d'Histoire naturelle, Paris, **190**: 211-229.

CUMBERLIDGE, N., S. A. E. MARIJNISSEN & J. THOMPSON, 2007. *Hydrothelphusa vencesi*, a new species of freshwater crab (Brachyura: Potamoidea: Potamonautidae) from southeastern Madagascar. Zootaxa, **1524**: 61-68.

CUMBERLIDGE, N. & K. S. MEYER, 2009. A new species of *Foza* Reed & Cumberlidge, 2006, from northern Madagascar (Decapoda, Brachyura, Potamoidea, Potamonautidae), with a redescription of *F. goudoti* (H. Milne Edwards, 1853) comb. n., and comments on *Skelosophusa prolixa* Ng & Takeda, 1994. ZooKeys, **18**: 77-89.

— — & — —, 2010. A new species of *Potamonautes* MacLeay, 1838, from southwestern Ethiopia (Decapoda, Brachyura, Potamonautidae). In: C. H. J. M. FRANSEN, S. DE GRAVE & P. K. L. NG (eds.), Studies on Malacostraca: Lipke Bijdeley Holthuis memorial volume. Crustaceana Monographs, **14**: 179-190.

— — & — —, 2011. A revision of the freshwater crabs of Lake Kivu, East Africa. Journal of Natural History, **45**: 1835-1857.

CUMBERLIDGE, N. & P. K. L. NG, 2009. Systematics, evolution, and biogeography of the freshwater crabs. In: J. W. MARTIN, K. A. CRANDALL & D. L. FELDER (eds.), Decapod crustacean phylogenetics. Crustacean Issues, **18**: 491-504. (CRC Press, Boca Raton, Florida).

CUMBERLIDGE, N., P. K. L. NG, D. C. J. YEO, C. MAGALHAES, M. R. CAMPOS, F. ALVAREZ, T. NARUSE, S. R. DANIELS, L. J. ESSER, F. Y. K. ATTIPOE, F.-L. CLOTILDE-BA, W. R. T. DARWALL, A. MCIVOR, M. RAM & B. COLLEN, 2009. Freshwater crabs and the biodiversity crisis: importance, threats, status, and conservation challenges. Biological Conservation, **142**: 1665-1673.

CUMBERLIDGE, N. & S. K. REED, 2004. *Erimetopus vandenbrandeni* (Balss, 1936) n. comb., with notes on the taxonomy of the genus *Erimetopus* Rathbun, 1894 (Brachyura: Potamoidea: Potamonautidae) from Central Africa. Zootaxa, **422**: 1-27.

CUMBERLIDGE, N., S. K. REED & C. B. BOYKO, 2004. Distribution patterns of the Malagasy freshwater crabs (Crustacea: Decapoda: Brachyura). Journal of Natural History, **38**: 1133-1157.

CUMBERLIDGE, N. & R. SACHS, 1991. Ecology, distribution, and growth in *Globonautes macropus* (Rathbun, 1898), a tree-living freshwater crab from the rain forests of Liberia (Parathelphusoidea, Gecarcinucidae). Crustaceana, **61**(1): 55-68.

CUMBERLIDGE, N. & R. V. STERNBERG, 2002. The freshwater crabs of Madagascar (Crustacea, Decapoda, Potamoidea). Zoosystema, **24**: 41-79.

CUMBERLIDGE, N., R. V. STERNBERG, I. R. BILLS & H. A. MARTIN, 1999. A revision of the genus *Platythelphusa* A. Milne-Edwards, 1887 from Lake Tanganyika, East Africa (Decapoda: Potamoidea: Platythelphusidae). Journal of Natural History, **33**: 1487-1512.

CUMBERLIDGE, N. & M. TAVARES, 2006. Remarks on the freshwater crabs of Angola, south Central Africa, with the description of *Potamonautes kensleyi*, new species (Brachyura: Potamoidea: Potamonautidae). Journal of Crustacean Biology, **26**: 248-257.

CUMBERLIDGE, N. & M. VANNINI, 2004. Ecology and taxonomy of a tree living freshwater crab (Brachyura: Potamoidea: Potamonautidae) from Kenya and Tanzania, East Africa. Journal of Natural History, **38**: 681-693.

CUMBERLIDGE, N. & W. WRANIK, 2002. A new genus and new species of freshwater crab (Potamoidea: Potamidae) from Socotra Island, Yemen. Journal of Natural History, **36**: 51-64.

DANIELS, S. R., 2011. Reconstructing the colonisation and diversification history of the endemic freshwater crab (*Seychellum alluaudi*) in the granitic and volcanic Seychelles Archipelago. Molecular Phylogenetics and Evolution, **61**: 534-542.

DANIELS, S. R. & J. BAYLISS, 2012. Neglected refugia of biodiversity: mountainous regions in Mozambique and Malawi yield three novel freshwater crab species (Potamonautidae: *Potamonautes*). Zoological Journal of the Linnean Society, **164**: 498-509.

DANIELS, S. R., N. CUMBERLIDGE & M. J. GIBBONS, 1998. *Potamonautes granularis* sp. nov. (Brachyura: Potamonautidae), a new cryptic species of river crab from the Olifants river system, South Africa. Crustaceana, **71**: 885-903.

DANIELS, S. R., N. CUMBERLIDGE, M. PÉREZ-LOSADA, S. A. E. MARIJNISSEN & K. A. CRANDALL, 2006. Evolution of Afrotropical freshwater crab lineages obscured by morphological convergence. Molecular Phylogenetics and Evolution, **40**: 227-235.

DANIELS, S. R., E. E. PHIRI & J. BAYLISS, 2014. Renewed sampling of inland aquatic habitats in southern Africa yields two novel freshwater crab species (Decapoda: Potamonautidae: *Potamonautes*). Zoological Journal of the Linnean Society, **171**(2): 356-369.

DANIELS, S. R., B. A. STEWART & L. BURMEISTER, 2001. Geographic patterns of genetic and morphological divergence amongst populations of a river crab (Decapoda, Potamonautidae) with the description of a new species from mountain streams in the Western Cape, South Africa. Zoologica Scripta, **30**: 181-197.

DANIELS, S. R., B. A. STEWART, G. GOUWS, M. CUNNINGHAM & C. A. MATTHEE, 2002. Phylogenetic relationships of the southern African freshwater crab fauna (Decapoda: Potamonautidae: *Potamonautes*) derived from multiple data sets reveal biogeographic patterning. Molecular Phylogenetics and Evolution, **25**: 511-523.

DARWALL, W. R. T., K. SMITH, T. LOWE & J.-C. VIÉ, 2005. The status and distribution of freshwater biodiversity in Eastern Africa. IUCN SSC Freshwater Biodiversity Assessment Programme: i-viii, 1-36. (IUCN, Gland, Switzerland and Cambridge, UK).

DARWALL, W. R. T., K. G. SMITH, D. J. ALLEN, R. A. HOLLAND, I. J. HARRISON & E. G. E. BROOKS (eds.), 2011. The diversity of life in African freshwaters: under water, under threat. An analysis of the status and distribution of freshwater species throughout mainland Africa: i-xiii, 1-347. (IUCN, Cambridge, UK and Gland, Switzerland).

DOBSON, M., 2004. Freshwater crabs in Africa. Freshwater Forum, **21**: 3-26.

DURIŠ, Z. & M. KOCH, 2010. *Potamonautes lipkei*, a new species of freshwater crab (Decapoda, Potamonautidae) from Senegal. In: C. H. J. M. FRANSEN, S. DE GRAVE & P. K. L. NG (eds.), Studies on Malacostraca: Lipke Bijdeley Holthuis memorial volume. Crustaceana Monographs, **14**: 219-225.

FLOWER, S. S., 1931. Notes on freshwater crabs in Egypt, Sinai and the Sudan. Proceedings of the Zoological Society of London, **1931**: 729-735.

GOUWS, G., B. A. STEWART & M. COKE, 2000. Evidence for a new species of river crab (Decapoda, Brachyura, Potamonautidae) from the Drakensberg, South Africa. Journal of Crustacean Biology, **20**: 743-758.

GOUWS, G., B. A. STEWART & P. REAVELL, 2001. A new species of freshwater crab (Decapoda, Potamonautidae) from the swamp forests of Kwazulu-Natal, South Africa: biochemical and morphological evidence. Crustaceana, **74**: 137-160.

HILGENDORF, F., 1869. Über eine neue Art der kurzschwänzigen Krebse aus den Sammlungen des Baron von der Decken, *Deckenia imitatrix*. Sitzungs-Bericht der Gesellschaft naturforschender Freunde zu Berlin, **1868**(1): 2.

IUCN, 2004. IUCN Red List of Threatened Species. www.iucnredlist.org.

KLAUS, S., D. BRANDIS, P. K. L. NG, D. C. J. YEO & C. D. SCHUBART, 2009. Phylogeny and biogeography of Asian freshwater crabs of the family Gecarcinucidae (Brachyura: Potamoidea). In: J. W. MARTIN, K. A. CRANDALL & D. L. FELDER (eds.), Decapod crustacean phylogenetics. Crustacean Issues, **18**: 509-531. (CRC Press, Boca Raton, Florida).

KLAUS, S., C. SCHUBART & D. BRANDIS, 2006. Phylogeny, biogeography and a new taxonomy for the Gecarcinucoidea Rathbun, 1904 (Decapoda: Brachyura). Organisms Diversity & Evolution, **6**: 199-217.

MACLEAY, W. S., 1838. Brachyurous decapod Crustacea. Illustrations of the Zoology of South Africa 5; being a portion of the objects of natural history chiefly collected during an expedition into the interior of South Africa, under the direction of Dr. Andrew Smith, in the years 1834, 1835, and 1836; fitted out by "The Cape of Good Hope Association for Exploring Central Africa". In: A. SMITH, Illustrations of the zoology of South Africa; consisting chiefly of figures and descriptions of the objects of natural history collected during an expedition into the interior of South Africa, in the years 1834, 1835, and 1836; fitted out by "The Cape of Good Hope Association for Exploring Central Africa", (Invertebrates), (1849): 1-75. (Smith, Elder and Co., London). [For dates of publication see Waterhouse, 1880: 489-491.]

MARIJNISSEN, S. A. E., F. SCHRAM, N. CUMBERLIDGE & E. MICHEL, 2004. Two new species of *Platythelphusa* A. Milne-Edwards, 1887 (Decapoda, Potamoidea, Platythelphusidae) and comments on the taxonomic position of *P. denticulata* Capart, 1952 from Lake Tanganyika, East Africa. Crustaceana, **77**(5): 513-532.

MARIJNISSEN, S. A. E., S. LANGE & N. CUMBERLIDGE, 2005. Revised distribution of the African freshwater crab genus *Deckenia* Hilgendorf, 1868 (Brachyura, Potamoidea, Deckeniidae). Crustaceana, **78**(7): 889-896.

MEYER, K. S. & N. CUMBERLIDGE, 2011. A revision of the freshwater crabs (Crustacea: Decapoda: Brachyura: Potamonautidae) of the Lake Kivu drainage basin in Central and East Africa. Zootaxa, **3011**: 45-58.

MILNE-EDWARDS, A. & E. L. BOUVIER, 1893. Sur une espèce nouvelle du genre *Deckenia* (Hilgendorf). Annales des Sciences naturelles, Zoologie Paris, **15**: 325-336.

MONOD, T., 1977. Sur quelques crustacés Décapodes africains (Sahel, Soudan). Bulletin de Muséum national d'Histoire naturelle Paris, (3) **500**: 1201-1236.

— —, 1980. Décapodes. In: J. R. DURAND & C. LÉVÊQUE (eds.), Flore et faune aquatiques de l'Afrique sahelo-soudanienne, **1**(44): 369-389. (ORSTOM, I.D.T., Paris).

NG, P. K. L., D. GUINOT & P. DAVIE, 2008. Systema Brachyuorum: Part I. An annotated checklist of extant brachyuran crabs of the world. Raffles Bulletin of Zoology, (Supplement) **17**: 1-286.

NG, P. K. L., Z. ŠTEVCIC & G. PRETZMANN, 1995. A revision of the family Deckeniidae Ortmann, 1897 (Crustacea, Decapoda, Brachyura, Potamoidea) from the Seychelles, Indian Ocean. Journal of Natural History, **29**: 581-600.

NG, P. K. L. & M. TAKEDA, 1994. *Skelosophusa* (Crustacea, Decapoda, Brachyura), a new genus of potamonautid freshwater crab from Madagascar, with descriptions of two new species. Bulletin of the National Science Museum, Tokyo, (A) (Zoology) **20**(4): 161-172.

NG, P. K. L. & L. W. H. TAN, 1984. A new freshwater crab, *Stoliczia* (*Johora*) *tiomanensis* n. sp. (Decapoda: Brachyura: Potamidae) from Pulau Tioman, Peninsular Malaysia. Malayan Nature Journal, **37**: 167-174.

ORTMANN, A. E., 1896. Das System der Decapoden-Krebse. Zoologische Jahrbücher, Abteilung für Systematik, Geographie und Biologie der Thiere, **9**: 409-453.

— —, 1897. Carcinologische Studien. Zoologische Jahrbücher, Abteilung für Systematik, Geographie und Biologie der Thiere, **10**: 256-372.

PHIRI, E. E. & S. R. DANIELS, 2013. Hidden in the highlands: the description and phylogenetic position of a novel endemic freshwater crab from Zimbabwe. Invertebrate Systematics, **27**: 530-539.

— — & — —, 2014. Disentangling the divergence and cladogenesis in the freshwater crab species (Potamonautidae: *Potamonautes perlatus* sensu lato) in the Cape Fold Mountains, South Africa, with the description of two novel cryptic lineages. Zoological Journal of the Linnean Society, **170**(2): 310-332.

RATHBUN, M. J., 1904. Les crabes d'eau douce (Potamonidae). Nouvelles Archives du Muséum d'Histoire naturelle, Paris, **6**(4): 255-312.

— —, 1905. Les crabes d'eau douce (Potamonidae). Nouvelles Archives du Muséum d'Histoire naturelle, Paris, **7**(4): 159-322.

— —, 1906. Les crabes d'eau douce (Potamonidae). Nouvelles Archives du Muséum d'Histoire naturelle, Paris, **8**(4): 106-124.

— —, 1921. Brachyuran Crabs of the Belgian Congo. Bulletin of the American Museum of Natural History, **43**: 379-474.

— —, 1935. II. Crustacea. No. 2. — Scientific results of an expedition to rain forest regions in eastern Africa. Bulletin of the Museum of Comparative Zoology at Harvard College, **79**: 23-28.

REED, S. K. & N. CUMBERLIDGE, 2004. Notes on the taxonomy of *Potamonautes obesus* (A. Milne-Edwards, 1868) and *Potamonautes calcaratus* (Gordon, 1929) (Brachyura: Potamoidea: Potamonautidae) from eastern and southern Africa. Zootaxa, **1262**: 1-139.

— — & — —, 2006. Taxonomy and biogeography of the freshwater crabs of Tanzania, East Africa (Brachyura: Potamoidea: Potamonautidae, Platythelphusidae, Deckeniidae). Zootaxa, **418**: 1-137.

SHIH, H.-T., D. C. J. YEO & P. K. L. NG, 2009. The collision of the Indian plate with Asia: molecular evidence for its impact on the phylogeny of freshwater crabs (Brachyura: Potamidae). Journal of Biogeography, **36**: 703-719.

SHY, J.-Y., P. K. L. NG & H.-P. YU, 1994. Crabs of the genus *Geothelphusa* Stimpson, 1858 (Crustacea: Decapoda: Brachyura: Potamidae) from Taiwan, with descriptions of 25 new species. Raffles Bulletin of Zoology, **42**(4): 781-846.

STERNBERG, R. V., N. CUMBERLIDGE & G. RODRIGUEZ, 1999. On the marine sister groups of the freshwater crabs (Crustacea: Decapoda). Journal of Zoological Systematics and Evolutionary Research, **37**: 19-38.

STEWART, B. A., 1997a. Morphological and genetic differentiation between populations of river crabs (Decapoda: Potamonautidae) from the Western Cape, South Africa, with a taxonomic reexamination of *Gecarcinautes brincki*. Zoological Journal of the Linnean Society, **99**: 1-21.

— —, 1997b. Biochemical and morphological evidence for a new species of river crab *Potamonautes parvispina* (Decapoda: Potamonautidae). Crustaceana, **70**: 737-753.

STEWART, B. A., M. COKE & P. A. COOK, 1995. *Potamonautes dentatus*, new species, a fresh-water crab (Brachyura: Potamoidea: Potamonautidae) from KwaZulu-Natal, South Africa. Journal of Crustacean Biology, **15**: 558-568.

STEWART, B. A. & P. A. COOK, 1998. Identification of a new species of river crab (Decapoda: Brachyura: Potamonautidae) from South Africa using morphological and genetic data. Journal of Crustacean Biology, **18**: 556-571.

YEO, D. C. J. & P. K. L. NG, 2003. Recognition of two subfamilies in the Potamidae Ortmann, 1896 (Brachyura, Potamidae), with a note on the genus *Potamon* Savigny. Crustaceana, **76**: 1219-1235.

YEO, D. C. J., P. K. L. NG, N. CUMBERLIDGE, C. MAGALHÃES, S. R. DANIELS & M. R. CAMPOS, 2008. Global diversity of crabs (Crustacea: Decapoda: Brachyura) in freshwater. Hydrobiologia, **595**: 275-286.

First received 28 June 2011.
Final version accepted 29 July 2013.

MORPHOLOGICAL AND MOLECULAR CHARACTERIZATION OF A NEW SPECIES OF *FREDIUS* (DECAPODA, PSEUDOTHELPHUSIDAE) FROM RONDÔNIA, SOUTHERN AMAZONIA, BRAZIL

BY

CÉLIO MAGALHÃES[1,4]), VITOR Q. A. SANCHES[2,5]), LEONARDO G. PILEGGI[3,6]) and FERNANDO L. MANTELATTO[3,7])

[1]) Instituto Nacional de Pesquisas da Amazônia (INPA), Caixa Postal 2223, 69080-971 Manaus, AM, Brazil

[2]) Instituto Federal de Educação, Ciência e Tecnologia do Mato Grosso do Sul (IFMS), Av. Júlio de Castilho, 4960 – Panamá, 79113-000 Campo Grande, MS, Brazil

[3]) Laboratório de Bioecologia e Sistemática de Crustáceos (LBSC), Programa de Pós-Graduação em Biologia Comparada, Departamento de Biologia, Faculdade de Filosofia, Ciências e Letras de Ribeirão Preto (FFCLRP), Universidade de São Paulo (USP), Av. Bandeirantes 3900, 14040-901, Ribeirão Preto, SP, Brazil

ABSTRACT

A new species of freshwater crab of the genus *Fredius* Pretzmann, 1967, from the state of Rondônia, Brazil, is described, illustrated, and characterized in terms of an mtDNA sequence (16S rRNA). *Fredius buritizatilis* sp. nov. is distinguished from its congeners by the following characters of the male first gonopod: a mesial lobe comprising a long, narrow, spear-like projection that is recurved upward; a cephalic lobe with a distinctly inflated proximal portion with a large patch of strong, corneous spines along its mesial and caudal surfaces; and an auxiliary lobe on the caudal surface that is inflated and shorter than cephalic lobe. The mt16S rRNA sequence of *Fredius buritizatilis* sp. nov. establishes its systematic position in relation to other species of *Fredius* included in the analysis. The occurrence of several species of Pseudothelphusidae in the state of Rondônia, Brazil, as well as the relationships of *Fredius buritizatilis* sp. nov. to other species of the genus, suggests that the affinities of some Brazilian pseudothelphusids are not as close as their geographical proximity might suggest.

[4]) Corresponding author; e-mail: celiomag@inpa.gov.br

[5]) e-mail: vitor.sanches@ifms.edu.br

[6]) e-mail: lpileggi@gmail.com

[7]) e-mail: flmantel@usp.br

 Advances in freshwater decapod systematics and biology: 101-114

RESUMO

Uma nova espécie de caranguejo de água doce do gênero *Fredius* Pretzmann, 1967, do estado de Rondônia, Brasil, é descrita e ilustrada, bem como contextualizada em termos de sequências de uma subunidade do DNA ribossomal. *Fredius buritizatilis* sp. nov. se distingue de suas congêneres pelos seguintes caracteres do primeiro gonópodo do macho: lobo mesial na forma de uma projeção longa, estreita, semelhante a uma lança recurvada para cima; lobo cefálico com a porção proximal claramente inflada, portando espinhos córneos robustos nas superfícies mesial e caudal; lobo auxiliar na superfície caudal, um tanto inflado, mais curto do que o lobo cefálico. As sequências do gene 16Smt de *Fredius buritizatilis* sp. nov. confirmaram seu posicionamento sistemático em relação às espécies próximas do gênero *Fredius* incluídas nas análises. A ocorrência de espécies de Pseudothelphusidae no estado de Rondônia, e as afinidades de *Fredius buritizatilis* sp. nov. com outras espécies do gênero, sugerem que a afinidade entre elas não são tão evidentes como a relativa proximidade geográfica poderia indicar.

INTRODUCTION

Freshwater crabs in the tropical environments of the Americas are a group of great relevance because of their high species diversity and their ecological importance (Ng et al., 2008; Cumberlidge & Ng, 2009; De Grave et al., 2009). Despite the species richness of this group there still are many areas of South America that are poorly surveyed for freshwater crabs. This is the case of the pseudothelphusid crab fauna in the Brazilian state of Rondônia in the southern Amazon basin. The only records available for pseudothelphusids in Rondônia are those of Magalhães (1986) who, 25 years ago, examined a series of female specimens from this region and assigned at least two of them to *Fredius reflexifrons* (Ortmann, 1897). Magalhães & Rodríguez (2002) later considered these records uncertain because the taxonomy of pseudothelphusids relies heavily on the morphology of the adult male gonopod to distinguish between species and genera. This means that there are no records of identified species of pseudothelphusids that are currently known from the state of Rondônia. In addition to those two works, the systematics and the taxonomy of the genus *Fredius* Pretzmann, 1967, was addressed by Rodríguez & Pereira (1992), Rodríguez & Campos (1998), Magalhães & Rodríguez (2002), Magalhães et al. (2005, 2006), and Magalhães (2009).

Recent collections from an urban area in the city of Ji-Paraná in east-central Rondônia, Brazil, found an undescribed species of the genus *Fredius*. Our knowledge of the diversity of the pseudothelphusids in the southern Amazon basin has increased recently and includes descriptions of species from adjacent areas including the Tapajós (Magalhães & Türkay, 1986), Xingu

(Magalhães, 2003), Tocantins (Magalhães, 2005), Madeira (Magalhães, 2009), and Aripuanã (Magalhães & Türkay, 2010) drainage basins. The discovery of this new species in Rondônia indicates that the southern Amazonian pseudothelphusid crab fauna is more diverse than previously thought. The new species is described, illustrated, and characterized in terms of a mtDNA sequence (16S rRNA). The new species is described by C. Magalhães and F. L. Mantelatto, who are the taxonomic authorities for *F. buritizatilis* sp. nov.

MATERIAL AND METHODS

Sampling was carried out on 21 May 2010 and on 13 October 2010, and all crabs were captured by hand, after 7:00 pm. Specimens were deposited in the crustacean collections of the Departamento de Biologia, Faculdade de Filosofia, Ciências e Letras de Ribeirão Preto, Universidade de São Paulo, Ribeirão Preto (CCDB), at the Instituto Nacional de Pesquisas da Amazônia, Manaus (INPA), and the Museu de Zoologia, Universidade de São Paulo, São Paulo (MZUSP). The morphological description is based on the male holotype. The following abbreviations were used: carapace breadth (cb), measured across the carapace at its widest point; carapace length (cl), measured along the midline, from the frontal to the posterior margin; carapace height (ch), the maximum height of the cephalothorax; frontal breadth (fb), the breadth of the frontal margin measured along the upper border. Measurements are in millimeters. Other abbreviations used are: P = pereiopod; s = thoracic sternite. The word gonopod, when used alone, refers to the first male gonopod; "igarapé" means small stream in Portuguese. Terminology for gonopod morphology follows Magalhães & Rodríguez (2002).

Both DNA extraction and amplification were made using fresh walking leg muscle tissue of *Fredius buritizatilis* sp. nov. DNA sequences from other species of *Fredius* were also obtained and compared to that of the new species (table I). Genetic vouchers of *F. buritizatilis* sp. nov., from which the tissue samples were obtained, are deposited in the CCDB (larger male paratype CCDB 342) and INPA (holotype INPA 1891) crustacean collections. Other specimens were acquired from the INPA crustacean collection. All specimens used in this study were identified by the first author. All DNA sequences used in this study were generated from our own extractions. The procedures followed Mantelatto et al. (2007, 2009a, b) and Pileggi and Mantelatto (2010), with appropriate modifications. A polymerase chain

TABLE I

Freshwater crab species used for the molecular comparison with collection site and catalogue number, and genetic database accession numbers (GenBank). CCDB = Coleção de Crustáceos, Departamento de Biologia, Faculdade de Filosofia, Ciências e Letras de Ribeirão Preto, Universidade de São Paulo, Ribeirão Preto; INPA = Instituto Nacional de Pesquisas da Amazônia, Manaus; MZUSP = Museu de Zoologia, Universidade de São Paulo, São Paulo; SMF = Forschungsinstitut und Naturmuseum Senckenberg, Frankfurt a. M., Germany

Species	Collection site	Catalogue no.	GenBank accession no.
Fredius buritizatilis sp. nov.	Ji-Paraná, Rondônia, Brazil	INPA 1891	JN402376
Fredius buritizatilis sp. nov.	Ji-Paraná, Rondônia, Brazil	CCDB 342	JN402377
Fredius denticulatus	Serra do Navio, Amapá, Brazil	INPA 582	JN402372
Fredius estevisi	Posto Indígena Parafuri, Roraima, Brazil	INPA 839	JN402379
Fredius fittkaui	Aldeia Balawa-ú, Amazonas, Brazil	INPA 1330	JN402373
Fredius platyacanthus	Comunidade Paapi-ú, Roraima, Brazil	INPA 841	Pending
Fredius reflexifrons	Rio Chumucuí, Bragança, Pará, Brazil	INPA 1512	JN402378
Fredius stenolobus	Rio Tawadu, Bolívar, Venezuela	INPA 833	JN402374
Fredius stenolobus	Aldeia Palimi-ú, Rio Uraricoera basin, Roraima, Brazil	INPA 848	JN402375
Trichodactylus dentatus	Bahia, Brazil	SMF 32763	FM208777

reaction (PCR) was performed in a Thermo® PxE 0.2 Thermal Cycler, using the universal primers: 16Sar (5′-CCGGTCTGAACTCAGATCACGT-3′) and 16Sbr (5′-CGCCTGTTTATCAAAAACAT-3′) (Palumbi et al., 1991) for the 16S rRNA (the large subunit of the ribosomal rRNA). PCR products were purified using a SureClean Plus kit, and were sequenced with the ABI Big Dye® Terminator Mix in an ABI Prism 3100 Genetic Analyzer®. All sequences were confirmed by sequencing both strands. The consensus sequence for the two strands was obtained using BioEdit Version 7.0.7.1 (Hall, 1999). Sequences were edited using BioEdit and aligned in Clustal W (Thompson et al., 1994) with interface in BioEdit, using default parameters. All sequences were submitted to GenBank (table I). A genetic distance analysis (minimum evolution) was conducted using MEGA 5.0 software (Tamura et al., 2011). Support for the nodes on the dendrogram was measured using a bootstrap method (1000 replicates), and only confidence values >50% were reported.

TAXONOMY

Family PSEUDOTHELPHUSIDAE Ortmann, 1893
Genus **Fredius** Pretzmann, 1967
Fredius buritizatilis Magalhães & Mantelatto sp. nov.
(fig. 1)

Material examined. — Holotype: male (genetic voucher: cb 39.7, cl 25.7, ch 17.2, fb 11.8), INPA 1891, Brazil, Rondônia, municipality of Ji-Paraná, buritizal no campus do Centro Universitário Luterano de Ji-Paraná (CEULJI-ULBRA), 10°51′50.6″N 61°57′30.9″W, coll. F. D. de Almeida, 21 May 2010. Paratypes: male (cb 31.4, cl 21.0), female (cb 38.4, cl 24.0), INPA 1892, same data as holotype; 2 males (cb 21.3, cl 14.2; genetic voucher: cb 31.6, cl 20.4), 1 female (cb 23.4, cl 15.6), 1 female with juveniles (cb 35.7, cl 23.0), CCDB 342, same data as holotype, 13 Oct. 2010; 1 male (cb 28.3, cl 18.9), 1 female (cb 22.0, cl 14.8), MZUSP 24441, same data as holotype, 13 Oct. 2010.

Other material examined. — *Fredius beccarii* (Coifmann, 1939), 1 male, Venezuela, Bolívar, Rio Uey, Serranía de Lema, trib. Rio Cuyuní, 06°02′23.5″N 61°30′26.4″W, 135 m elevation, coll. C. Lasso et al., 23 Jan. 2008. *Fredius denticulatus* (H. Milne Edwards, 1853): 1 male (genetic voucher), INPA 582, Brazil, Amapá, Serra do Navio, coll. Projeto Diversitas Neotropica, 7 May 1994. *Fredius estevisi* (Rodríguez, 1966): 1 male (genetic voucher), INPA 839, Brasil, Roraima, igarapé Inajá, Posto Indígena Parafuri, coll. Victor Py-Daniel et al., 22 Apr. 1994. *Fredius fittkaui* (Bott, 1967): 1 male (genetic voucher: cb 69.0, cl 43.2), INPA 1330, Brasil, Amazonas, unnamed stream, Rio Demini watershed, Balawa-ú Yanomami village, 01°47′N 63°46′W, coll. U. C. Barbosa, 5 Sep. 2003. *Fredius platyacanthus* Rodríguez & Pereira, 1992: 2 males, INPA 1328, Brasil, Amazonas, surroundings of Balawa-ú village, igarapé Loahik, affluent Rio Balawa-ú, Rio Demini watershed, 01°47′N 63°46′W, coll. J. Yanomami, 12 Nov. 2003; 1 male, INPA 841, Brasil, Roraima, Comunidade Paapi-ú, 02°39′N 63°09′W, igarapé unnamed, coll. U. C. Barbosa et al., 28 Apr. 1994. *Fredius*

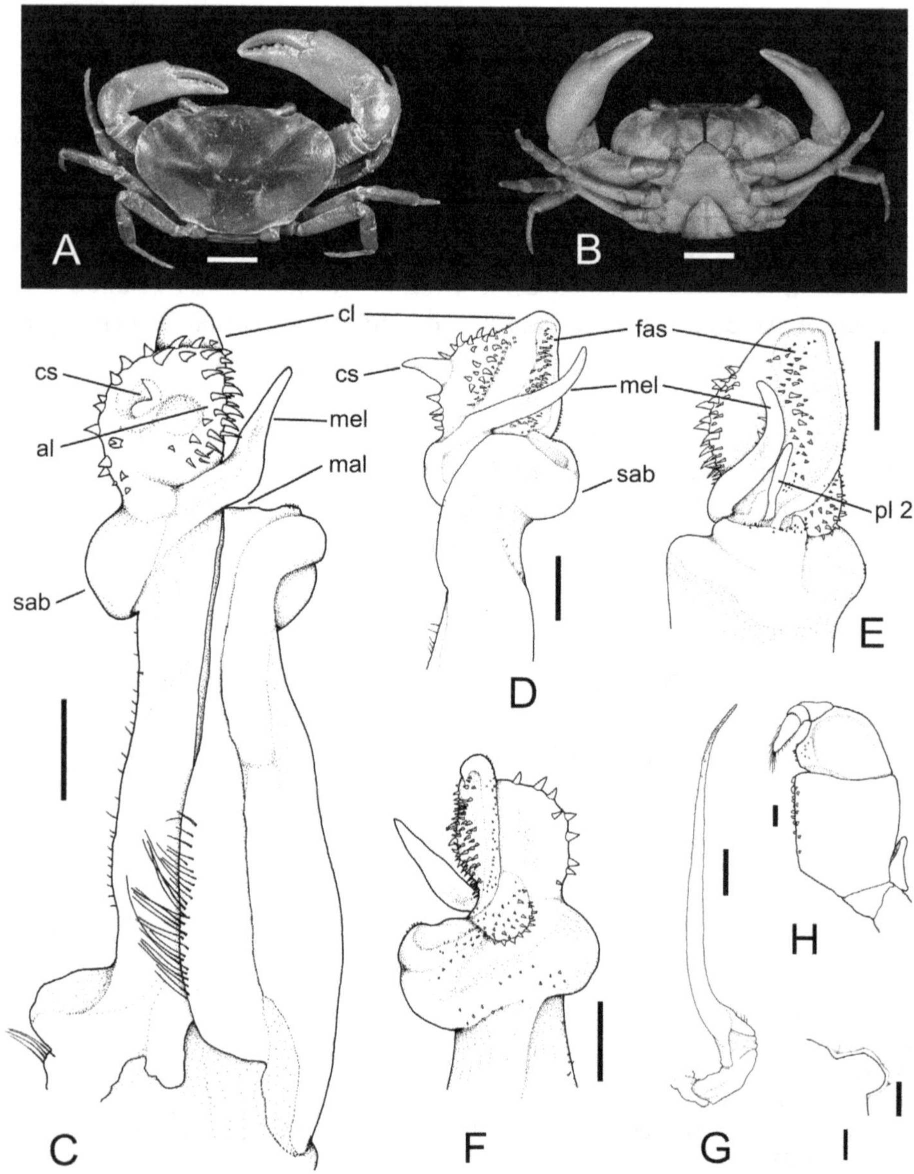

Fig. 1. *Fredius buritizatilis* sp. nov., male holotype (cb 39.7, cl 25.7), INPA 1891. A, dorsal view; B, ventral view; C, left first gonopod, whole limb, mesiocaudal view; D, distal part of left first gonopod, caudal view; E, distal part of left first gonopod, lateral view; F, distal part of left first gonopod, laterocephalic view; G, left second gonopod, whole limb, mesiocaudal view; H, left third maxilliped, external view; I, left aperture of efferent branchial channel. Abbreviations: al = auxiliary lobe; cl = cephalic lobe; cs = cephalic spine; fas = field of apical spines; mal = marginal lobe; mel = mesial lobe; pl 2 = left second gonopod (distal part only); sab = subapical bulge. Scales: A, B = 10 mm; C-I = 1 mm.

reflexifrons (Ortmann, 1897): 1 male (genetic voucher), INPA 1512, Brazil, Pará, Bragança, Rio Chumucuí, coll. S. Alves, 12 Nov. 2004. *Fredius stenolobus* Rodríguez & Suárez, 1994: 1 male (genetic voucher: cb 65.6, cl 42.0), INPA 833, Venezuela, Bolívar, Rio Tawadu, 06°20.147′N 65°02.057′W, coll. C. Magalhães & G. Pereira, 5 Dec. 2000; 1 male (genetic voucher: cb 74.0, cl 48.7), INPA 848, Brasil, Roraima, Rio Uraricoera, Palimi-ú Yanomami village, coll. Victor Py-Daniel et al., 22 May 1995. *Fredius ykaa* Magalhães, 2009: male, holotype, INPA 1465, Brazil, Amazonas, municipality of Maués, Rio Marau, tributary of Rio Maués-Açu, Vila Nova I, 03°44′48″S 57°11′23″W, coll. M. A. V. Correa, 14-28 Jul. 2004.

Diagnosis. — Mesial lobe of male first gonopod long, narrow, spear-like projection recurved upward; cephalic lobe with proximal portion distinctly inflated, bearing large patch of strong, corneous spines along mesial, caudal surfaces. Auxiliary lobe on caudal surface inflated, shorter than cephalic lobe.

Description of the holotype. — Carapace outline ellipsoid (fig. 1A), widest in middle (cb/cl average: 1.54); dorsal surface smooth, convex, regions poorly defined. Pair of distinct gastric pits very close to each other on metagastric region. Cervical grooves narrow, deep, nearly straight, extremities barely reaching anterolateral margin. Postfrontal lobules small, oval; median groove indistinct between postfrontal lobules. Carapace surface between front, postfrontal lobules smooth, sloping gently downwards. Front with distinct upper border; upper border slightly convex in dorsal view, marked with row of very faint papillae; lower border carinate, barely visible in dorsal view, slightly sinuous in frontal view. Upper, lower orbital margins marked by row of very faint papillae; exorbital angle obtuse. Anterolateral margins of carapace with poorly defined notch just behind exorbital angle, fringed by 3 tubercles before cervical groove, followed by row of minute, tuberculiform teeth; posterolateral margins smooth, rounded, marked by faint suture. Epistome narrow; epistomial tooth triangular, deflexed, borders carinate. Suborbital, subhepatic regions of carapace sidewall smooth; pterygostomial regions covered by rather thin pubescence around mouthparts, otherwise smooth.

Endopod of third maxilliped with outer margin of ischium slightly convex, inner margin straight. Exopod of third maxilliped short, approximately 0.31 times length of outer margin of ischium (fig. 1H). Orifice of efferent branchial channel wide, subquadrate (fig. 1I).

First pereiopods moderately heterochelous, holotype with right cheliped larger than left. Major cheliped with merus subtriangular in cross section; upper crest rounded, with irregular longitudinal row of tubercles; internal lower crest with irregular row of conical, blunt teeth increasing in size distally; external lower crest delimited by regular row of low tubercles. Carpus with row of small tubercles and prominent spine on inner side; outer side rounded, smooth. Palm slightly swollen (length/breadth 1.1), smooth on both sides; upper, lower borders rounded, smooth. Fingers slightly gaping when closed (not gaping in smaller cheliped), tips not crossing; both fingers with large rounded teeth, smaller distally; smaller teeth sometimes interspersed with larger ones. Pereiopods 2-5 slender, dactyli 1.42-1.68 times as long as propodi, with 5 longitudinal rows of corneous spines diminishing in size proximally.

Thoracic sternites (fig. 1B) of third maxillipeds, first pereiopods completely fused, except for small notches at lateral edges of sternum. Sternal sulci s4/s5, s5/s6, s6/s7 distinct, faint in middle, deeper near midline of thoracic sternum, ending just before reaching midline; sternal sulcus s7/s8 deep throughout length, reaching midline. Midline of thoracic sternum marked by deep groove in sternites VII-VIII; indistinct in sternites V-VI. Sternoabdominal cavity densely pilose, especially in sternites V-VI.

All abdominal segments free (fig. 1B). Lateral margins of male telson parallel in line with midline axis, tip rounded.

First gonopod robust, widest at base, narrowest approximately 2/3 along length; subapical bulge well developed around lateral, cephalic sides. Marginal suture straight, situated on mesial side (fig. 1C). Marginal lobe large, short, distal border nearly straight, with scattered patch of minute teeth (fig. 1C); marginal lobe separated from caudal surface by shallow groove. Mesial lobe extremely developed as long, narrow, sinuous projection recurved upwards, directed laterally, distally (fig. 1C-F). Cephalic spine strong, recurved, conical, with acuminate tip pointing in mesocaudal direction (fig. 1C, D). Cephalic lobe enlarged, rounded distally, its proximal portion distinctly inflated, bearing large patch of strong, corneous spines along mesial, caudal sides (fig. 1C, D, F). Auxiliary lobe on caudal side, rather inflated, slightly shorter than cephalic lobe (fig. 1C-E); lateral channel indistinct. Field of apical spines well developed as elongated, dense patch of strong spines along lateral side, delimited by cephalic, caudal borders of apex (fig. 1D-F). Second gonopod (fig. 1G) slightly shorter than first, very slender, tapering, distal part rather flattened, with dense patch of minute teeth.

Distribution. — *Fredius buritizatilis* sp. nov. is currently known only from the type locality, which is the drainage area of the middle course of the Rio Machado, a tributary of the Rio Madeira, in turn, a right bank tributary of the Amazon River. Magalhães (1986) previously assigned to *Fredius reflexifrons* two ovigerous females (MZUSP 7047) from Santa Cruz da Serra (in Rio Jaru watershed, an affluent of Rio Machado) and a female (MZUSP 6382) from Nova Esperança, also in the Rio Machado drainage. However, these records of female specimens were later considered uncertain (Magalhães & Rodríguez, 2002) because the taxonomy of these crabs relies almost entirely on the morphology of the male first gonopod. Magalhães (1986) also reported two other records of undetermined pseudothelphusid specimens from Rondônia: one female (MZUSP 6392 — the male symbol was a printing mistake) from the Rio Jaru watershed, and a juvenile (INPA) from the headwaters of Rio Formoso, in the Rio Jaciparaná drainage. Therefore *Fredius buritizalis* is the first identified pseudothelphusid species to be recorded from the state of Rondônia. It is possible that the females of uncertain identity that were recorded earlier from localities in the Rio Machado drainage basin might also belong to *Fredius buritizatilis* sp. nov., but without gonopod characters or DNA sequences their identity cannot be confirmed. More comprehensive surveys of the decapod fauna of Rondônia are needed in order to establish the actual distribution of this new species.

Fredius buritizatilis sp. nov. is the southernmost record of any species of *Fredius* known to date. Apart from a few records of *F. reflexifrons* from the state of Ceará (Brazil) and from the main axis of the Amazon River (Magalhães et al., 2005), and *F. ykaa* from the lower Rio Madeira basin (Magalhães, 2009). All other species of the genus are found north of this in northern Brazil, French Guiana, Suriname, Guiana, Colombia, and Venezuela (Rodríguez, 1982; Rodríguez & Campos, 1998; Magalhães & Rodríguez, 2002; Magalhães et al., 2005, 2006; Magalhães, 2009). Both *F. buritizatilis* sp. nov. and *F. ykaa* are found in the Rio Madeira basin. However the latter species occurs in the lower part of the basin and was found in "terra firme" (highlands never inundated by the periodical floods of the main rivers of the Amazon basin) forest streams, while the former occurs a little more than 7 degrees latitude southwards in a different kind of habitat (see below).

Habitat. — Specimens of the new species were collected in a "buritizal" swamp of the Centro Universitário Luterano de Ji-Paraná (CEULJI-ULBRA), within the urban area of the city of Ji-Paraná. A "buritizal" is a "forest" of basically the *Mauritia flexuosa* palm popularly known as "buriti" which tends to occur in moist depressions with poorly drained soils (Ribeiro & Walter, 2008). The sampling area was about 195 m long and 80 m wide and is currently impacted by human activities, invasion by domestic fauna, and domestic sewage discharge. According to the Secretaria de Estado do Desenvolvimento Ambiental (Rondônia, 2009), the temperature during the year varies between 21.6 and 31.5°C (mean of 26.5°C), the humidity ranges from 35 to 84%, and the mean precipitation is 1711.2 mm/year (varying from 1.8 mm in July to

337 mm in January). The capture of crabs is easier during the dry season when lower water levels expose the entrances of the crab burrows. Male and female crabs were collected from the same burrow.

Etymology. — The species is named after its habitat and the name is formed from the word "buritizal" (forest of buritis) and the Latin suffix, *-atilis*, meaning place of growth.

Molecular aspects. — A 543 base sequence of the 16S rRNA gene from a number of species of *Fredius* was aligned, and the sequence included insertions and deletions. The sequence divergence rates (i.e., uncorrected pairwise differences) estimated among all species of *Fredius* included in the analysis ranged from 1.4-20.9% for the 16S rRNA gene; sequence divergence rates for intraspecific individuals were 0% for 16S rRNA gene. Sequence divergence between *F. buritizatilis* sp. nov. and a number of closely related congeneric species is shown in the Minimum Evolution dendrogram (fig. 2).

Remarks. — The new species is included in the genus *Fredius* based on characters of the male first gonopod such as the open, ear-shaped field of apical spines, the well developed subapical bulge around the lateral and cephalic surfaces ("bourrelet subapical", after Rodríguez, 1982), and the simple marginal process. The molecular analysis positions this taxon within the group formed by the other species of the genus *Fredius* used in this study (table I, fig. 2).

Comparisons between *F. buritizatilis* sp. nov. and *F. ykaa* (whose distribution is closest to that of the new species), suggest that their affinities are not

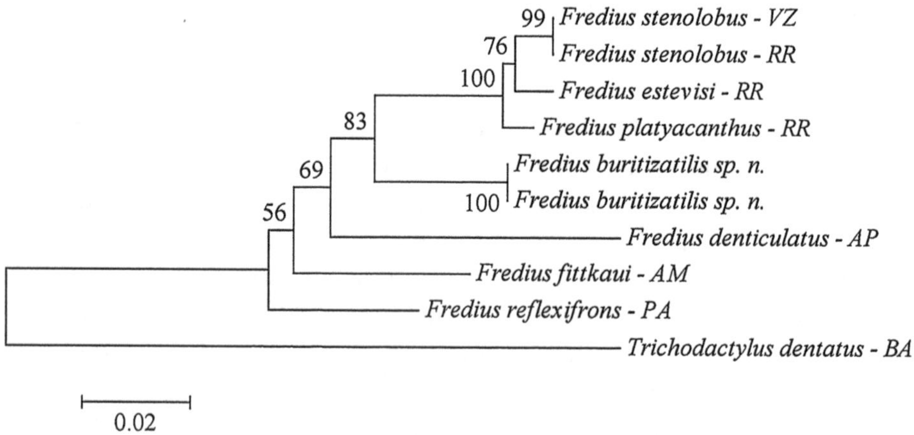

Fig. 2. Dendrogram of distance analysis among some species of *Fredius*, based on the 16S rRNA gene.

as close as their relative geographical proximity might suggest. The first gonopod of *F. buritizatilis* sp. nov. bears a very long, spear-like, distally recurved mesial lobe while in *F. ykaa* this lobe is very small, and is reduced to a small triangular spine pointing in the caudal direction. Other differences between the gonopod of these two taxa can be seen in the cephalic lobe (proximal portion inflated and bearing patch of strong, corneous spines in *F. buritizatilis* sp. nov. vs. flat, with a patch of very small spines in *F. ykaa*), and the cephalic spine (conical, with sharp tip pointing in mesiocaudal direction in of *F. buritizatilis* sp. nov. vs. broadly triangular, with a blunt tip pointing laterally in *F. ykaa*). In addition, the position of the mesial lobe of the gonopod in relation to the cephalic spine, and the situation of the auxiliary lobe are very distinct between both species.

The new species seems to be morphologically most closely related to *F. platyacanthus* Rodríguez & Pereira, 1992, *F. estevisi*, and *F. stenolobus*, based on similar morphologies and positions of the mesial lobe and cephalic spine of the adult male gonopod, in spite of differences in the shape of these processes. Similarly, *F. buritizatilis* sp. nov., *F. platyacanthus* and *F. estevisi* all have a gonopod that has a mesial lobe that is much larger than the cephalic spine (see Rodríguez & Pereira, 1992: 306, fig. 4A-E), and this character is a little larger than that of both *F. stenolobus* (see Rodríguez & Campos, 1998: 766, fig. 2A, B) and *F. adpressus* Rodríguez & Pereira, 1992 (see Rodríguez & Pereira, 1992: 306, fig. 4H-J). The mesial lobe of the male gonopod is equally as developed as the cephalic spine in *F. granulatus* Rodríguez & Campos, 1998 (see Rodríguez & Campos, 1998: 766, fig. 2C, D), *F. chaffanjoni* (Rathbun, 1905) (see Rodríguez & Pereira, 1992: 306, fig. 4F, G), and *F. beccarii* (see Rodríguez & Pereira, 1992: 306, fig. 4FM, N). Other species of *Fredius* [e.g., *F. ykaa*, *F. reflexifrons*, *F. fittkaui* (Bott, 1967), and *F. denticulatus* (H. Milne Edwards, 1853)] have a gonopod with a cephalic spine that is more developed than the mesial lobe (see Magalhães & Rodríguez, 2002: 679, fig. 1; 683, fig. 2, respectively; Rodríguez & Campos, 1998: 766, fig. 2O, P).

These affinities are supported by the dendrogram based on the mt16S rRNA gene (fig. 2). Although the analysis is incomplete due to the unavailability of samples for all species of the genus, *F. buritizatilis* sp. nov. lies closer to *F. platyacanthus*, *F. estevisi* and *F. stenolobus* than to the other species that have a weakly developed mesial lobe. Integration of molecular taxonomy and comparative morphology of pending available species will provide a more robust insight into the systematics of this genus in South America.

ACKNOWLEDGEMENTS

CM and FLM are grateful to the Conselho Nacional de Desenvolvimento Científico e Tecnológico (CNPq) for an ongoing Research Grant (304468/2009-6 and 302748/2010-5, respectively) and LGP is supported by an ongoing Post-Doctoral fellowship CAPES (02630/2009-5). Additional support for this project was provided by FAPESP (Biota Proc. 2010/50188-8; Coleções Cientificas Proc. 2009/54931-0), CNPq (Proc. 491490/2004-6; 490353/2007-0; 471011/2011-8) and CAPES/DAAD (Proc. 315/09) to FLM. We are deeply grateful to Francisco Dutra de Almeida for help in collecting the new species, and to Jose Christopher E. Mendoza and an anonymous reviewer for valuable comments that greatly improved the text. The authors thank Barbara Robertson for correcting the English text, and Darren Yeo and Sebastian Klaus for the invitation to contribute to this special volume on freshwater decapods.

REFERENCES

BOTT, R., 1967. Flusskrabben aus Brasilien und benachbarter Gebiete. *Potamocarcinus* (*Kingsleya*) Ortmann 1987 (Crustacea, Decapoda). Senckenbergiana biologica, **48**(4): 301-312.

COIFMANN, I., 1939. Potamonidi dell R. Museo Zoologico di Torino. Archivo Zoologico Italiano, **27**: 93-116.

CUMBERLIDGE, N. & P. K. L. NG, 2009. Systematics, evolution, and biogeography of freshwater crabs. In: J. W. MARTIN, K. A. CRANDALL & D. L. FELDER (eds.), Decapod crustacean phylogenetics. Crustacean Issues, **18**: 245-260. (Taylor & Francis/CRC Press, Boca Raton).

DE GRAVE, S., N. D. PENTCHEFF, S. T. AHYONG, T. Y. CHAN, K. A. CRANDALL, P. C. DWORSCHAK, D. L. FELDER, R. M. FELDMANN, C. H. J. M. FRANSEN, L. Y. D. GOULDING, R. LEMAITRE, M. E. Y. LOW, J. W. MARTIN, P. K. L. NG, C. E. SCHWEITZER, S. H. TAN, D. TSHUDY & R. WETZER, 2009. A classification of living and fossil genera of decapod crustaceans. Raffles Bulletin of Zoology, (Supplement) **21**: 1-109.

HALL, T. A., 1999. BioEdit: a user-friendly biological sequence alignment editor and analysis program for Windows 95/98/ NT. Nucleic Acids Symposium Series, **41**: 95-98.

MAGALHÃES, C., 1986. Revisão taxonômica dos caranguejos de água doce brasileiros da família Pseudothelphusidae (Crustacea, Decapoda). Amazoniana, **9**(4): 609-636.

— —, 2003. The occurrence of freshwater crabs (Crustacea: Decapoda: Pseudothelphusidae, Trichodactylidae) in the Rio Xingu, Amazon Region, Brazil, with description of a new species of Pseudothelphusidae. Amazoniana, **17**(3/4): 377-386.

— —, 2005. A new species of freshwater crab (Crustacea: Decapoda: Pseudothelphusidae) from the southeastern Amazon Basin. Nauplius, **12**(2): 99-107 [2004].

— —, 2009. A new species of freshwater crab of the genus *Fredius* Pretzmann, 1967 from the middle Amazon River basin, Brazil (Crustacea: Decapoda: Pseudothelphusidae). Proceedings of the Biological Society of Washington, **112**(1): 81-86.

MAGALHÃES, C., F. A. ABRUNHOSA, M. O. PEREIRA & M. A. MELO, 2005. New records of *Fredius denticulatus* (H. Milne Edwards, 1853) and *F. reflexifrons* (Ortmann, 1897), and the eastern limits of the distribution of the pseudothelphusid crabs (Crustacea: Decapoda) in Brazil. Acta Amazonica, **35**(1): 93-96.

MAGALHÃES, C., U. C. BARBOSA & V. PY-DANIEL, 2006. Decapod crustaceans used as food by the Yanomami Indians of the Balawa-ú village, State of Amazonas, Brazil. Acta Amazonica, **36**(3): 369-374.

MAGALHÃES, C. & G. RODRÍGUEZ, 2002. The systematic and biogeographical status of *Fredius reflexifrons* (Ortmann, 1897) and *Fredius fittkaui* (Bott, 1967) (Crustacea: Brachyura: Pseudothelphusidae) from the Amazon and Atlantic Guianas river basins. Acta Amazonica, **34**(2): 677-689.

MAGALHÃES, C. & M. TÜRKAY, 1986. *Brasiliothelphusa*, a new Brazilian freshwater-crab genus (Crustacea: Decapoda: Pseudothelphusidae). Senckenbergiana biologica, **66**(4/6): 371-376.

— — & — —, 2010. A new freshwater crab of the genus *Brasiliothelphusa* Magalhães & Türkay, 1986 from Rio Aripuanã, southern Amazon Region, Brazil (Crustacea: Decapoda: Pseudothelphusidae). Nauplius, **18**(2): 103-108.

MANTELATTO, F. L., L. M. PARDO, L. G. PILEGGI & D. L. FELDER, 2009b. Taxonomic re-examination of the hermit crab species *Pagurus forceps* and *Pagurus comptus* (Decapoda: Paguridae) by molecular analysis. Zootaxa, **2133**: 20-32.

MANTELATTO, F. L., R. ROBLES & D. L. FELDER, 2007. Molecular phylogeny and taxonomic approach of the crab genus *Portunus* (Crustacea, Portunidae) from Western Atlantic. Zoological Journal of the Linnean Society, **150**(1): 211-220.

MANTELATTO, F. L., R. ROBLES, C. D. SCHUBART & D. L. FELDER, 2009a. Molecular phylogeny of the genus *Cronius* Stimpson, 1860, with reassignment of *C. tumidulus* and several American species of *Portunus* to the genus *Achelous* De Haan, 1833 (Brachyura: Portunidae). In: J. W. MARTIN, K. A. CRANDALL & D. L. FELDER (eds.), Decapod crustacean phylogenetics. Crustacean Issues, **18**: 567-579. (Taylor & Francis/CRC Press, Boca Raton).

MILNE EDWARDS, H., 1853. Mémoire sur la famille des Ocypodien. Annales des Sciences Naturelles, Zoologie, 3e série, **20**: 163-228.

NG, P. K. L., D. GUINOT & P. J. F. DAVIE, 2008. Systema Brachyurorum: Part I. An annotated checklist of extant brachyuran crabs of the world. Raffles Bulletin of Zoology, (Supplement) **17**: 1-286.

ORTMANN, A., 1893. Die Decapoden-Krebse des Straßburger Museum, mit besonderer Berücksichtigung der von Herrn Dr. Döderlein bei Japan und bei den Liu-Kiu Inseln gesammelten und zur Zeit im Straßburger Museum aufbewahrten Formen. VII. Theil. Abtheilung: Brachyura (Brachyura genuina Boas) II. Unterabtheilung: Cancroidea, 2. Section: Cancrinea, 1. Gruppe: Cyclometopa. Zoologischen Jahrbücher, Abtheilung für Systematik, Geographie und Biologie der Thiere, **7**: 411-495, pl. 17.

— —, 1897. Carcinologische Studien. Zoologische Jahrbücher, Abteilung für Systematik, Geographie und Biologie der Tiere, **10**: 258-372, pl. 17.

PALUMBI, S., A. MARTIN, S. ROMANO, W. O. MCMILLAN, L. STICE & G. GRABOWSKI, 1991. The simple fool's guide to PCR: 1-46. (University of Hawaii, Department of Zoology, Honolulu).

PILEGGI, L. G. & F. L. MANTELATTO, 2010. Molecular phylogeny of the freshwater prawn genus *Macrobrachium* (Decapoda, Palaemonidae), with emphasis on the relationships among selected American species. Invertebrate Systematics, **24**(2): 194-208.

PRETZMANN, G., 1967. Über einige südamerikanische Süßwasserkrabben (Pseudothelphusidae). Vorläufige Mitteilung. Entomologische Nachrichtenblatt, **14**(2): 23-26.

RATHBUN, M. J., 1905. Les Crabes d'eau douce (Potamonidae). Nouvelles archives du Muséum national d'Histoire naturelle, **7**(4): 159-321.

RIBEIRO, J. F. & B. M. T. WALTER, 2008. As principais fitofisionomias do cerrado. In: S. M. SANO, S. P. DE ALMEIDA & J. F. RIBEIRO (eds.), Cerrado: ecologia e flora, **1**: 151-212. (EMBRAPA Informação Tecnológica, Planaltina).

RODRÍGUEZ, G., 1966. Three new species of *Pseudothelphusa* from Venezuela. Zoologische Mededelingen, **41**: 259-267.

— —, 1982. Les crabes d'eau douce d'Amerique. Famille des Pseudothelphusidae. Faune Tropicale, **22**: 1-224. (ORSTOM, Paris).

RODRÍGUEZ, G. & M. CAMPOS, 1998. A cladistic revision of the genus *Fredius* (Crustacea: Decapoda: Pseudothelphusidae) and its significance to the biogeography of the Guianan lowlands of South America. Journal of Natural History, **32**(5): 763-775.

RODRÍGUEZ, G. & G. PEREIRA, 1992. New species, cladistic relationships and biogeography of the genus *Fredius* (Crustacea: Decapoda: Pseudothelphusidae) from South America. Journal of Crustacean Biology, **12**(2): 298-311.

RODRÍGUEZ, G. & H. SUÁREZ, 1994. *Fredius stenolobus*, a new species of freshwater crab (Decapoda: Brachyura: Pseudothelphusidae) from the Venezuelan Guiana. Proceedings of the Biological Society of Washington, **107**(1): 132-136.

RONDÔNIA, 2009. Secretaria de Estado do Desenvolvimento Ambiental (SEDAM). Boletim Climatológico de Rondônia, **2007**: 1-36. (SEDAM, Porto Velho).

TAMURA, K., D. PETERSON, N. PETERSON, G. STECHER, M. NEI & S. KUMAR, 2011. MEGA5: molecular evolutionary genetics analysis using maximum likelihood, evolutionary distance, and maximum parsimony methods. Molecular Biology and Evolution, **28**(10): 2731-2739.

THOMPSON, J. D., D. G. HIGGING & T. J. GIBSON, 1994. CLUSTAL W: improving the sensitivity of progressive multiple sequence alignment through sequence weighting specific gap penalties and weight matrix choice. Nucleic Acids Research, **22**: 4673-4680.

First received 23 July 2011.
Final version accepted 29 July 2013.

DESCRIPTION OF A NEW FRESHWATER CRAB SPECIES OF THE GENUS *POTAMON* (DECAPODA, BRACHYURA, POTAMIDAE) FROM IRAN, BASED ON MORPHOLOGICAL AND GENETIC CHARACTERS

BY

ALIREZA KEIKHOSRAVI[1,2,3]) and CHRISTOPH D. SCHUBART[1,4])

[1]) Biologie 1, Institut für Zoologie, Universität Regensburg, 93040 Regensburg, Germany

[2]) Department of Biology, Faculty of Science, University of Hakim Sabzevari, Sabzevar, Iran

ABSTRACT

A new freshwater crab species of the genus *Potamon* Savigny, 1816, from Iran is described. *Potamon ilam* sp. nov. differs from the closely related species *Potamon persicum* Pretzmann, 1962 and *P. mesopotamicum* Brandis et al., 1998, by the shape of the first gonopod and carapace characters. Other differences between these taxa are evident from nuclear 28S rRNA and mitochondrial 16S rRNA gene sequences. The phylogenetic data also show noticeable geographic variation within *P. persicum* from different parts of its range. The present work increases the number of species in the genus *Potamon* to twenty-two.

INTRODUCTION

Potamon Savigny, 1816, is a well-known freshwater crab genus with a distribution ranging from the Middle East to southern Europe and North Africa. Pretzmann (1962, 1965, 1966, 1971, 1976a, b) revised the taxonomy of the freshwater crabs of the Middle East and described a number of new species and subspecies of *Potamon*. However, Pretzmann's collections did not cover the entire range of *Potamon* in Iran and left large areas unexplored. Brandis et al. (2000) revised the entire genus *Potamon* and described *Potamon bilobatum* Brandis, Storch & Türkay, 2000 from Iran, and synonymized a number of Pretzmann's subspecies and infraspecific taxa. A total of twenty-one species were thus included in the genus *Potamon* (Jesse et al., 2010, 2011) of which six

[3]) Corresponding author; e-mail: A.Keikhosravi@hsu.ac.ir

[4]) e-mail: christoph.schubart@biologie.uni-regensburg.de

species, i.e., *P. bilobatum*; *P. persicum* Pretzmann, 1962; *P. ruttneri* Pretzmann, 1962; *P. strouhali* Pretzmann, 1962; *P. transcaspicum* Pretzmann, 1962; and *P. ibericum* (Bieberstein, 1809) are found in Iran.

Taxonomic revisions of *Potamon* focused on the Middle East are important for two reasons. First, it has been suggested that the European freshwater crab fauna originated in the Middle East (Pretzmann, 1987; Klaus & Gross, 2010; Jesse et al., 2011) and thus a better resolution of the taxonomy and phylogeny of *Potamon* in this area promises to provide important insights into the biogeography and diversity within this genus (Brandis et al., 2000). Second, the Middle East was recently a center of orogenic activities (viz., the current structure of Zagros Mountains) that may have resulted in the isolation of river systems in the past (Kinzelbach, 1980; Banarescu, 1991; Rangzan & Iqbaluddin, 1995).

The present paper describes a new crab species from the Zagros Mountains of Iran based on morphological and molecular evidence. This is the seventh species of *Potamon* from Iran.

MATERIAL AND METHODS

Specimens were collected during three field trips in 2009 and 2010 (fig. 1, table I) and preserved in 70% ethanol for morphological studies. One walking leg from each individual was removed and preserved in 100% ethanol for molecular studies. Two specimens of *P. mesopotamicum* (SMF 23315) from Syria and *P. transcaspicum* (ZUTC Pot.1088) from Iran (see table I) were used for tissue and DNA extraction for phylogenetic analyses. The studied specimens are deposited in the following collections: Zoology Museum, University of Tehran (ZUTC); Senckenberg Research Institute and Natural History Museum, Frankfurt am Main (SMF); Natural History Museum, London (NHM); Naturhistorisches Museum, Wien (NHMW); Zoological Reference Collection of the Raffles Museum of Biodiversity Research, National University of Singapore (ZRC); and Naturalis Biodiversity Center, Leiden (RMNH).

DNA was isolated from the muscle tissue extracted from walking legs using a modified Puregene method (Gentra Systems). One mitochondrial (16S rRNA) and one nuclear (28S rRNA) ribosomal gene were partially amplified using the following primer combinations: 16L29 (5′-CATATTATCTGCC AAAATAG-3′) and 16HLeu (5′-YGCCTGTTTATCAAAAACAT-3′) (Schu-

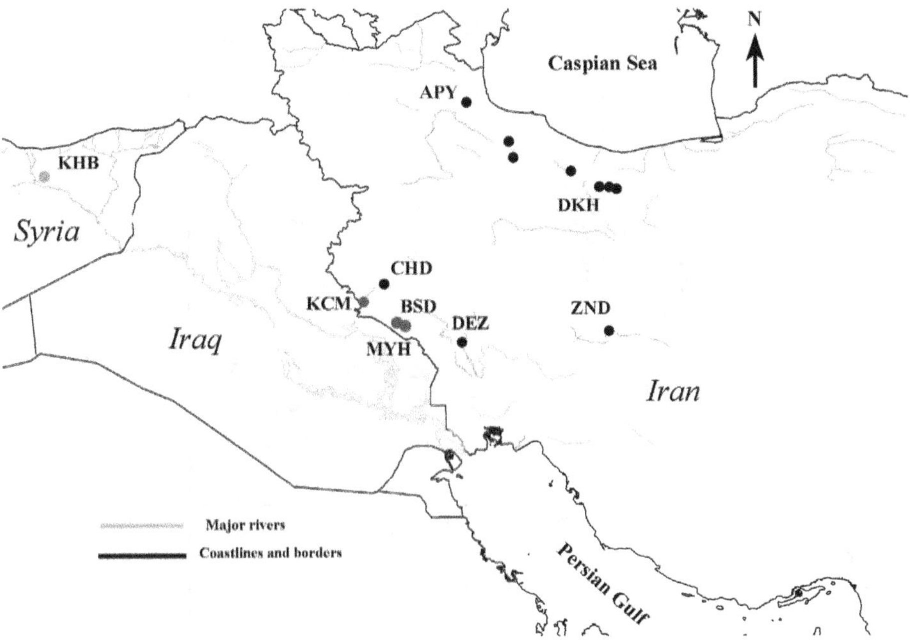

Fig. 1. Map of Iran showing the principal rivers. Black circles represent the sampling sites for *Potamon persicum* Pretzmann, 1962, dark gray circles represent sampling sites for *P. ilam* sp. nov., and, light gray circles represent sampling sites for *P. mesopotamicum* Brandis, Storch & Türkay, 1998. For abbreviations see table I.

bart, 2009) for the 16S rRNA, 28L9 (5′-GACCCGTCTTGAAACACGG-3′) (newly designed) and 28Sb (5′-TCGGAAGGAACCAGCTACTA-3′) (Whiting et al., 1997) for the 28S rRNA. Polymerase chain reaction (PCR) was carried out under the following conditions: 94°C 45 s / 48°C 1 min / 72°C 1 min (40 cycles) for 16S rRNA (16S) and 97°C 45 s / 58°C 1 min / 72°C 1 min (40 cycles) for 28S rRNA (28S) (denaturing / annealing / extension, respectively). The sequences were obtained with an ABI Prism 310 Genetic Analyser (Applied Biosystem, Foster City, USA) or by outsourcing to LGC Genomics. The sequences were corrected manually with BioEdit (version 5.09; Hall, 1999), aligned with Mafft (version 6; Katoh et al., 2002), and deposited at the European Molecular Database EMBL (accession numbers in table I). In addition, a sequence of *P. persicum* from GenBank (FM180116) (Klaus et al., 2009) was included in the analyses.

A phylogeny based on a 644 basepair alignment of the 16S mitochondrial DNA (mtDNA) was inferred by Bayesian Inference (BI) and performed with MrBayes (version 3.1.2; Huelsenbeck & Ronquist, 2001). The appropriate substitution model was previously evaluated using MrModeltest (version 2.3; Ny-

TABLE I

Specimens used for DNA sequencing with collection number, locality, coordinates and GenBank accession numbers

Species	Voucher number	Locality	Coordinates	Label	Accession no. of 16S	Accession no. of 28S
P. ibericum	ZUTC Pot.1089	Iran, Gorgan, Naharkhoran	36°45.68′N 54°28.74′E	–	HE963848	–
P. ilam sp. nov.	SMF 39028	Iran, Ilam Prv., Meymeh R.	32°44.60′N 47°9.366′E	MYH	HE963847	HE970761
P. ilam sp. nov.	ZUTC Pot.1078	Iran, Ilam Prv., Meymeh R.	32°44.60′N 47°9.366′E	MYH	–	HE970762
P. ilam sp. nov.	SMF 39026	Iran, Ilam Prv., Bishehderaz R.	32°49.41′N 46°58.34′E	BSD	HE963846	HE970759
P. ilam sp. nov.	SMF 39025	Iran, Ilam Prv., Bishehderaz R.	32°49.41′N 46°58.34′E	BSD	–	HE970760
P. ilam sp. nov.	SMF 39027	Iran, Ilam Prv., Konjacham R.	33°17.35′N 46°15.13′E	KCM	HE963845	–
P. ilam sp. nov.	SMF 39027	Iran, Ilam Prv., Konjacham R.	33°17.35′N 46°15.13′E	KCM	–	HE970758

TABLE I
(Continued)

Species	Voucher number	Locality	Coordinates	Label	Accession no. of 16S	Accession no. of 28S
P. persicum	SMF 39030	Iran, Khuzestan Prv., Dezful, Dez R.	32°23.28′N 48°23.67′E	DEZ	HE963843	HE970753
P. persicum	SMF 39029	Iran, Ilam Prv., Chardavol R.	33°40.17′N 46°41.48′E	CHD	HE963842	HE970752
P. persicum	SMF 39031	Iran, Isfahan, Zayandehrood R.	32°38.32′N 51°35.97′E	ZND	HE963844	HE970754
P. persicum	SMF 39038	Iran, Khalkhal, Arpachay R.	37°40.92′N 48°30.23′E	APY	–	HE970757
P. persicum	SMF 39032	Iran, Tehran, Darakeh R.	35°49.09′N 51°22.85′E	DKH	–	HE970755
P. persicum	ZUTC Pot.1082	Iran, Tehran, Darakeh R.	35°49.09′N 51°22.85′E	DKH	–	HE970756
P. mesopotamicum	SMF 23315	Syria, Al-Hassakah Prv., Al-Khabur R.	36°42′N 40°11′E	KHB	HE963849	HE970763
P. transcaspicum	ZUTC Pot.1088	Iran, Sabzevar, Zardkoohi R.	36°17.29′N 57°23.34′E	–	HE963841	–

lander, 2004). The best model was selected by the Akaike information criterion (AIC). Four Montecarlo Markov chains (MCMC) were run for 2 000 000 generations, saving trees every 500 generations. The − ln L converged on a stable value between 5000 and 10 000 generations ('burn-in phase'). Consequently, the first 20 000 generations were excluded from the analysis. The topology and posterior probabilities of the phylogeny were determined by constructing a 50% majority-rule consensus tree using the sumpt option in MrBayes. *Potamon ibericum* (ZUTC Pot.1089) from Iran served as outgroup for phylogenetic reconstructions, because it belongs to a distinct morphological group (a former subgenus).

A statistical parsimony network analysis was constructed using the software TCS version 1.21 (Clement et al., 2000) for twelve nuclear 28S rRNA sequences (accession numbers in table I) with an alignment length of 646 bases, in order to represent the relationships among the closely related sequences (table I, fig. 6).

Characters of the first gonopod, median tooth of gastric mill, cheliped, walking legs, carapace and abdomen were observed under light microscopy. Figures were prepared using a microscope fitted with a camera lucida. The gastric mill terminology follows that used by Naderloo et al. (2010).

TAXONOMY

Family POTAMIDAE Ortmann, 1896
Genus **Potamon** Savigny, 1816
Potamon ilam sp. nov.
(figs. 2-4)

Material examined. — Holotype: 1 ♂ (38.27 × 31.01 carapace width × carapace length in mm) (SMF 39025), Iran: Ilam Province, 82 km southeast of Mehran, road to Dehloran, Bishehderaz River, 361 m a.s.l., 32°49.412′N 46°58.346′E, leg. A. Keikhosravi, B. Fathinia & M. Moradmand, 17 Apr. 2009. Paratypes: 1 ♂, 1 ♀, 1 j (j) (44.64 × 35.20, 46.10 × 37.26, 22.69 × 18.15) (SMF 39026), 2 ♂♂, 1 ♀, 1 j (35.13 × 28.02, 35.90 × 28.34, 46 × 36.17, 26.50 × 21.69) (NHMW 25425), 1 ♂, 1 ♀, 1 j (42.80 × 33.81, 38.05 × 30.46, 21.46 × 17.18) (RMNH.CRUS.D.54892), 1 ♂, 1 ♀, 2 j (31.25 × 24.24, 37.74 × 30.36, 18.50 × 14.73, 16.85 × 13.63) (ZUTC Pot.1077), same collection data as holotype; 1 ♂, 4 ♀♀, 6 j (34.19 × 27.30, 40.98 × 32.77, 33.40 × 26.73, 35.42 × 28.20, 34.24 × 27.31, 18.86 × 15.38, 28.80 × 22.78, 19.35 × 15.41, 28.90 × 23.10, 23.90 × 18.65, 27.80 × 21.53) (SMF 39027), Ilam Province, 43 km southwest of Ilam, road to Mehran, Konjacham River, 33°17.350′N 46°15.130′E, leg. A. Keikhosravi, B. Fathinia & M. Moradmand, 16 Apr. 2009; 1 ♂, 1 ♀, 5 j (41.74 × 32.76, 46.30 × 36.36, 27.08 × 21.91, 25.40 × 20.21, 26.74 × 21.74, 17.61 × 14.06, 15.68 × 12.66) (SMF 39028), 2 ♂♂, 1 ♀, 4 j (37.60 × 29.48, 29.10 × 23.74, 34.50 × 27.93, 26.31 × 21.47, 19.80 ×

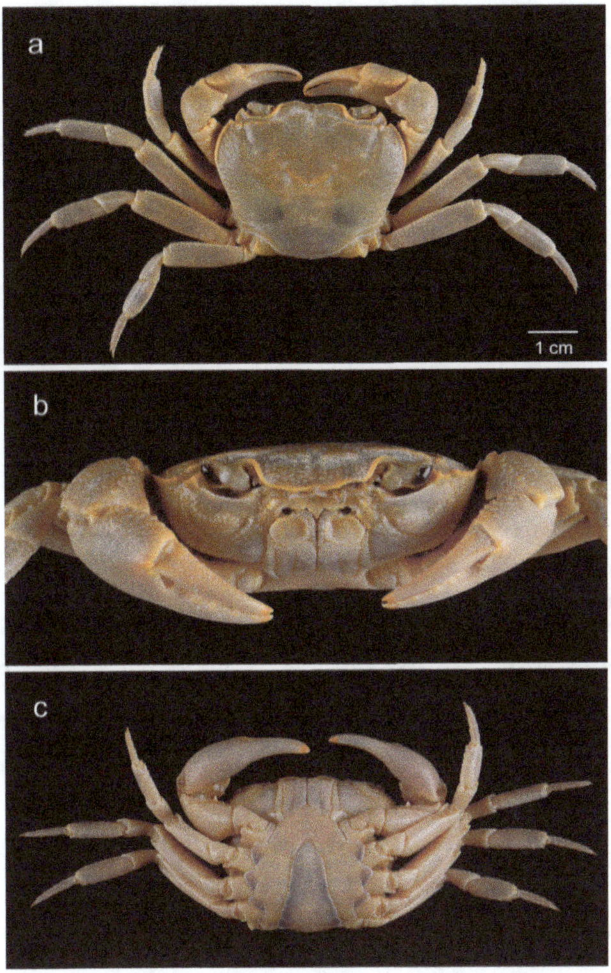

Fig. 2. *Potamon ilam* sp. nov., holotype (SMF 39025). a, dorsal view; b, frontal view; c, ventral view.

16.13, 16.75 × 13.44, 17.32 × 14.03) (ZRC 2012.0162), 1 ♂, 3 j (37.89 × 30.80, 24.65 × 19.83, 26.76 × 21.45, 16.40 × 13.11) (ZUTC Pot.1078), Ilam Province, 12 km northwest of Dehloran, near to Meymeh, Meymeh River, 32°44.6′N 47°9.366′E, leg. A. Keikhosravi, B. Fathinia & M. Moradmand, 17 Apr. 2009.

Diagnosis. — Posterolateral region of the carapace gently flexed downward. Terminal segment of the first gonopod (G1) conical in shape, mesial part straight, slight distal depression, lateral part slightly convex (fig. 3a-c).

Description. — Carapace distinctly broader than long (cw/cl = 1.25); upper surface of carapace convex (ch/cw = 0.4), glabrous, smooth; anterolateral region granular; posterolateral region with curved finely granulated ridges

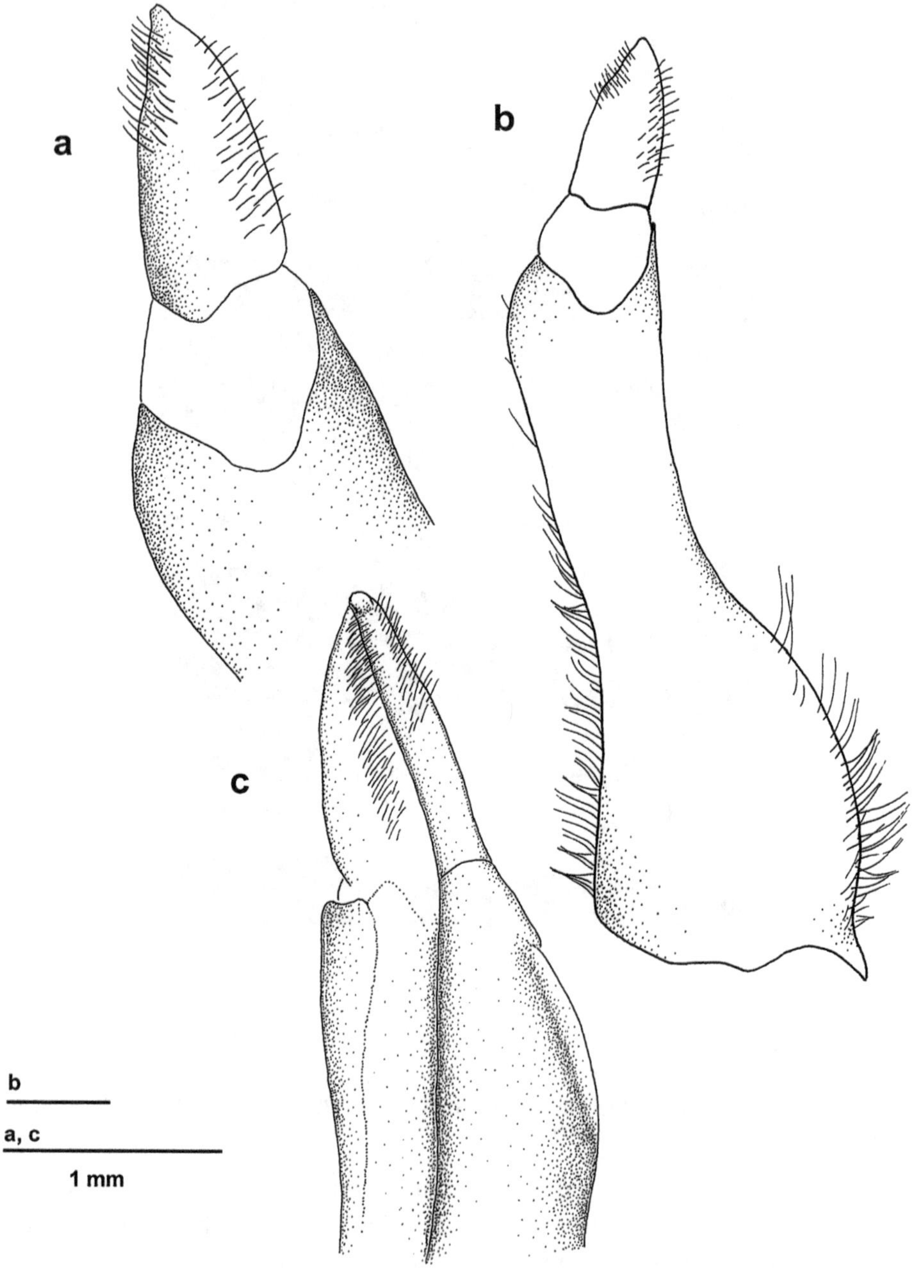

Fig. 3. *Potamon ilam* sp. nov., holotype (SMF 39025). a-b, right G1 (dorsal view); c, ventral view of terminal article.

(carinae), nearly parallel, variably sized, last one longest, nearly parallel with posterior margin; posterolateral margins strongly converging posteriorly.

Cervical groove distinct, deeper anteriorly; H-shaped depression shallow, distinct. Cephalothoracic region slightly convex; mesogastric region distinct, almost smooth; mesogastric grooves well defined; epigastric region visible, separated from postorbital crest by shallow groove. Frontal ridge low, granulated; frontal region flexed downward, strongly granulated, depressed medially, almost bilobed, with field of short, faint setae; frontal margin gently sinuous, covered by blunt, hardly visible granules; postfrontal crest sharp, very prominent and detached from anterolateral region by deep cervical groove. Inner supraorbital margin smooth, outer margin finely tuberculated under magnification; infraorbital margin finely tuberculated. Exorbital tooth triangular, distinct, blunt, lateral side slightly serrate. Anterolateral margin arched, bending inward posteriorly; first epibranchial tooth short, blunt; epibranchial denticles (31-43) short, slightly serrate. Posterolateral region gently flexed downward, with short and hardly visible setae sparsely covering it. Epistome smooth laterally with some short setae; slightly granulated medially.

Third maxilliped ischium with deep median sulcus; exopod with flagellum nearly as long as merus. Anterior part of suborbital lobe smooth, glabrous; posterior region with curved lines of tubercles.

Thoracic sternites smooth, glabrous, pilose only at margins of segments. ♂ pleon long, triangular, smooth; lateral margins with short, dense setae; abdominal segments 2-6 progressively longer; telson longer than abdominal segment 6, tip rounded.

Male chelipeds medium-sized, unequal. Merus with upper margin serrate; inner-ventral margin granular, granules with wide base, progressively larger from proximal to distal; outer-ventral margin granular, progressively larger from proximal to distal; outer surface smooth, becoming serrated near upper margin; inner surface smooth; ventral surface with distinct spine-shaped tooth on distal part near inner-ventral margin. Carpus with large acute medial spine on upper-inner margin, 1-2 smaller spines at base of large spine; outer surface with transverse granulated rows, slightly depressed distally. Palm with outer surface slightly swollen, nearly smooth, transverse rows of small granules on median portion, becoming larger dorsally; ventral margin completely smooth; dorsal margin denticulate; inner surface of palm smooth. Movable dactyl as long as propodal pollex, dorsal longitudinal row of small granules, disappearing at distal half; cutting edges with distinct row of conical teeth.

Pereiopods 2-5 (P2-P5) relatively long, P3 longest (about 1.4 times as long as CW), P5 shortest; dorsal margin of merus, carpus and propodus of P2-P5 slightly serrate, dorsal margin of merus with subdistal notch; carpus with longitudinal carinae, medially on posterior surface, carpus of P5 without carinae; propodus with two rows of spines along ventral margin, spines larger distally; propodus of P2 with three rows of spines; dorsal margin of dactylus with two rows of 4-8 sharp spines, proximal to distal progressively larger, ventral margin with two rows of 2-4 spines, proximal to distal progressively larger; spines of dorsal margin slightly larger than those of ventral margin; three rows of short setae along dorsal and ventral margins.

Male first gonopod (G1) sinuous, gently curving outward; terminal segment conical in shape, mesial margin nearly straight, slightly concave distally, lateral part slightly convex. Flexible zone well developed. Subterminal segment slender; mesial edge subapically swollen, laterally bent, proximally covered by setae, disappearing distally; lateral margin strongly curved (sinus shape), with scattered setae near base (fig. 3a-c).

Gastric mill plate elongated, goblet shaped, slender at base; median tooth with three teeth; first one largest with bilobed appearance; second tooth slightly bilobed, laterally extended; third one very small, hardly discernible (fig. 4).

Remarks. — *Potamon ilam* sp. nov. is most similar to *P. persicum* and *P. mesopotamicum* in terms of G1 morphology. The new species is easily distinguished from these and other species of *Potamon* by characters of the carapace and G1. The terminal article of the G1 in *P. ilam* is conical, with the mesial margin slightly concave distally (fig. 3a-c), whereas that of *P. persicum* and *P. mesopotamicum* is nearly triangular, with straight mesial margins in the proximal half, becoming sharply convergent distally (Brandis et al., 2000: figs. 11c-d, 12c-d). The posterolateral carapace region in *P. ilam* is gently flexed downward, with relatively long curving parallel tuberculate ridges (fig. 2a), whereas in *P. persicum* and *P. mesopotamicum* this margin is arc-shaped, with only short tuberculations (Brandis et al., 2000: figs. 11a-b, 12a-b). The anterolateral margin of the carapace of *P. persicum* and *P. ilam* is slightly cristate, whereas that of *P. mesopotamicum* is distinctly cristate. These characters, together with others that distinguish between *P. persicum*, *P. mesopotamicum* and *P. ilam* are summarized in table II.

Size. — *Potamon ilam* sp. nov. is a medium-sized species, the largest ♂ encountered measured 35.2 mm in carapace length (CL) and 44.64 mm in carapace width (CW), and the largest ♀ 36.4 mm CL and 46.3 mm CW.

Etymology. — The species name is derived from the type locality in Ilam Province in the west of Iran whose capital city is Ilam.

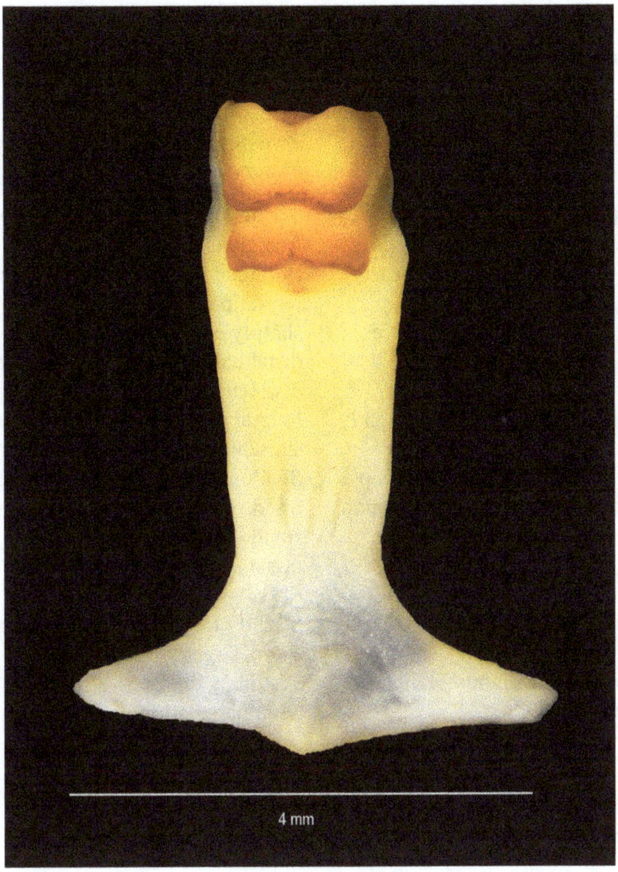

Fig. 4. *Potamon ilam* sp. nov., paratype (SMF 39028). Ventral surface of median tooth plate of gastric mill.

Colour. — Carapace and walking legs dark gray/blue. The outer surfaces of the fingers of the chelipeds (dactylus, propodus), and of the carpus and merus are all bright orange.

Distribution. — *Potamon ilam* n. sp. is found in most of the rivers in Ilam Province that drain into the Tigris River (except for some rivers in the eastern part), but it is not found in the Tigris River itself where *P. mesopotamicum*, *P. persicum*, and *P. magnum* occur (Pretzmann, 1962).

Molecular results. — The tree based on mtDNA (16S rRNA gene) had a total alignment length of 644 basepairs, and GTR + I (with proportion of invariable sites = 0.8358) was chosen as the best evolutionary model of substitution by MrModeltest. The model was used to construct a phylogenetic tree with ten specimens with MrBayes.

TABLE II

Distinct morphological differences between *Potamon ilam* sp. nov., *P. persicum* Pretzmann, 1962, and *P. mesopotamicum* Brandis, Storch & Türkay, 1998

Character	*Potamon ilam* sp. nov.	*Potamon mesopotamicum*	*Potamon persicum*
G1 (terminal article)	Conical	Triangular or nearly triangular	Nearly triangular
	Mesial part nearly straight, only slightly concave distally	Proximal half of mesial part straight, sharply convergent in distal half or distinctly triangular	Proximal half of mesial part straight, sharply convergent in distal half
	Lateral part slightly convex	Lateral part almost straight	Lateral part gently straight
Cervical groove	Distinct and deep	Shallow	Very shallow
Anterolateral surface	Long rows of small granules (>12), transversely arranged	Short or long rows of small granules, transversely arranged	Many short rows of larger granules (>6), irregularly arranged
Anterolateral margin teeth	Short and blunt	Slightly long and sharp, slightly upward	Short and blunt
Anterolateral margin	Slightly granular	Distinctly cristate	Slightly cristate
Posterolateral surface ridge	Curved tuberculation, relatively long, nearly parallel	No tuberculation	Short tuberculation
Posterolateral region	Gently flexed downward	Arc shaped	Arc shaped
Mesogastric sulcus	Shallow and short	Deep and more elongated	Deep and more elongated
Notch between exorbital and first anterolateral teeth	Small, almost contiguous view	Large U-shaped notch	Small, almost contiguous view
Colour	Dark gray bluish	Gray-brown or gray-blue	Dark green
Largest ♂ body size (carapace width × carapace length in mm)	44.64 × 35.2 mm	55.2 × 45.2 mm	79.05 × 64.64 mm

Two main clades can be distinguished (fig. 5), one that unites *P. ilam* sp. nov. and *P. mesopotamicum*, and a second one that unites different populations of *P. persicum*. These clades are well supported by posterior probability values and group taxa that are found in the eastern and western slopes of the Zagros Mountains. The clade that includes *P. persicum* is subdivided into two groups

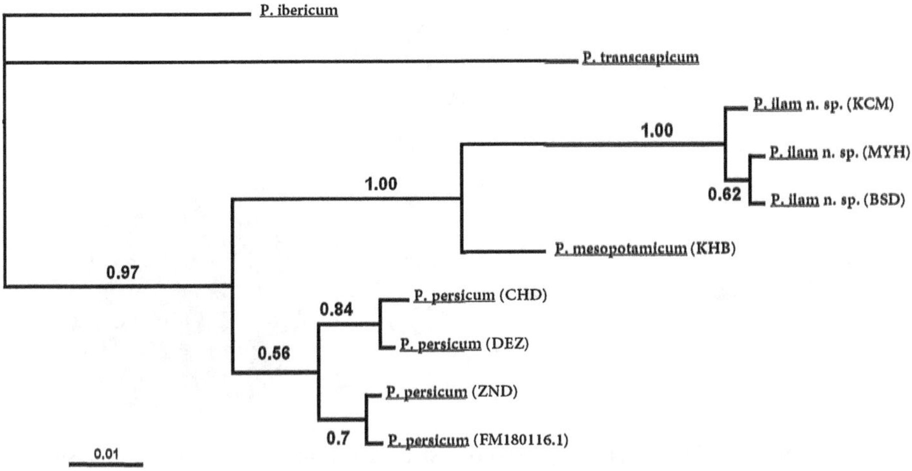

Fig. 5. Phylogenetic relationships of *Potamon ilam* sp. nov. and related species according to a Bayesian analysis (GTR + I) of a 644 basepair alignment of 16S ribosomal mtDNA. Posterior probability values show support for the corresponding clades. Terminal taxa labels are provided in fig. 1 and table I.

(that each have weak support) that corresponds well with the geographical distribution of the taxa. The remarkably low genetic distances (based on 16S rRNA) between the two clades of *P. persicum* (0.008 to 0.01, mean = 0.009) are unexpected and warrant further investigations of the genetic and morphometric structure of the populations over the entire range of this species (Keikhosravi & Schubart, 2014).

A 646 basepair DNA fragment of the nuclear 28S rRNA was obtained for a subset of specimens (see table I). The conservative nature of the 28S rRNA locus means that it has only limited resolution power in this context and only four genotypes were identified by this marker. The genotypes were separated by less than six substitutions from neighboring genotypes, and the most frequent genotype (six individuals from five populations) was found in *P. persicum*. Two genotypes were found among three populations of *P. ilam* separated by 1-3 mutation steps from *P. persicum* and 4-6 mutation steps from *P. mesopotamicum* (see fig. 6).

Comparative material. — *Potamon persicum*, Iran: 3 ♂♂, 1 j (SMF 39029), 3 ♂♂, 1 ♀ (NHMW 25426), 2 ♂♂, 1 j (ZUTC Pot.1079), Ilam Province, 25 km West of Ilam, Chardavol River, Shemsheh Strait, 33°40.177′N 46°41.489′E, leg. A. Keikhosravi, B. Fathinia & M. Moradmand, 15 Apr. 2009; 2 ♂♂, 1 j (NHMW 25427), 2 ♂♂, 1 j (SMF 39030), 2 ♂♂ (ZUTC Pot.1080), Khuzestan Province, Dezful, Dez River, 32°23.282′N 48°23.679′E, leg. A. Keikhosravi

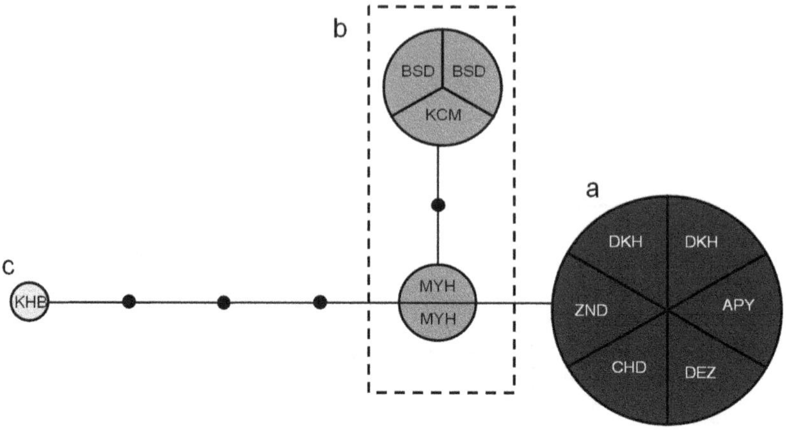

Fig. 6. Maximum parsimony spanning network constructed with TCS of *Potamon ilam* sp. nov. and related species, based on a 646 basepair alignment of 28S ribosomal nuclear DNA. The size of the circle is proportional to the frequency of the genotypes; each line represents one substitution and dots on lines indicate additional substitutions. a, *P. persicum* Pretzmann, 1962, from five different rivers in Iran; b, *P. ilam* sp. nov. collected from three adjacent rivers in Iran; c, *P. mesopotamicum* Pretzmann, 1962 from Syria. See fig. 1 and table I for population labels.

& M. Moradmand, 18 Apr. 2009; 4 ♂♂, 1 ♀ (SMF 39031), 3 ♂♂ (NHMW 25428), 3 ♂♂, 1 j (ZUTC Pot.1081), Isfahan Province, Isfahan, Zayandehrood River, 32°38.322′N 51°35.977′E, leg. A. Keikhosravi & M. Moradmand, 19 Apr. 2009; 2 ♂♂ (SMF 39032), 2 ♂♂ (RMNH.CRUS.D.54893), 3 ♂♂ (NHMW 25429), 2 ♂♂, 1 j (ZUTC Pot.1082), Tehran Province, Tehran, Darakeh, Darakeh River, 35°49.092′N 51°22.855′E, leg. A. Keikhosravi, 28 Apr. 2009; 2 ♂♂, 1 ♀ (RMNH.CRUS.D.54894), 2 ♂♂ (NHMW 25430), 3 ♂♂ (SMF 39033), 2 ♂♂, 1 ♀ (ZUTC Pot.1087), Guilan Province, Roodbar, Sepidrood River, 36°49.060′N 49°25.256′E, leg. A. Keikhosravi & M. Houshmand, 23 Dec. 2009; 2 ♂♂, 2 j (SMF 39034), 1 ♂, 1 j (RMNH.CRUS.D.54895), 2 ♂♂, 2 j (ZUTC Pot.1083), Alborz Province, Taleghan, Thaleghan River, 36°10′8.22″N 50°45′45.45″E, leg. F. A. Houshmand, 18 Aug. 2009; 3 ♂♂, 2 ♀♀ (SMF 39035), 4 ♂♂, 1 ♀ (NHMW 25431), 3 ♂♂, 2 j (RMNH.CRUS.D.54896), 2 ♂♂, 1 ♀, 1 j (ZUTC Pot.1084), northwest of Qazvin, 55 km to Qazvin, Molaali River, Trib to Shahrood R., 36°27.742′N 49°30.868′E, leg. Y. A. Houshmand & A. Keikhosravi, 03 Sep. 2010; 1 ♀, 1 ♂, 1 j (SMF 39036), 1 ♂, 2 j (ZUTC Pot.1085), Tehran, Lavasan, 3 km west of Latyan Dam 35°48′46.44″N 51°36′03.54″E, leg. Y. Houshmand & A. Keikhosravi, 13 Nov. 2009; 2 ♂♂, 3 ♀♀ (SMF 39037), 3 ♂♂, 3 ♀♀, 1 j (NHMW 25432), 30 km northeast of Tehran, Kamard, Trib to Jajrood River, 35°46′20.98″N 51°45′15.80″E, leg. A. Keikhosravi, Y. A. Houshmand & H.

Heydari, 1 Oct. 2009; 1 ♂, 2 ♀♀ (SMF 39038), 2 ♀♀ (ZUTC Pot.1086), 5 km north of Khalkhal, Arpachay River, 37°40.926′N 48°30.237′E, leg. Y. Houshmand & A. Keikhosravi, 2 Sep. 2010; 1 ♂ (NHM 1920.2.5.3-4), northwest of Qazvin, leg. P. A. Buxton, R. Gurney, 1920; 3 ♂♂, 2 ♀♀, 13 j (NHM 1899.10.6.1-5), Azarbayejan-e Gharbi Province, near Orumiyeh, Northwest of Iran, leg. R. J. Gunther, 1899; 1 ♂ (ZMUC CRU-9522), Kermanshah, no further locality information, leg. S. W. Kaiser, 1937; 1 ♂ (SMF 2640), Tehran Province, Tehran, leg. F. Bruhns; 1 ♂, 1 ♀ (SMF 4157), road of Abu-Ali, 30 km off Tehran, leg. J. Theodorides, 1959. Iraq: 1 ♂ (NHM 1934.8.29.1-6), stream on Kirkuk road, 16 km west of Sulaymaniyah, 760 m, 35°33′N 45°16′E, leg. MacFadyen, 1933. Turkey: 2 ♂♂, 1 ♀ (SMF 5881), 2 ♂♂, 2 ♀♀ (SMF 5882), Van, Mengene Daglari, SE of Van Gölu, Baskale, leg. Lampe, 1972.

Potamon mesopotamicum, Syria: Holotype: 1 ♂ (SMF 23311), Source of Al-Khabur River, Ras Al-Ain, 36°51′N 40°04′E, leg. F. Krupp, D. Kock & H. Martens, 1989; paratypes: 13 ♂♂, 13 ♀♀, 14 j (SMF 23312), same data as holotype; 1 ♂ (SMF 23315), Al-Hassakah Province, Al-Khabur River, Tell Atash, 36°42′N 40°11′E, leg. F. Krupp, D. Kock & G. Eppler, 1988; 2 ♂♂ (SMF 23318), Al-Khabur River, Ras Al-Ain, 36°51′N 40°04′E, leg. F. Krupp, D. Kock & G. Eppler, 1988; 1 ♂ (SMF 23320), Al-Khabur River, Tell Shaikh Hamed, 35°37′N 40°45′E, leg. F. Krupp & H. Martens, 1989; 1 ♂ (SMF 23326), Al-Khabur River, Bahrat Khatuniya, 36°24′N 41°13′E, leg. F. Krupp & W. Schneider, 1986; 3 ♂♂ (SMF 23331), Euphrates at Halabiyeh, leg. R. Kinzelbach, 1978; 1 ♂ (SMF 23332), Euphrates tributary N of Maadan, leg. R. Kinzelbach, 1978. Iraq: 1 ♂, 1 ♀ (ZRC 2009.0693), Al-Huwaizah Marshes.

DISCUSSION

The phylogeny and taxonomy of freshwater crabs in general, and particularly within the genus *Potamon* in the Middle East, has only recently been studied in detail by specialists. Formerly, five subgenera were recognized within *Potamon* based on G1 morphology: *P.* (*Potamon*) Savigny, 1816; *P.* (*Euthelphusa*) Pretzmann, 1962; *P.* (*Pontipotamon*) Pretzmann, 1962; *P.* (*Orientopotamon*) Pretzmann, 1962; *P.* (*Centropotamon*) Pretzmann, 1962. Brandis et al. (2000) synonymised the subgenus *Centropotamon* under *Orientopotamon* giving priority to the latter. Later, Ng et al. (2008) omitted all the subgenera, but a phylogenetic study on evolution of *Potamon* in the Aegean region supported the recognition of these subgenera that were also congruent with geographic distributions and past geological events (Jesse et al., 2011).

All rivers populated by *P. ilam* rise on the westernmost slopes of the Zagros Mountains in western Iran. These rivers belong to the Euphrates-Tigris system and drain westward into the Tigris River in eastern Iraq. *Potamon ilam* and *P. mesopotamicum* are phylogenetically closer to each other than both are to *P. persicum* in terms of mtDNA, and both are distributed beyond the Zagros Mountains. These taxa share a large drainage system: *P. mesopotamicum* occurs in the Khabur River, the Euphrates River (in the region from the mouth of the Khabur River upstream) and in the Al-Huwaizah marshes, Iraq (downstream of Tigris River) (Brandis et al., 1998; Naser, 2009). The presence of *P. magnum* and *P. persicum* in the headwaters of the Tigris River suggests that *P. ilam* may have limits to its dispersal in this direction. The fact that three different species of freshwater crabs have evolved to co-exist in a single drainage system emphasizes their abilities to adapt to different local conditions.

The average 16S rRNA nucleotide divergence between the clades of *P. persicum* and *P. ilam* plus *P. mesopotamicum* is 2.3%. This represents a separation of about 3.6 Mya based on the substitution rate proposed by Schubart et al. (1998) (0.65% per million years for non-endemic Jamaican freshwater crab species). The separation time between these taxa also corresponds to isolation events associated with the orogenic events occurring in the Zagros Mountains at that time. The Zagros Mountains formed when the Arabian and Eurasian plates collided during the Miocene (Takin, 1972; Agard et al., 2011), but the current uplift (Zagros Simply Folded Belt) of the Zagros Mountains took place 4-5 Mya, concomitant with the opening of the Red Sea (Rangzan & Iqbaluddin, 1995; Agard et al., 2011). The generally poor dispersal abilities of freshwater crabs imply that vicariance events such as mountain formation could have resulted in the reproductive isolation of crab populations (Banarescu, 1991; Sternberg et al., 1999; Shih et al., 2006, 2009). The formation of the Zagros Mountains in the recent past may have blocked gene flow between populations of *Potamon* that consequently speciated, resulting in the distinct species *P. ilam* and *P. mesopotamicum*.

The present study brings the number of species in the genus *Potamon* to 22. This number may increase further when the results of studies on the *P. persicum* complex have been completed, because this taxon may comprise one or more cryptic species (Keikhosravi & Schubart, 2014). The vast range of this species in Iran (about 518 000 km^2), which includes a diverse topography and climate zones, requires characterization of the evolutionary diversity within the *P. persicum* complex. Future morphological and molecular studies based on larger sampling are needed to resolve the phylogenetic relationships within the *P. persicum* and *P. mesopotamicum* complexes, and to reveal the true diversity of Iran's freshwater crabs.

ACKNOWLEDGEMENTS

The first author would like to express his special gratitude to the people who helped to collect specimens: Mr. Yadollah Houshmand, Fatollah Houshmand, Behzad Fathinia and Madjid Moradmand. We are grateful to Dr. Jørgen Olesen (ZMUC), Dr. Paul Clark (NHM), Prof. Michael Türkay (SMF), Dr. Charles H. J. M. Fransen (RMNH) and Prof. Peter K. L. Ng (ZRC) for loaning us material from their collections. We extend our gratitude to Reza Naderloo for his useful comments and excellent illustrations, and to Nicole Rivera and Richard Landstorfer for their help with TCS and MrBayes. Prof. Jürgen Heinze (University of Regensburg) and his staff are also thanked for the use of their facilities and for their continuing support throughout this project.

REFERENCES

AGARD, P., J. OMRANI, L. JOLIVET, H. WITHECHURCH, B. VRIELYNCK, W. SPAKMAN, P. MONIE, B. MEYER & R. WORTEL, 2011. Zagros orogeny: a subduction-dominated process. Geological Magazine, **148**: 1-34.

BANARESCU, P., 1991. Zoogeography of freshwaters, Volume 1. General distribution and dispersal of freshwater animals: 1-511. (AULA-Verlag, Wiesbaden).

BIEBERSTEIN, M. DE, 1809. Notice sur quelques Insectes su Caucase. Mémoires de la Société Impériale des Naturalistes de Moscou, **2**: 3-5.

BRANDIS, D., V. STORCH & M. TÜRKAY, 1998. The status of the freshwater crab populations of the Khabur River (Syria). Journal of Natural History, **32**: 1439-1445.

— —, — — & — —, 2000. Taxonomy and zoogeography of the freshwater crabs of Europe, North Africa, and the Middle East. Senckenbergiana biologica, **80**: 5-56.

CLEMENT, M., D. POSADA & K. A. CRANDALL, 2000. TCS: a computer program to estimate gene genealogies. Molecular Ecology, **9**: 1657-1659.

HALL, T. A., 1999. Bioedit: a user-friendly biological sequence alignment editor and analysis program for Windows 95/98/NT. Nucleic Acids Symposium Series, **41**: 95-98.

HUELSENBECK, J. P. & F. RONQUIST, 2001. MrBayes: Bayesian inference of phylogenetic trees. Bioinformatics, **17**: 754-755.

JESSE, R., M. GRUDINSKI, S. KLAUS, B. STREIT & M. PFENNINGER, 2011. Evolution of freshwater crab diversity in the Aegean region (Crustacea: Brachyura: Potamidae). Molecular Phylogenetics and Evolution, **59**: 23-33.

JESSE, R., C. D. SCHUBART & S. KLAUS, 2010. Identification of a cryptic lineage within *Potamon fluviatile* (Herbst) (Crustacea: Brachyura: Potamidae). Invertebrate Systematics, **24**: 348-356.

KATOH, K., K. MISAWA, K. KUMA & T. MIYATA, 2002. MAFFT: a novel method for rapid multiple sequence alignment based on fast Fourier transform. Nucleic Acids Research, **30**: 3059-3066.

KEIKHOSRAVI, A. & C. D. SCHUBART, 2014. Revalidation and redescription of *Potamon elbursi* Pretzmann, 1976 (Brachyura, Potamidae) from Iran, based on morphology and genetics. Central European Journal of Biology, **9**(1): 114-123.

KINZELBACH, R. K., 1980. Hydrobiologie am Orontes. Natur und Museum, Frankfurt am Main, **110**(1): 9-18.

KLAUS, S., D. BRANDIS, P. K. L. NG, D. C. J. YEO & C. D. SCHUBART, 2009. Phylogeny and biogeography of Asian freshwater crabs of the family Gecarcinucidae (Brachyura Potamidae). In: J. W. MARTIN, K. A. CRANDALL & D. L. FELDER (eds.), Decapod crustaceans phylogenetics. Crustacean Issues, **18**: 509-531. (Taylor & Francis/CRC Press, Boca Raton, Florida).

KLAUS, S. & M. GROSS, 2010. Synopsis of the fossil freshwater crabs of Europe (Brachyura: Potamoidea: Potamidae). Neues Jahrbuch für Geologie und Paläontologie, Abhandlungen, **256**: 39-59.

NADERLOO, R., M. TÜRKAY & H.-L. CHEN, 2010. Taxonomic revision of the wide-front fiddler crabs of the *Uca lactea* group (Crustacea: Decapoda: Brachyura: Ocypodidae) in the Indo-West Pacific. Zootaxa, **2500**: 1-38.

NASER, M. D., 2009. First record of the freshwater crab, *Potamon mesopotamicum* Brandis, Storch & Türkay, 1998 (Decapoda, Brachyura, Potamidae) from the Al-Huwaizah marshes, Iraq. Crustaceana, **82**(12): 1599-1602.

NG, P. K. L., D. GUINOT & P. J. F. DAVIE, 2008. Systema Brachyurorum: Part 1. An annotated checklist of extant brachyuran crabs of the world. Raffles Bulletin of Zoology, (Supplement) **17**: 1-286.

NYLANDER, J. A. A., 2004. MrModeltest v2. Evolutionary Biology Centre, Uppsala University, Uppsala, Sweden. http://www.abc.se/~nylander/mrmodeltest2/mrmodeltest2.html.

ORTMANN, A., 1896. Das System der Decapoden-Krebse. Zoologische Jahrbücher, (Systematik) **9**(13): 409-453.

PRETZMANN, G., 1962. Die mediterranen und vorderasiatischen Süßwasserkrabben. Annalen des Naturhistorischen Museums Wien, **65**: 205-240.

— —, 1965. Die Süßwasserkrabben des Mittelmeers und Vorderasiens des British Museums of Natural History, London. Annalen des Naturhistorischen Museums Wien, **68**: 519-525.

— —, 1966. Potamoniden (Crustacea Brachyura) aus Afghanistan. Annalen des Naturhistorischen Museums Wien, **69**: 297-298.

— —, 1971. Ergebnisse einiger Sammelreisen nach Vorderasien: 1. Teil: Zwei neue Unterarten von Süßwasserkrabben. Annalen des Naturhistorischen Museums Wien, 75: 473-475.

— —, 1976a. Ergebnisse einiger Sammelreisen nach Vorderasien: 6. Die Süßwasserkrabben Persiens. Annalen des Naturhistorischen Museums Wien, **80**: 457-472.

— —, 1976b. Ergebnisse einiger Sammelreisen nach Vorderasien: 5. Weitere neue Süßwasserkrabben. Annalen des Naturhistorischen Museums Wien, **80**: 453-456.

— —, 1987. Versuch einer historischen Deutung des Verbreitungsbildes der mediterranen und europäischen Süßwasserdecapoden. Sitzungsbericht der Österreichischen Akademie der Wissenschaften Mathematisch-Naturwissenschaftlich Klasse, **196**: 1-4.

RANGZAN, K. & IQBALUDDIN, 1995. Morpho-tectonic study of Zagros Structural Belt of SW Iran using remote sensing techniques. Journal of the Indian Society of Remote Sensing, **23**(4): 211-224.

SAVIGNY, J. C., 1816. Mémoires sur les animaux sans vertèbres: Première Partie: Description et classification des animaux invertébrés et articulés, connus sous les noms de Crustacés, d'Insectes, d'Annélides, etc.: i-v, 1-118, pls. 1-8. (G. Dufour, Paris).

SCHUBART, C. D., 2009. Mitochondrial DNA and decapod phylogenies; the importance of pseudogenes and primer optimization. In: J. W. MARTIN, K. A. CRANDALL & D. L. FELDER (eds.), Decapod crustaceans phylogenetics. Crustacean Issues, **18**: 47-65. (Taylor & Francis/CRC Press, Boca Raton, Florida).

SCHUBART, C. D., R. DIESEL & S. B. HEDGES, 1998. Rapid evolution in terrestrial life in Jamaican crabs. Nature, **393**: 363- 365.

SHIH, H.-T., H.-C. HUNG, C. D. SCHUBART, C. A. CHEN & H.-W. CHANG, 2006. Intraspecific genetic diversity of the endemic freshwater crab *Candidiopotamon rathbunae* (Decapoda, Brachyura, Potamidae) reflects five million years of geological history of Taiwan. Journal of Biogeography, **33**: 980-989.

SHIH, H.-T., D. C. J. YEO & P. K. L. NG, 2009. The collision of the Indian plate with Asia: molecular evidence for its impact on the phylogeny of freshwater crabs (Brachyura: Potamidae). Journal of Biogeography, **36**: 703-719.

STERNBERG, R. V., N. CUMBERLIDGE & G. RODRIGUEZ, 1999. On the marine sister groups of the freshwater crabs (Crustacea: Decapoda: Brachyura). Journal of Zoological Systematics and Evolutionary Research, **37**: 19-38.

TAKIN, M., 1972. Iranian geology and continental drift in the Middle East. Nature, **235**: 147-150.

WHITING, M. F., J. C. CARPENTER, Q. D. WHEELER & W. C. WHEELER, 1997. The Strepsiptera problem: phylogeny of the holometabolous insect orders inferred from 18S and 28S ribosomal DNA sequences and morphology. Systematic Biology, **46**: 1-68.

First received 19 October 2011.
Final version accepted 29 July 2013.

A NEW SPECIES OF *ISOLAPOTAMON* BOTT, 1968 (DECAPODA, BRACHYURA, POTAMIDAE) FROM MINDANAO, WITH NOTES ON THE PHILIPPINE *ISOLAPOTAMON* SPECIES

BY

JOSE C. E. MENDOZA[1,2]) and DARREN C. J. YEO[1,3])

[1]) Department of Biological Sciences, Faculty of Science, National University of Singapore, 14 Science Drive 4, 117543 Singapore

ABSTRACT

The genus *Isolapotamon* Bott, 1968 (Brachyura: Potamidae) is represented in the Philippines by four poorly known species, all found on the large southern island of Mindanao. *Isolapotamon mindanaoense* (Rathbun, 1904) is re-described based on newly collected material, and an update on its distribution is also provided. A fifth species, *I. maranao* sp. nov., is here described from the northwestern region of Mindanao. It differs from its congeners primarily in the morphology of the male first pleopods, although minor external differences are also observed in the form of the carapace, ambulatory legs and of the male thoracic sternum and abdomen. A key to the Philippine species of *Isolapotamon* Bott, 1968, is provided.

INTRODUCTION

The Philippines is in a part of the world that has high freshwater crab diversity including a great potential for the discovery of new species (Chia & Ng, 2006; Yeo et al., 2008; Mendoza & Naruse, 2010). The primary freshwater crabs of the Philippines are represented by two Old-World families, Gecarcinucidae Rathbun, 1904 (with three genera: *Mainitia* Bott, 1969, *Parathelphusa* H. Milne Edwards, 1853, and *Sundathelphusa* Bott, 1969) and Potamidae Ortmann, 1896 (with five genera: *Isolapotamon* Bott, 1968, *Ovitamon* Ng & Takeda, 1992, *Insulamon* Ng & Takeda, 1992, *Mindoron* Ng & Takeda, 1992, and *Carpomon* Tan & Ng, 1998). To date, 54 species, all endemic, have been described from the Philippines (Ng et al., 2008; Husana et al., 2009; Mendoza & Naruse, 2010; Ng, 2010; Freitag, 2012; Husana et al., 2013; Manuel-Santos

[2]) Corresponding author; e-mail: jcmendoza@nus.edu.sg

[3]) e-mail: dbsyeod@nus.edu.sg

& Ng, 2013). Considering the archipelagic nature and geological history of this country and the surrounding region (Hall, 1998, 2002, 2009; Lohman et al., 2011) and the likelihood of allopatric speciation driven by the limited dispersal ability of freshwater crabs (Ng & Rodriguez, 1995; Daniels et al., 2002; Daniels, 2003), more species discoveries are expected. It is also important to state that many areas where freshwater crabs are found are under threat due to habitat alteration and other environmental changes (Cumberlidge et al., 2009; Freitag, 2012), thus increasing the urgency for such discoveries and species descriptions.

The genus *Isolapotamon* was established by Bott (1968) for several species, and included three subgenera, *Isolapotamon*, *Nanhaipotamon* and *Malayopotamon*. Bott (1970a) subsequently raised all three subgenera to full genus rank under a new family, Isolapotamidae, recognizing eight species in the type genus *Isolapotamon* (type species: *Potamon anomalus* Chace, 1938). Ng & Tan (1998) provided a thorough systematic revision of the genus, where they considered Isolapotamidae Bott, 1968, to be a junior synonym of Potamidae Ortmann, 1896 (also previously mooted by Ng & Yang, 1985, 1986), and confirmed Bott's (1970a) action in raising the aforementioned subgenera to full genus rank, citing differences in the morphology of the third maxillipeds and first gonopod (G1). However, Brandis (2002) considered the Isolapotamidae to be a valid taxon and distinguishable from the Potamidae on the basis of the anatomy of the second gonopod (G2). Ng & Tan (1998), in their review of the genus, discussed other taxonomic actions concerning *Isolapotamon* and recognized 18 species from Borneo and the Philippines, including two new Bornean species. In the Philippines, four species are currently recognized, and all are from the southern island of Mindanao; these are: *I. sinuatifrons* (H. Milne Edwards, 1853), *I. mindanaoense* (Rathbun, 1904), *I. spatha* Ng & Takeda, 1992, and *I. danielae* Manuel-Santos, 2010. These species are still not very well known, as only a few specimens from a few localities have been sampled to date. The exact type localities of the earlier described species, e.g., *I. mindanaoense* and *I. sinuatifrons*, remain unknown. Furthermore, the range of *Isolapotamon* in the Philippines may prove to be more extensive given the geological history and the biogeography of the region surrounding Mindanao. For instance, the genus may also occur in the islands to the northeast of Mindanao, particularly Samar and Leyte, as these were connected to Mindanao as recently as the late Pleistocene and share some similarity with Mindanao in their floral and faunal assemblages (Heaney, 1986; Heaney & Regalado, 1998; Ong et al., 2002; Catibog-Sinha & Heaney, 2006; Vallejo, 2011).

Recently, a series of freshwater crabs, collected during herpetological surveys in the Philippines, was sent for study to the authors by Dr. Arvin C. Diesmos, curator for Herpetology at the National Museum of the Philippines in Manila. Among these were specimens of *I. mindanaoense* collected from different localities in Mindanao, which form the basis for the present redescription of this species. A new species from Lanao del Sur province in western Mindanao is described. Other unreported specimens in the collection of the Raffles Museum of Biodiversity Research, National University of Singapore, including the other known species of Philippine *Isolapotamon*, are also evaluated in the present work for a complete account.

For material examined, all measurements are of the carapace maximum width by median length, in millimeters. The terminology follows that of Ng (1988), and the classification follows Ng et al. (2008). Specimens are deposited in the Museum National d'Histoire Naturelle (MNHN), Paris; the National Museum of the Philippines (NMCR), Manila; the National Museum of Science & Technology (NMST), Tokyo; and the Zoological Reference Collection (ZRC) of the Raffles Museum of Biodiversity Research, National University of Singapore.

Photographs of the types of *Isolapotamon mindanaoense* (Rathbun, 1904) and *I. sinuatifrons* (H. Milne Edwards, 1853), deposited in the MNHN, were kindly provided by Dr. Tohru Naruse (University of the Ryukyus, Japan), and photographs of the holotype of *I. spatha* Ng & Takeda, 1992, deposited in the NMST, were kindly provided by Dr. Hironori Komatsu (NMST).

The following abbreviations are used: G1 and G2 = first and second male pleopods, respectively; P1-P5 = first to fifth pereopods, P1 being the chelipeds and P2-P5 being the ambulatory legs; coll. = collected by; imm. = immature; Bgy. = "barangay" or village, the smallest geo-political unit in the Philippines, several of which make up a municipality or city; and Mt. = Mount, as used in the proper names for mountains. Where available, altitude above sea level is indicated in meters.

SYSTEMATIC ACCOUNT

Family POTAMIDAE Ortmann, 1896

Subfamily POTAMISCINAE Bott, 1970

Isolapotamon Bott, 1968

Isolapotamon danielae Manuel-Santos, 2010

Telphusa sinuatifrons, Bürger, 1894: 2 (in part). Not *Thelphusa sinuatifrons* H. Milne Edwards, 1853.

Isolapotamon danielae Manuel-Santos, 2010: 198, figs. 1, 2.

Description. — See Manuel-Santos (2010: 198).

Remarks. — This species was recently described by Manuel-Santos (2010), who provided a comprehensive description and diagnosis based on three specimens (two males, one female). Bürger (1894: 2) reported three specimens, which he identified as "*Telphusa sinuatifrons* M.-Edwards", from Carl Semper's Philippine collection, and particularly from the localities "Zamboanga" (two specimens) and "Rio Jibon" (one specimen). Following the usage at that time, the first of the two names is unmistakably referring to the port of Zamboanga (= Zamboanga City) in western Mindanao, whereas the latter name most probably refers to a river called Jibon or a river in or near a place called Jibon. Unfortunately, this name is apparently no longer in current usage. However, an entry for "Jibón" was found in a colonial-era gazetteer of the Philippines (Bureau of Insular Affairs, 1902: 564), where it was defined as a "summit overlooking E. bank of Agusan Riv., central Surigao". This area is in the northern part of present-day Surigao del Sur Province, in northeastern Mindanao. This places the two localities mentioned by Bürger (1894) on opposite sides of the island of Mindanao, and it is very unlikely that the specimens from there are conspecific. Bürger's (1894) "Rio Jibon" specimen is, therefore, most likely to be *Isolapotamon danielae* Manuel-Santos, 2010, which closely resembles *I. sinuatifrons* (H. Milne Edwards, 1853), and has its type locality in neighboring Agusan del Norte province ("Maguinca, Sumili, Dagandang Salt Spring") (cf. Manuel-Santos, 2010: 198, fig. 1), and it is provisionally considered so in this report.

We reserve judgement, however, on the conspecificity of the paratype male, cw 44.7 by cl 36.2 mm (ZRC 2009.0583), from Sebu, South Cotabato, mentioned by Manuel-Santos (2010). The locality from which this specimen was collected is more than 300 km away from where the holotype was collected (Agusan del Norte), does not share a common freshwater drainage, and is actually closer to the type locality of *I. spatha* Ng & Takeda, 1992 (Palimbang, Sultan Kudarat). This specimen was not illustrated by Manuel-Santos (2010), and while it has a ZRC catalogue number, it has not been found by the authors in the collection of the museum and, therefore, could not be reexamined. For this reason, this specimen is provisionally considered incerta sedis.

Size. — This is a large species, adult between cw 40 and 53 mm.

Distribution. — Dagandang Salt Spring, Butuan City, Agusan del Norte (type locality) (Manuel-Santos, 2010); Río Jibón, Surigao del Sur (Bürger, 1894) (see fig. 9).

Isolapotamon maranao sp. nov.
(figs. 1, 2A, 5A-E)

Potamon mindanaoensis, Balss, 1937: 162, figs. 22, 23. Not *Potamon* (*Potamon*) *mindanaoensis* Rathbun, 1904.

Material examined. — Holotype, male, 32.8 by 25.9 mm (NMCR 39071), Lake Dapao, Lanao del Sur Province, Mindanao, coll. A. C. Diesmos, Mar. 2007.

Paratype: 1 female (damaged carapace), 26.3 by 20.1 mm (NMCR 39072), same data as holotype.

Diagnosis. — Carapace relatively narrow, compared to most congeners; mesobranchial, intestinal regions inflated; posterolateral margins straight rather than concave. Dactylus of P5 short, stout. G1 short, stout, terminal segment distinctly shorter than subterminal segment; apical lobe subcircular, not prominently expanded.

Description. — Carapace (fig. 1A, C) subquadrate, cordiform in outline, wider than long, width about 1.3 times length; dorsal surface depressed, mostly smooth except for granular frontal region; branchial region lightly striated; regions poorly defined; epigastric, suborbital cristae low, but distinct, not confluent; front bilobed, not advanced beyond orbits, ventrally deflexed, anterior margin sinuous, cristate; orbits cristate, supraorbital margin more-or-less straight, transverse; anterolateral margin convex, cristate, granular; exorbital tooth triangular, apex acute, lateral margin longer than mesial margin; epibranchial tooth small, broadly triangular, separated from exorbital tooth by shallow, V-shaped notch; cervical groove indistinct, central H-shaped gastric groove distinct, deep; mesobranchial, intestinal regions slightly inflated.

Eyes (fig. 1C) well developed, corneas large. Posterior margin of epistome with well produced, central, triangular projection, straight to slightly sinuous laterally; endostome with distinct, complete ridges laterally. Third maxilliped smooth; merus subquadrate, length about half that of ischium; ischium with distinct submedian sulcus; exopod tapering toward distal end, lateral, mesial margins gently convex, tip reaching up to level of midlength of merus, flagellum slightly longer than width of merus (although both broken in holotype; one intact, one broken in paratype).

Thoracic sternum (fig. 1B) long, narrow, smooth; sternites 1, 2, sternites 3, 4 completely fused, deep transverse sulcus separating two fused units; sternites 5-7 distinct externally; sterno-abdominal cavity moderately deep, setose, thoracic press-button distinct, visible at posterior half of sternite 5 median line visible only between sternites 7, 8.

Chelipeds (fig. 1A) unequal, surfaces smooth to slightly rugose; merus short, barely exceeding lateral margins of carapace, distal end of flexor surface

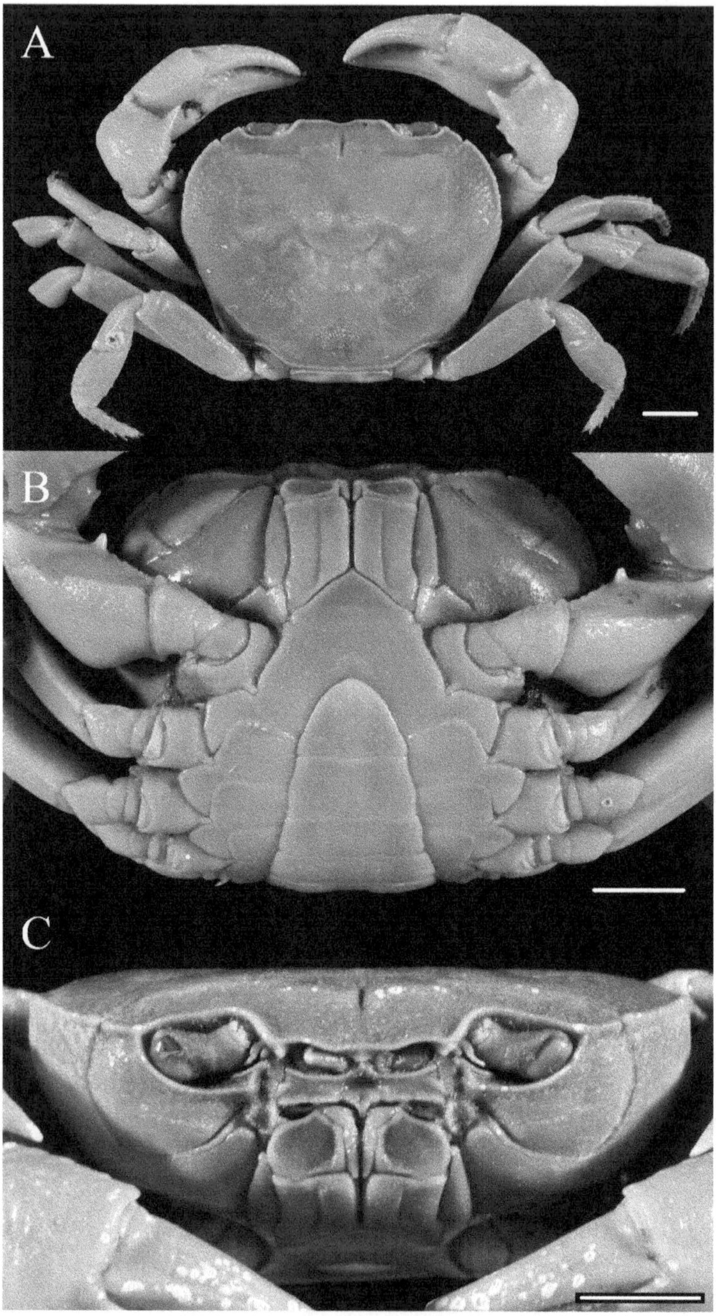

Fig. 1. *Isolapotamon maranao* sp. nov., holotype, male, 32.8 by 25.9 mm (NMCR 39071), Lake Dapao, Marawi City, Lanao del Sur Province, Mindanao. A, dorsal view; B, ventral view; C, anterior view. Scale bars: 5.0 mm.

with small but distinct conical tubercle; carpus inner angle marked by long, acute spine, with second similar but smaller spine posteriorly; palm inflated, smooth except for rugose supero-external surface, length subequal to that of dactylus; both fingers with rounded keel running along length in both external, internal surfaces, inner margins armed with variably sized teeth, tips pointed, incurved.

Ambulatory legs (figs. 1A, 2A) mostly smooth, glabrous, moderate in length, P3, P4, longest, about 1.4 times carapace width; meri subrectangular, thrice longer than broad, cristate on anterior, posterior margins; carpi with low longitudinal ridge on dorsal surface; propodi with shallow longitudinal sulcus on dorsal surface; anterior, posterior margins spinulose; dactyli stout, spinose, slightly curved, slightly longer than propodus.

Male abdomen (fig. 1B) broad, gradually tapering distally towards telson, with six freely articulated somites; somite 1 strip-like, with transverse crest; somite 2 trapezoidal; somites 3-6 subrectangular, somite 6 more subquadrate, wider than long, lateral margins slightly convex; telson subtriangular, apex rounded, width of base about 1.4 times median length, median length of somite 6 about 0.8 times that of telson. Female abdomen wider, longer, almost completely covering thoracic sternum.

G1 (fig. 5A, B, D, E) moderately long, relatively thick; subterminal segment distinctly longer than terminal segment, with prominent shelf proximally on external margin; apex with lobiform, subcircular, lateral extension angled about 45° from the vertical axis. G2 (fig. 5C) filiform; basal segment cylindrical for most of length; distal segment flattened, curved, about 0.5 times as long as basal segment.

Etymology. — The specific epithet for the new species comes from "Maranao", the name of the largest ethnic group residing in Lanao del Sur province (and roughly translated as "the people of the Lake"), where the type locality is located. Used as a noun in apposition.

Remarks. — *Isolapotamon maranao* sp. nov. most closely resembles *I. spatha* Ng & Takeda, 1992, in morphology, particularly in the form of the carapace and of the G1. Both species have a narrow carapace, and a short and stout dactylus of P5, compared to the other species known from Mindanao, and both have a G1 with a much less extended, subcircular, apical lobe, though similar in orientation to that of *I. mindanaoense* (Rathbun, 1904). Furthermore, both species are also found on the western region of Mindanao, with *I. maranao* sp. nov. occurring in the northern part and *I. spatha* occurring in the southern part, though it is not yet clear whether they are sympatric or not. However, *I. maranao* sp. nov. differs from *I. spatha* in the following

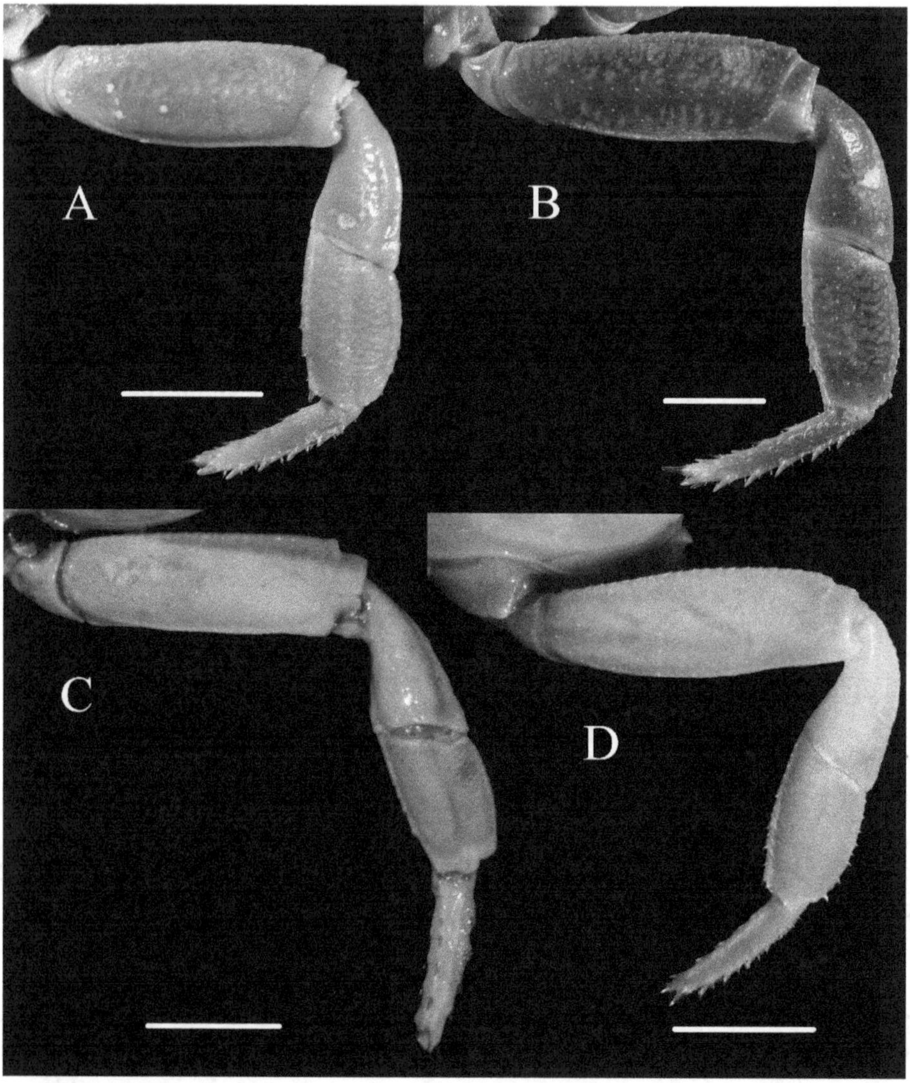

Fig. 2. Fifth pereopods (P5), dorsal view: A, *Isolapotamon maranao* sp. nov., holotype, male (NMCR 39071); B, *I. mindanaoense* (Rathbun, 1904), female (NMCR 39073); C, *I. sinuatifrons* (H. Milne Edwards, 1853), lectotype, male (MNHN-BP4353); D, *I. spatha* Ng & Takeda, 1992, holotype, male (NMST-Cr 11225). Scale bars: 5.0 mm.

morphological features: 1) the dorsal surface of the carapace is more inflated, particularly in the branchial regions (fig. 1C) (vs. branchial regions more depressed; fig. 8C); 2) the supraorbital margin is smooth (fig. 1A, C) (vs. supraorbital margin granulate; fig. 8A, C); 3) the exorbital tooth is more obtuse (fig. 1A) (vs. exorbital tooth more acute; fig. 8A); 4) the epibranchial tooth is less distinct, with the notch separating it from the exorbital tooth shallower (fig. 1A) (vs. epibranchial tooth more distinct, notch deeper; fig. 8A); 5) the male telson and abdominal somite 6 are proportionately shorter (fig. 1B) (vs. male telson and abdominal somite 6 proportionately longer; fig. 8B); 6) the G1 is relatively shorter and stouter, with the terminal segment relatively shorter (fig. 5A, B, D, E) (vs. G1 longer and more slender, with terminal segment relatively longer); and 7) the distal segment of the G2 is relatively longer, about 0.5 times the length of basal segment (fig. 5C) (vs. distal segment of G2 relatively shorter, 0.4 times length of basal segment) (see also Ng & Takeda, 1992: fig. 7; Ng & Tan, 1998: fig. 12A-D).

Isolapotamon maranao sp. nov. differs from *I. mindanaoense* in having a stouter, relatively shorter G1, with the apical lobe less extended and subcircular in shape (vs. G1 longer and more slender, apical lobe longer and oblongate; figs. 3C, 5F-K); more inflated branchial regions when viewed anteriorly (vs. branchial regions depressed; fig. 4C); less convex carapace anterolateral margins (vs. carapace anterolateral regions more convex; figs. 3A, 4A); straight carapace posterolateral margins (vs. carapace posterolateral margins gently concave; figs. 3A, 4A); lateral regions of the posterior margin of the epistome straight (vs. lateral regions of the posterior margin of the epistome concave; fig. 4C); and the P5 with a relatively shorter dactylus (fig. 2A) (vs. dactylus of P5 relatively longer; fig. 2B).

Balss (1937) reported *Potamon mindanaoense* also from Lake Dapao; his specimens consisted of one adult male (25.6 by 21.0 mm) and one juvenile female (20.0 by 16.5 mm). He also provided a photograph of the male specimen and an illustration of its G1 (Balss, 1937: figs. 22, 23). Based on the photograph, this male specimen more closely resembles *I. maranao* sp. nov. in the form of the carapace and proportion of the pereopods rather than *I. mindanaoense*. Furthermore, Balss' (1937) specimens were also collected from Lake Dapao, the type locality of *I. maranao* sp. nov. However, there is an appreciable difference in the form of the G1 of the holotype and that of Balss' (1937) specimen as illustrated, in that the holotype has a less extended apical lobe, although the proportions of the terminal and subterminal segments appear to be the same in the holotype and in Balss' specimen. Also, while from their synonymy list, Ng & Tan (1998) considered Balss' specimens as *I. sinuatifrons*, it is clear now, based on a stricter definition of this species' G1 morphology, that they are not conspecific (see remarks on *I. sinuatifrons*). For the reasons stated above, Balss' specimens are now considered as conspecific with *I. maranao* sp. nov.

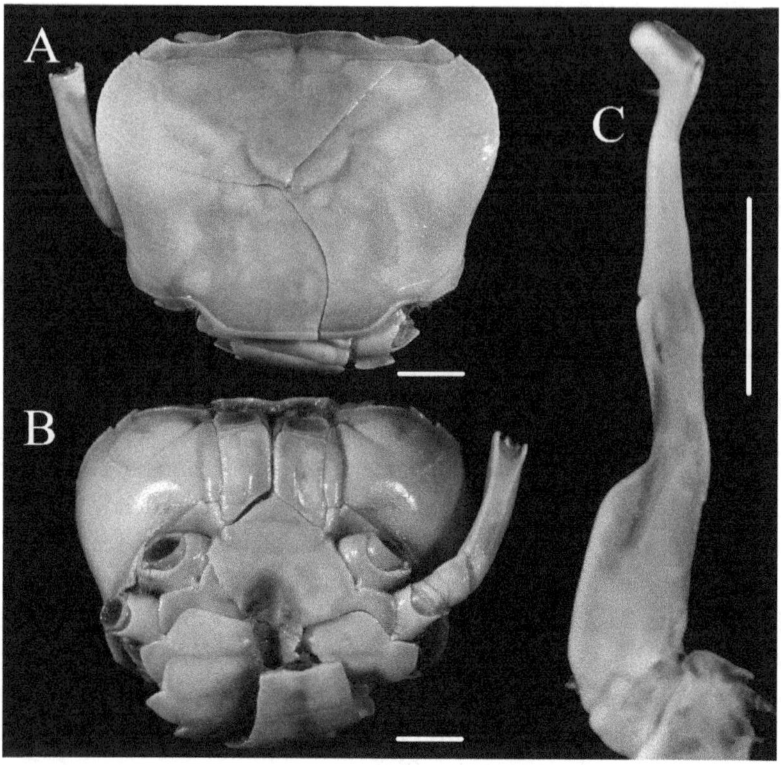

Fig. 3. *Isolapotamon mindanaoense* (Rathbun, 1904), holotype, male, 30.5 by 24.5 mm (MNHN-B5297), Mindanao (no specific locality). A, dorsal view; B, ventral view, left G1, internal view. Scale bars: 5.0 mm. Photos by Tohru Naruse.

Size. — The female paratype specimen (cw 26.3 mm) has a fully expanded abdomen; hence this is a small species, adult between cw 26 and 32 mm.

Distribution. — Lake Dapao, Lanao del Sur (type locality) (Balss, 1937; present record) (see fig. 9).

Isolapotamon mindanaoense (Rathbun, 1904) (figs. 2B, 3, 4, 5F-L)

Potamon (*Potamon*) *mindanaoensis* Rathbun, 1904: 268, fig. 8, pl. 10 fig. 5.
Isolapotamon (*Isolapotamon*) *mindanaoense*, Bott, 1968: 121, fig. 5.
Isolapotamon mindanaoense, Bott, 1970b: 192, pl. 41 fig. 79, pl. 56 fig. 79. — Ng & Tan, 1998: 72, fig. 10A-D. — Ng et al., 2008: 163 (list).

Material examined. — 2 males, 22.7 by 18.2 mm, 44.1 by 34.6 mm (ZRC 2012.1128), Mt. Hamiguitan, Davao Oriental Province, Mindanao, coll. A. C. Diesmos, May 2005; 1 male, 36.2 by 29.1 mm, 5 females, 26.5 by 22.2 mm, 27.3 by 22.5 mm, 28.5 by 23.8 mm, 33.7 by 27.6 mm, 42.5 by 33.9 mm (NMCR 39073) Bgy. Kimlawis, Kiblawan, Davao del Sur Province,

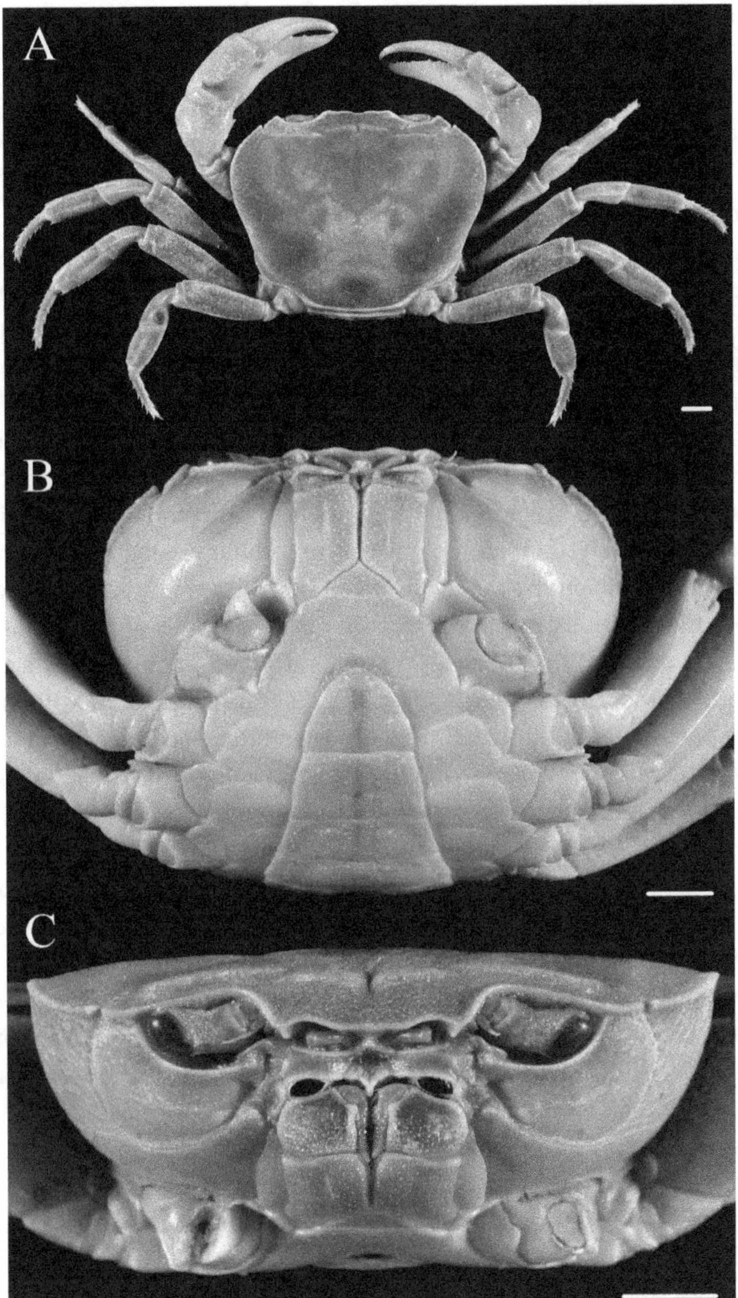

Fig. 4. *Isolapotamon mindanaoense* (Rathbun, 1904), A, female, 42.5 by 33.9 mm; B, C, male, 36.2 by 29.1 mm (NMCR 39073), Bgy. Kimlawis, Kiblawan, Davao del Sur Province, Mindanao. A, dorsal view; B, ventral view; C, anterior view. Scale bars: 5.0 mm.

Mindanao, coll. A. C. Diesmos, 17 Oct. 2009; 1 male, 30.1 by 24.5 mm, 1 female, 39.5 by 30.8 mm (ZRC 2012.1129), Mt. Kimlawis, Kiblawan, Davao del Sur Province, Mindanao, coll. A. C. Diesmos, Nov.-Dec. 2009; 1 male, 39.4 by 31.9 mm, 1 female, 33.9 by 27.1 mm, 5 juveniles, 10.3 by 8.5 mm, 12.5 by 10.2 mm, 13.9 by 11.5 mm, 15.2 by 12.5 mm, 17.8 by 14.7 mm (ZRC 2012.1130), around base camp, Mt. Magdiwata, San Francisco, Agusan del Sur Province, Mindanao, coll. A. C. Diesmos, Oct. 2008.

Diagnosis. — Carapace relatively narrow; mesobranchial, intestinal regions depressed dorsally; posterolateral margins gently concave. Dactylus of P5 long, slender, gently curving. G1 relatively long, slender, terminal segment and subterminal segment subequal in length; apical lobe prominently expanded, oblongate, angled about 45° from vertical axis.

Description (re-description based on newly collected material as well as recent photographs of holotype). — Carapace subquadrate, cordiform in outline, wider than long, width about 1.2-1.3 times length; dorsal surface depressed, mostly smooth, frontal region slightly granular, epibranchial region with faint striae; regions poorly defined; epigastric, suborbital cristae low, but distinct, not confluent; front bilobed, not advanced beyond orbits, ventrally deflexed, anterior margin sinuous, cristate; orbits cristate, granulate, supraorbital margin more-or-less straight, transverse; anterolateral margin convex, cristate, serrulate; exorbital tooth triangular, apex acute, lateral margin longer than mesial margin; epibranchial tooth small, broadly triangular, separated from exorbital tooth by moderately deep, V-shaped notch; cervical groove indistinct, central H-shaped groove distinct, deep; mesobranchial, intestinal regions slightly depressed.

Eyes well developed, corneas large. Posterior margin of epistome with well produced, central, triangular projection, straight to slightly sinuous laterally; endostome with distinct, complete ridges laterally. Third maxilliped smooth; merus subquadrate, length about half that of ischium; ischium with distinct submedian sulcus; exopod tapering toward distal end, lateral, mesial margins gently convex, tip reaching up to level of midlength of merus, flagellum slightly longer than width of merus.

Thoracic sternum long, narrow, smooth; sternites 1, 2, sternites 3, 4 completely fused, deep transverse sulcus separating two fused units; sternites 5-7 distinct externally; sterno-abdominal cavity moderately deep, setose, thoracic press-button distinct, visible at posterior half of sternite 5 median line visible only between sternites 7, 8.

Chelipeds unequal, surfaces smooth to slightly rugose; merus short, barely exceeding lateral margins of carapace, distal end of flexor surface with small but distinct, conical tubercle; carpus inner angle marked by long, acute spine, with second similar but smaller spine posteriorly; palm inflated, smooth except for rugose supero-external surface, length subequal to that of dactylus; both fingers with rounded keel running along length in both external, internal surfaces, internal margins well armed with more-or-less uniform teeth, tips pointed, incurved.

Ambulatory legs mostly smooth, glabrous, moderate in length, P3, P4, longest, about 1.5 times carapace width; meri subrectangular, thrice longer than broad, cristate on anterior, posterior margins; carpi with low longitudinal ridge on dorsal surface; propodi with shallow longitudinal sulcus on dorsal surface; anterior, posterior margins spinulose; dactyli slender, spinose, slightly curved, slightly longer than propodus.

Male abdomen broad, gradually tapering distally towards telson, with six freely articulated somites; somite 1 strip-like, with transverse crest; somite 2 trapezoidal; somites 3-6 subrectangular, somite 6 more subquadrate, wider than long, lateral margins straight to slightly convex; telson subtriangular, apex rounded, basal width about 1.3 times median length, median length of somite 6 about 0.8 times that of telson. Female abdomen wider, longer, almost completely covering thoracic sternum.

G1 long, relatively slender; terminal and subterminal segment subequal, with prominent shelf on external margin; apex with lobiform, oblongate, lateral extension angled about 45° from the vertical axis, with distinct triangular fold near terminal opening. G2 filiform; basal segment cylindrical for most of length; distal segment flattened, curved, about 0.6 times as long as basal segment.

Remarks. — Mary J. Rathbun wrote extensively on the taxonomy of the freshwater crabs of the world, describing several new species (Rathbun, 1904, 1905, 1906). Rathbun (1904) described *Potamon* (*Potamon*) *mindanaoensis* based on a single male specimen (30.5 by 24.5 mm), collected by Mr. Montano, and deposited in the MNHN in Paris (MNHN-B5297) (fig. 3). As to the provenance of this specimen, no more specific locality was provided by Rathbun besides "Mindanao". Bott (1968) transferred this species to the genus *Isolapotamon* on account of its distinctive G1 morphology. All subsequent treatments of the species (Bott, 1968, 1970b) were based on the holotype until Ng & Tan (1998) reported additional specimens, some of which had a specific locality indicated — "East Mindanao: Gulf of Davao, Tibuan River". No other record of "Tibuan River" could be found, but there is in current usage the name "Tiblawan River", which is a small river in Davao Oriental province, and which also drains into the Gulf of Davao. There are no other rivers draining into the Davao Gulf with a similar-sounding name. In any case, the consensus appears to be that *I. mindanaoense* is found in the southern region of Mindanao (Manuel-Santos, 2010). There are only a few illustrations of this species available in the literature, and those only of the holotype, which is devoid of most of its pereopods.

Additional material from other localities in southern and southeastern Mindanao have now been examined in this study, and some appreciable variation has now been noted, particularly in the morphology of the G1. The new material generally agrees well with the published description and illustrations of the holotype of *I. mindanaoense*. This was corroborated with further comparison with more recent photographs of the holotype (fig. 3). The variations in the G1 are mainly in the size of the apical lobe of the G1, with the G1 of the male from Mt. Hamiguitan, Davao Oriental (ZRC 2012.1128) (fig. 5F-J) most closely approximating that of the holotype (fig. 3C; also Bott, 1970b: pl. 41 fig. 79; Ng & Tan, 1998: fig. 10A-D). By contrast, the males from Kiblawan, Davao del Sur (NMCR 39073) (fig. 5L), and Mt. Magdiwata, Agusan del Sur (ZRC 2012.1130) (fig. 5K), tend to have larger apical lobes with a more prominent triangular fold near the terminal opening of the G1. The size of the apical lobe of the male from Mt. Magdiwata recalls that of *I. danielae*, the type locality of which is about 80 km away to the north (Manuel-Santos, 2010: fig. 2). The main difference between the two is in the angle at which the apical lobe joins the main axis of the G1, with that of *I. danielae* more bent away from the vertical axis, and in the shape of the lobe, where that of *I. danielae* is proportionately longer and narrower. The shape and proportions of the abdomen of the holotype (fig. 3B; also Bott, 1970b: pl. 56 fig. 79) are more similar to those of the males from Kiblawan (fig. 4B) and Mt. Magdiwata, with more concave lateral margins than that of the adult male from Mt. Hamiguitan, which tend to be straight. These minor differences are here considered part of the intraspecific variation one would observe within this group of freshwater crabs, however, and not enough on their own to warrant consideration of these forms as distinct species.

Isolapotamon mindanaoense is most similar to *I. danielae* from northeastern Mindanao, and *I. sinuatifrons* from the Zamboanga Peninsula, in the general appearance of the carapace and pereopods. It is also quite similar to these species in the general form of the G1 (see comments on G1 of *I. sinuatifrons* below), and can only be reliably distinguished from either species by the way the apical lobe is angled from the main axis of the G1 — about 45° in *I. mindanaoense* versus more than 45° in *I. danielae* (which also has a longer and narrower apical lobe) and about 90° in *I. sinuatifrons*.

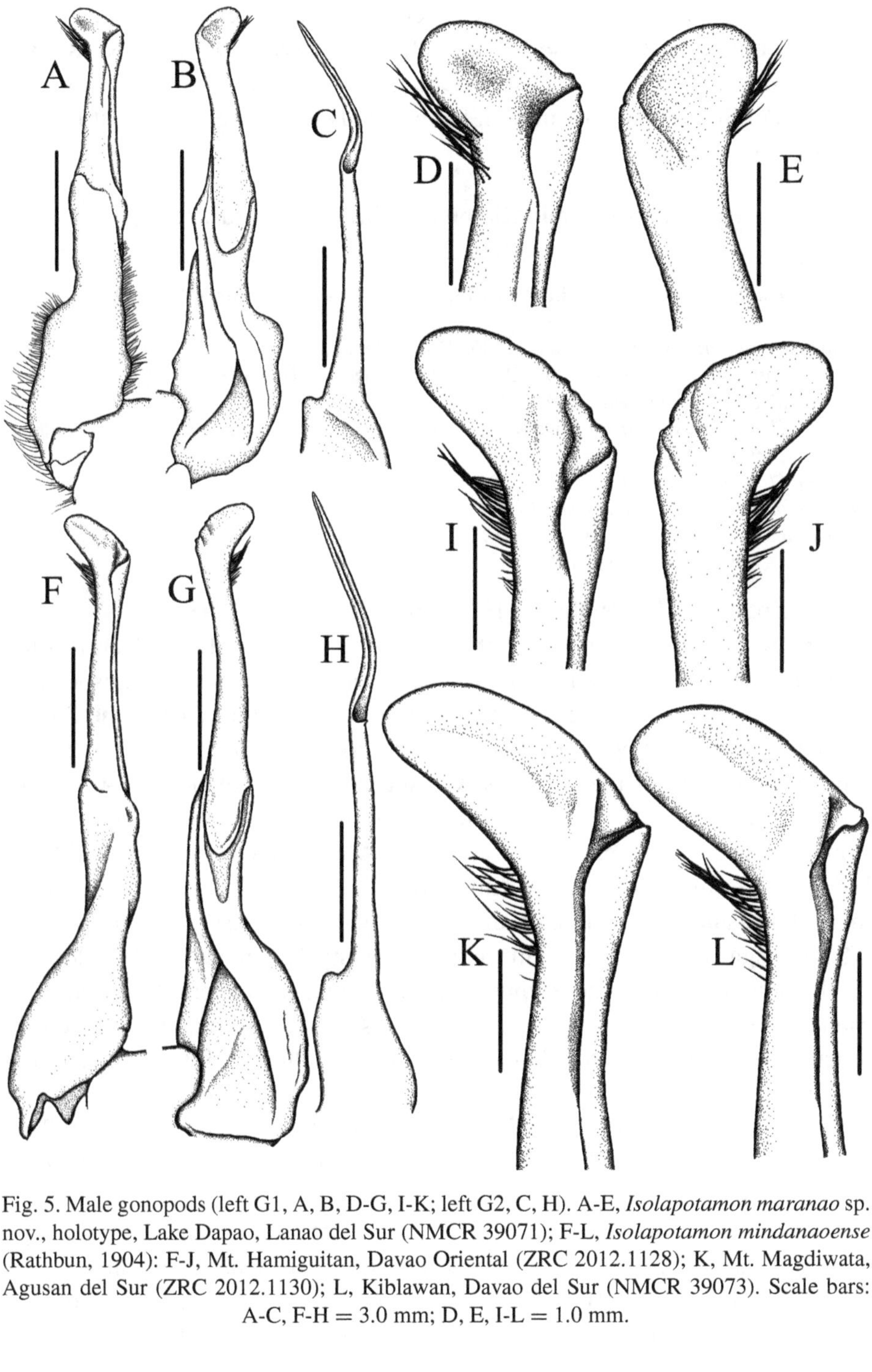

Fig. 5. Male gonopods (left G1, A, B, D-G, I-K; left G2, C, H). A-E, *Isolapotamon maranao* sp. nov., holotype, Lake Dapao, Lanao del Sur (NMCR 39071); F-L, *Isolapotamon mindanaoense* (Rathbun, 1904): F-J, Mt. Hamiguitan, Davao Oriental (ZRC 2012.1128); K, Mt. Magdiwata, Agusan del Sur (ZRC 2012.1130); L, Kiblawan, Davao del Sur (NMCR 39073). Scale bars: A-C, F-H = 3.0 mm; D, E, I-L = 1.0 mm.

Size. — This is a medium to large species, adult between cw 30 and 44 mm.

Distribution. — Tibuan (= Tiblawan?) River, Davao Oriental (Ng & Tan, 1998); Mt. Hamiguitan, Davao Oriental, Mt. Magdiwata, San Francisco, Agusan del Sur, and Mt. Kimlawis, Kiblawan, Davao del Sur (present record) (see fig. 9).

Isolapotamon sinuatifrons (H. Milne Edwards, 1853)
(figs. 2C, 6, 7)

Thelphusa sinuatifrons H. Milne Edwards, 1853: 211. — A. Milne-Edwards, 1869: 167, pl. 10 fig. 2.

Thelphusa sinuatifrons (?) var., Miers, 1886: 214, pl. 18 fig. 1.

Telphusa sinuatifrons, De Man, 1892: 296. — Bürger, 1894: 2 (in part).

Potamon sinuatifrons, De Man, 1898: 404.

Potamon (*Potamon*) *sinuatifrons*, De Man, 1899: 92, 100, pl. 8 fig. 9, pl. 9 fig. 9. — Rathbun, 1904: 266, pl. 10 fig. 9.

Potamon (*Telphusa*) *sinuatifrons*, Estampador, 1937: 532; 1959: 87.

Isolapotamon (*Isolapotamon*) *sinuatifrons*, Bott, 1968: 121, fig. 6.

Isolapotamon sinuatifrons, Bott, 1970b: 192, pl. 41 fig. 83. — Ng & Tan, 1998: 75, figs. 10M-P, 11. — Ng et al., 2008: 163 (list).

Description. — See Rathbun (1904: 266); Ng & Tan (1998: 75).

Remarks. — The taxonomy of this species is rather confused. Henri Milne Edwards (1853) described *Thelphusa sinuatifrons* based on an unspecified number of specimens collected during the voyage of the Zelée to the Pacific (collected by Mr. Legillou), but the type locality was not specified. An illustration of one of the syntypes was subsequently provided by A. Milne-Edwards (1869: pl. 10 fig. 2). Subsequent specimens, however, were collected from "Pasananca" (Miers, 1886) and from "Zamboanga" (Bürger, 1894: 2). Using the terminology of the time, Miers' "Pasananca" can be taken to mean Pasonanca, a district in the port city of Zamboanga, and Bürger's "Zamboanga" can be taken to mean the same city. Zamboanga City is located on the southern tip of the Zamboanga Peninsula, in western Mindanao. De Man (1899) reported having examined three syntypes (all male), and a fourth, female, specimen collected from Zamboanga. Rathbun (1904) also reported examining the three male syntypes. Bott (1968) examined two males and one female, and the larger male ("MPa 2371-41") he designated as lectotype. This lectotype, measuring 53.0 by 39.0 mm, is deposited in the MNHN (MNHN-BP4353) (fig. 6).

Bott (1968, 1970b) was the first to provide illustrations of the G1 of the type material, although he did not specify which specimen he based the illustrations on. These illustrations (Bott, 1968: fig. 6a, b; 1970b: pl. 41 fig. 83) show the apical lobe positioned at a right angle to the long axis of the G1, and somewhat

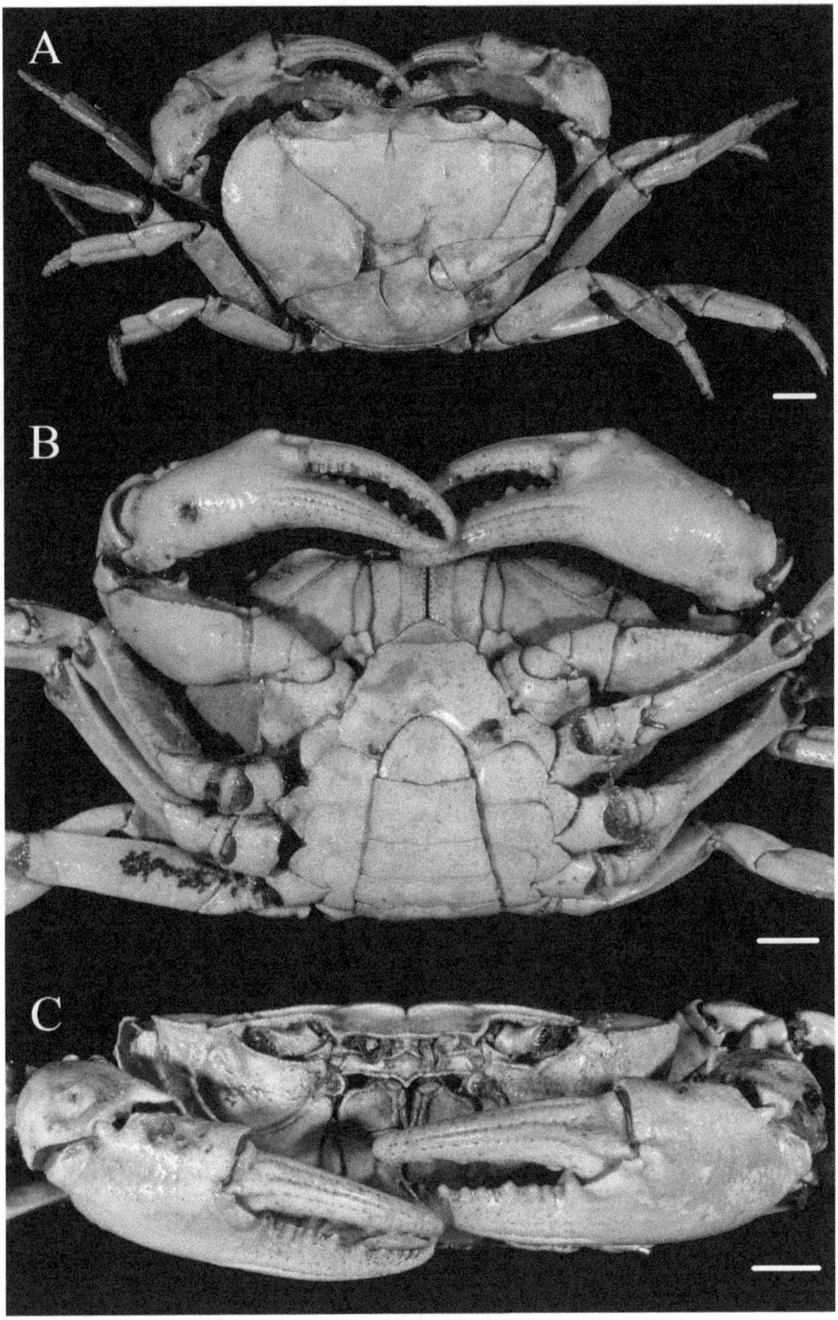

Fig. 6. *Isolapotamon sinuatifrons* (H. Milne Edwards, 1853), lectotype, male, 53.0 by 39.0 mm (MNHN-BP4353), locality unknown. A, dorsal view; B, ventral view; C, anterior view. Scale bars: 5.0 mm. Photos by Tohru Naruse.

oblongate in outline, with the tip more acute than in *I. mindanaoense*; they also show that the terminal segment is not twisted along the longitudinal axis of the G1. However, Ng & Tan (1998), illustrated the actual G1 of the lectotype (Ng & Tan, 1998: fig. 10M-P), where the terminal segment is clearly twisted and the apical lobe is much narrower and more pointed at the tip (also fig. 7A, B). They also mention that a syntype male, also deposited in the MNHN, differs from the lectotype in the following features: "the tip of the G1 terminal segment lies flat on the sternum and not twisted, whereas it is twisted and facing towards the sternum in the lectotype; the syntype male has a proportionately shorter ambulatory merus; the left ambulatory leg propodus of the syntype male is proportinately half as short as that on the lectotype; there are four low spines on the posterior margin of the ambulatory propodus of the syntype male but it is smooth in the lectotype; the syntype male has relatively flatter branchial regions" (Ng & Tan, 1998: 76). They then concluded that although it is possible that the syntype and lectotype are not conspecific, there was not enough evidence to conclusively say so. It is very likely that this syntype male (MNHN-BP3845) was the one whose G1 was figured by Bott (1968, 1970b). It is also quite similar to the G1 illustrated by Balss (1937: fig. 22) from a specimen from Lake Dapao in Lanao del Sur province, which he identified as *Potamon mindanaoensis*, but which is now considered to be conspecific with *I. maranao* sp. nov. (see remarks for *I. maranao* sp. nov.). Ng & Tan (1998) re-examined the material reported by Miers (1886) from Pasonanca, Zamboanga, now deposited in the Natural History Museum, London (NHM 1884:31), and confirmed those to be *I. sinuatifrons*, with the terminal segment of the G1 exhibiting the diagnostic twist, although the tip of the apical lobe was observed to be more rounded rather than pointed. The taxonomy of *I. sinuatifrons* is far from settled and can only be conclusively resolved with additional material from western Mindanao, preferably specimens similar in size to the lectotype.

Size. — This is a large species, adult at cw 53 mm.

Distribution. — Zamboanga City, Zamboanga Peninsula (Miers, 1886; Bürger, 1894; Ng & Tan, 1998) (see fig. 9).

Isolapotamon spatha Ng & Takeda, 1992
(figs. 2D, 8)

Isolapotamon spatha Ng & Takeda, 1992: 163, fig. 7. — Ng & Tan, 1998: 76, fig. 12A-D. — Ng et al., 2008: 163 (list).

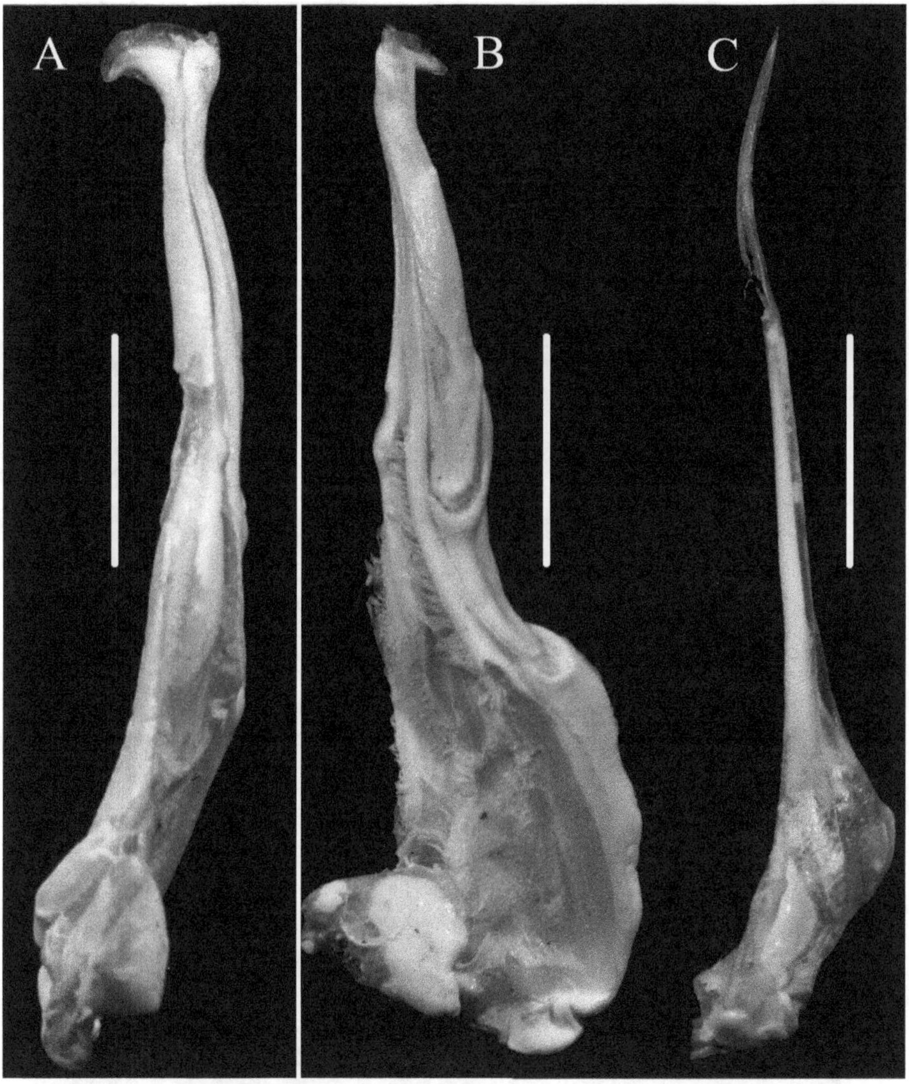

Fig. 7. *Isolapotamon sinuatifrons* (H. Milne Edwards, 1853), lectotype, male (MNHN-BP4353), locality unknown. A, left G1, mesio-internal view; B, left G1, external view; C, left G2, internal view. Scale bars: 5.0 mm. Photos by Tohru Naruse.

Material examined. — Paratypes, 1 male, 18.8 by 15.4 mm, 1 female, 25.3 by 19.9 mm (ZRC 1999.0070-0071), Bgy. Kraan, Palimbang, Sultan Kudarat Province, Mindanao, coll. Y. Nishikawa, 12 Aug. 1985.

Description. — See Ng & Takeda (1992: 163); Ng & Tan (1998: 76).

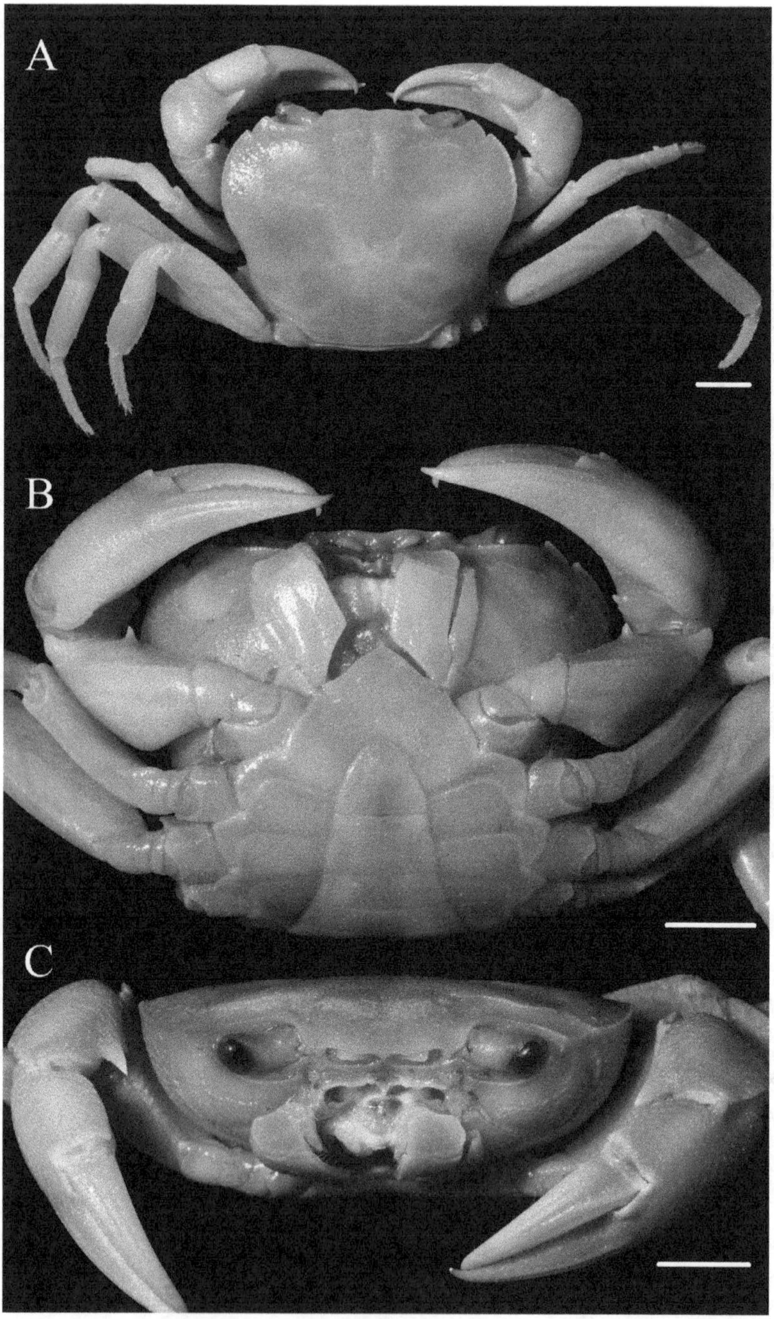

Fig. 8. *Isolapotamon spatha* Ng & Takeda, 1992, holotype, male, 28.0 by 22.7 mm (NMST-Cr 11225), Palimbang, Sultan Kudarat province, Mindanao. A, dorsal view; B, ventral view; C, anterior view. Scale bars: 5.0 mm. Photos by Hironori Komatsu.

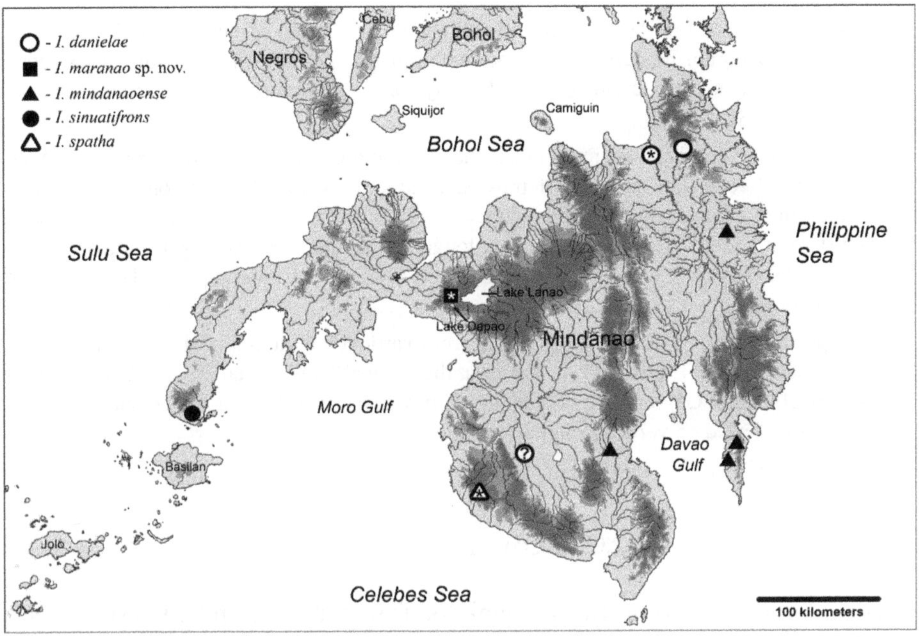

Fig. 9. Map of Mindanao, showing known collecting localities for Philippine *Isolapotamon* spp. Dark grey areas represent elevated regions (500-2500 m a.s.l.). Legend: *Isolapotamon danielae* — white circles; *I. maranao* sp. nov. — black square; *I. mindanaoense* — black triangles; *I. sinuatifrons* — black circle; *I. spatha* — white triangle; question mark — uncertain identification; asterisk — type locality.

Remarks. — Ng & Takeda (1992) described *Isolapotamon spatha* from one adult male and two subadult (male and female) specimens collected from Kraan, Palimbang, Sultan Kudarat province, near the southwestern coast of Mindanao, bordering the Moro Gulf. This poorly known species has not been reported from any other locality in Mindanao. Its closest congener is *I. maranao* sp. nov. (see remarks on *I. maranao*).

Size. — This is a small species, adult between cw 28 and 30 mm.

Distribution. — Palimbang, Sultan Kudarat (type locality) (Ng & Takeda, 1992; Ng & Tan, 1998) (see fig. 9).

KEY TO THE PHILIPPINE SPECIES OF *ISOLAPOTAMON*

1. Apical lobe of G1 tilted at about 45° from the vertical/longitudinal axis 2
– Apical lobe of G1 bent at more than 45° from the vertical/longitudinal axis 4

2. Apical lobe of the G1 short, subcircular, about as long as wide; carapace intestinal region depressed; external margin of exopod of third maxilliped straight; dactylus of P5 much shorter and stouter (about 4.8 times as long as wide) 3

– Apical lobe of G1 relatively longer, oblongate, distinctly longer than wide; carapace intestinal region moderately inflated; external margin of exopod of third maxilliped distinctly convex; dactylus of P5 relatively longer and more slender (about 6.0 times as long as wide) . . . . *I. mindanaoense*

3. G1 slender; G2 distal segment much shorter, about 0.4 times as long as basal segment; branchial region of carapace depressed in anterior view; striations on posterolateral margin of carapace present and distinct . . . . *I. spatha*

– G1 stout; G2 distal segment relatively longer, about 0.5 times as long as basal segment; branchial region of carapace inflated in anterior view; striations on posterolateral margin of carapace absent . . . . *I. maranao* sp. nov.

4. Apical lobe of G1 bent at 90° or more from vertical/longitudinal axis; distal portion of G1 twisted by 90° counter-clockwise on the vertical/longitudinal axis . . . *I. sinuatifrons*

– Apical lobe of G1 bent at less than 90° from vertical/longitudinal axis; distal portion of G1 not twisted on vertical/longitudinal axis . . . . *I. danielae*

ACKNOWLEDGEMENTS

The authors wish to profusely thank the following: Arvin C. Diesmos (National Museum of the Philippines, Manila) for kindly sharing his collection of Philippine freshwater crabs with them; the Mindanao State University (MSU), in Marawi City, for its support during fieldwork; Tohru Naruse (University of the Ryukyus) & Ng Ngan Kee (National University of Singapore) for their help with the examination and photography of the types of *Isolapotamon mindanaoense* and *I. sinuatifrons*; Hironori Komatsu (NMST) for providing photos of the holotype of *I. spatha*; Marivene Manuel-Santos (NMCR) and Tan Siong Kiat (ZRC) for specimen cataloguing; and Peter Ng (NUS) for valuable discussions on this topic. The authors are also very grateful to Sebastian Klaus and Neil Cumberlidge for their critical review of this article. Support from the National University of Singapore to DCJY (grant number R-154-000-465-133) is acknowledged.

REFERENCES

BALSS, H., 1937. Potamiden (Dekapoda Brachyura) der Philippinen und des Malayischen Archipels. Internationale Revue der gesamten Hydrobiologie und Hydrographie, **34**(3-5): 143-187.

BOTT, R., 1968. Potamiden aus Süd-Asien (Crustacea, Decapoda). Senckenbergiana biologica, **49**: 119-130.

— —, 1969. Flußkrabben aus Asien und ihre Klassifikation (Crustacea, Decapoda). Senckenbergiana biologica, **50**: 359-366.

— —, 1970a. Betrachtungen über die Entwicklungsgeschichte und Verbreitung der Süßwasserkrabben nach der Sammlung des Naturhistorischen Museums in Genf/Schweiz. Revue Suisse de Zoologie, **77**(24): 327-344.

— —, 1970b. Die Süßwasserkrabben von Europa, Asien, Australien und ihre Stammesgeschichte. Eine Revision der Potamoidea und Parathelphusoidea (Crustacea, Decapoda). Abhandlungen der Senckenbergischen Naturforschenden Gesellschaft, **526**: 1-338.

BRANDIS, D., 2002. On the taxonomic status and biogeography of the Isolapotamidae Bott, 1970 (Decapoda, Brachyura). Journal of Natural History, **36**(11): 1291-1339.

BUREAU OF INSULAR AFFAIRS, 1902. Pronouncing gazetteer and geographical dictionary of the Philippine Islands, United States of America, with maps, charts and illustrations, also the law of government in the Philippine Islands passed by Congress and approved by the President, July 1, 1902, with a complete index: 933 pp. (Bureau of Insular Affairs, War Department, & Government Printing Office, Washington).

BÜRGER, O., 1894. Beiträge zur Kenntnis der Gattung *Telphusa*. Zoologische Jahrbücher, (Systematik) **8**: 1-7.

CATIBOG-SINHA, C. & L. R. HEANEY, 2006. Philippine biodiversity: principles and practice: 1-495. (Haribon Foundation for the Conservation of Nature, Quezon City).

CUMBERLIDGE, N., P. K. L. NG, D. C. J. YEO, C. MAGALHÃES, M. R. CAMPOS, F. ALVAREZ, T. NARUSE, S. R. DANIELS, L. J. ESSER, F. Y. K. ATTIPOE, F.-L. CLOTILDE-BA, W. DARWALL, A. MCIVOR, J. E. M. BAILLIE, B. COLLEN & M. RAM, 2009. Freshwater crabs and the biodiversity crisis: importance, threats, status, and conservation challenges. Biological Conservation, **142**(8): 1665-1673.

DANIELS, S. R., 2003. Examining the genetic structure among populations of the common Cape river crab *Potamonautes perlatus* from river systems in South Africa reveals hydrographic boundaries. Journal of Crustacean Biology, **23**: 936-950.

DANIELS, S. R., B. A. STEWART, G. GOUWS, M. CUNNINGHAM & C. A. MATTHEE, 2002. Phylogenetic relationships of the southern African freshwater crab fauna (Decapoda: Potamonautidae: *Potamonautes*) derived from multiple data sets reveal biogeographic patterning. Molecular Phylogenetics and Evolution, **25**: 511-523.

ESTAMPADOR, E., 1937. A check list of Philippine crustacean decapods. Philippine Journal of Science, **62**: 465-559.

— —, 1959. Revised check list of Philippine crustacean decapods. Natural & Applied Science Bulletin, **17**: 1-125.

FREITAG, H., 2012. Revision of the genus *Insulamon* Ng & Takeda, 1992 (Crustacea: Decapoda: Potamidae) with description of four new species. Raffles Bulletin of Zoology, **60**(1): 37-55.

HALL, R., 1998. The plate tectonics of Cenozoic SE Asia and the distribution of land and sea. In: R. HALL & J. D. HOLLOWAY (eds.), Biogeography and geological evolution of South East Asia: 99-131. (Backhuys Publishers, Leiden).

— —, 2002. Cenozoic geological and plate tectonic evolution of SE Asia and the SW Pacific: computer-based reconstructions and animations. Journal of Asian Earth Sciences, **20**: 353-434.

— —, 2009. Southeast Asia's changing palaeogeography. Blumea, **54**: 148-161.

HEANEY, L. R., 1986. Biogeography of mammals in SE Asia: estimates of rates of colonization, extinction and speciation. Biological Journal of the Linnaean Society, **28**: 127-165.

HEANEY, L. R. & J. C. REGALADO, 1998. Vanishing treasures of the Philippine rainforest: 1-88. (The Field Museum of Natural History, Chicago).

HUSANA, D. E. M., T. KASE & P. K. L. NG, 2013. On a new species of freshwater crab of the genus *Ovitamon* Ng & Takeda, 1992 (Crustacea: Brachyura: Potamidae) from Panay Island, Philippines. Raffles Bulletin of Zoology, **61**(2): 651-655.

HUSANA, D. E. M., T. NARUSE & T. KASE, 2009. Two new cavernicolous species of the genus *Sundathelphusa* from western Samar, Philippines (Decapoda: Brachyura: Parathelphusidae). Journal of Crustacean Biology, **29**(3): 419-427.

LOHMAN, D. J., M. DE BRUYN, T. PAGE, K. VON RINTELEN, R. HALL, P. K. L. NG, H.-T. SHIH, G. R. CARVALHO & T. VON RINTELEN, 2011. Biogeography of the Indo-Australian Archipelago. Annual Review of Ecology, Evolution and Systematics, **42**: 205-226.

MAN, J. G. DE, 1892. Decapoden des Indischen Archipels. Zoologische Ergebnisse einer Reise in Niederländisch Ost-Indien, **2**: 265-567.

— —, 1898. Viaggo di Leonardo Fea in Birmanis. Annali del Museo Civico di Storia Naturale di Genova, **2**(19): 284-440.

— —, 1899. Zoological results of the Dutch Scientific Expedition to Central Borneo. The Crustacea. Part II, Brachyura. Notes from the Leyden Museum, **21**: 53-144.

MANUEL-SANTOS, M. R., 2010. A new species of *Isolapotamon* Bott, 1968 (Brachyura, Potamidae) from Mindanao, Philippines. In: P. CASTRO, P. J. F. DAVIE, P. K. L. NG & B. RICHER DE FORGES (eds.), Studies on Brachyura: a homage to Danièle Guinot. Crustaceana Monographs, **11**: 197-202.

MANUEL-SANTOS, M. R. & P. K. L. NG, 2013. Two new species of *Ovitamon* Ng & Takeda, 1992 (Crustacea: Brachyura: Potamidae) from the Philippines. Zootaxa, **3619**(3): 394-400.

MENDOZA, J. C. E. & T. NARUSE, 2010. A new species of riverine crab of the genus *Sundathelphusa* Bott, 1969 (Crustacea: Brachyura: Gecarcinucidae) from northeastern Luzon, Philippines. Philippine Journal of Science, **139**(1): 61-70.

MIERS, E. J., 1886. Report on the Brachyura collected by H.M.S. Challenger during the years 1873-1876. Reports of the Voyage of the Challenger, **17**(2): 1-362, pls. 1-29.

MILNE-EDWARDS, A., 1869. Revision du genre Thelphuse et description de quelques espèces nouvelles faisant partie de la collection du museum. Nouvelles Archives du Muséum d'Histoire Naturelle, **5**: 161-191, pls. 8-11.

MILNE EDWARDS, H., 1853. Memoire sur la famille des Ocypodiens. Annales du Science Naturelles, Zoologie, (3) **20**: 163-228.

NG, P. K. L., 1988. The freshwater crabs of Peninsular Malaysia and Singapore: 1-156, pls. 1-4. (Department of Zoology, National University of the Singapore and Shinglee Press, Singapore).

— —, 2010. On the identity of *Para-Bary-Thelphusa grapsoides longipes* Balss, 1937, with description of a new species from the Philippines (Brachyura, Gecarcinucidae). In: C. H. J. M. FRANSEN, S. DE GRAVE & P. K. L. NG (eds.), Studies on Malacostraca: Lipke Bijdeley Holthuis memorial volume. Crustaceana Monographs, **14**: 561-571.

NG, P. K. L., D. GUINOT & P. J. F. DAVIE, 2008. Systema Brachyuorum. Part I. An annotated checklist of extant brachyuran crabs of the world. Raffles Bulletin of Zoology, (Supplement) **17**: 1-286.

NG, P. K. L. & G. RODRÍGUEZ, 1995. Freshwater crabs as poor zoogeographical indicators: a critique of Bănărescu (1990). Crustaceana, **68**: 636-645.

NG, P. K. L. & M. TAKEDA, 1992. The freshwater crab fauna (Crustacea, Brachyura) of the Philippines. I. The family Potamidae Ortmann, 1896. Bulletin of the National Science Museum, **18**: 149-166.

NG, P. K. L. & S. H. TAN, 1998. A revision of the southeast Asian freshwater crabs of the genus *Isolapotamon* Bott, 1968 (Crustacea: Decapoda: Brachyura: Potamidae). Proceedings of the Biological Society of Washington, **111**(1): 52-80.

NG, P. K. L. & C. M. YANG, 1985. On three new species of freshwater crabs (Crustacea, Decapoda, Brachyura) from Singapore and Peninsular Malaysia. Malayan Nature Journal, **39**: 57-73.

— — & — —, 1986. A new freshwater crab of the genus *Isolapotamon* Bott, 1968 (Decapoda, Brachyura, Potamidae) from Sarawak, Borneo. Indo-Malayan Zoology, **3**: 15-18.

ONG, P. S., L. E. AFUANG & R. G. ROSELL-AMBAL (eds.), 2002. Philippine biodiversity conservation priorities: a second iteration of the National Biodiversity Strategy and Action Plan: 1-113. (Department of Environment and Natural Resources, Conservation International, University of the Philippines, and Foundation for the Philippine Environment, Quezon City).

ORTMANN, A., 1896. Das System der Decapoden-Krebse. Zoologische Jahrbücher, (Systematik) **9**: 409-453.

RATHBUN, M. J., 1904. Les crabes d'eau douce. Nouvelles Archives du Muséum d'Histoire Naturelle, **4**(6): 225-312.

TAN, S. H. & P. K. L. NG, 1998. The freshwater crab fauna (Crustacea: Brachyura) of the Philippines. V. On a new genus and species of potamid from Palawan Island, Philippines. Hydrobiologia, **379**: 93-96.

VALLEJO, B. JR., 2011. The Philippines in Wallacea. In: D. TELNOV (ed.), Biodiversity, biogeography and nature conservation in Wallacea and New Guinea, **1**: 1-450. (Entomological Society of Latvia, Riga).

YEO, D. C. J., P. K. L. NG, N. CUMBERLIDGE, C. MAGALHÃES, S. R. DANIELS & M. R. CAMPOS, 2008. Global diversity of crabs (Crustacea: Decapoda: Brachyura) in freshwater. Hydrobiologia, **595**: 275-286.

First received 13 July 2012.
Final version accepted 29 July 2013.

NEW OCCURRENCE OF MIOCENE FRESHWATER CRABS (BRACHYURA, POTAMIDAE) IN THE NORTH ALPINE FORELAND BASIN, GERMANY, WITH A NOTE ON FOSSIL *POTAMON* TO CALIBRATE MOLECULAR CLOCKS

BY

SEBASTIAN KLAUS[1,2,5]) and JÉRÔME PRIETO[3,4,6])

[1]) Department of Ecology and Evolution, J.W. Goethe-Universität, Max-von-Laue-Straße 13, 60438 Frankfurt am Main, Germany

[2]) Department of Biological Sciences, National University of Singapore, 14 Science Drive 4, Singapore 117543, Republic of Singapore

[3]) Senckenberg Centre for Human Evolution and Palaeoecology (HEP), Eberhard-Karls University, Institute for Geoscience, Sigwartstraße 10, 72076 Tübingen, Germany

[4]) Department for Earth and Environmental Sciences, Ludwig-Maximilians-University Munich & Bavarian State Collections for Palaeontology and Geology, Richard-Wagner-Straße 10, 80333 Munich, Germany

ABSTRACT

Miocene freshwater crabs remain a rarity in the North Alpine Foreland basin despite reports in the palaeontological literature dating back to the beginning of the XIXth century. We report on the presence of potamid freshwater crabs *Potamon* cf. *speciosus* (Von Meyer, 1844) (Decapoda, Potamidae) from the Middle Miocene locality at Giggenhausen near Freising, Germany. The biogeography of fossil *Potamon* is discussed and the use of the first occurrence date of *Potamon* in molecular phylogenetic studies as a calibration point for molecular clocks is evaluated. We propose that the earliest fossil from this genus, *Potamon quenstedti* (Zittel, 1885), belongs to the stem group of extant *Potamon*, but probably does not represent its direct ancestor because biogeographic and phylogenetic evidence indicates that *Potamon* originated in the Near/Middle East. This taxonomic uncertainty should be taken into account when calibrating molecular clocks with fossil *Potamon*.

ZUSAMMENFASSUNG

Obwohl in der paläontologischen Literatur bereits am Anfang des XIX. Jhdts. erwähnt, bleiben miozäne Süßwasserkrabben eine Seltenheit im nordalpinen Molassetrog. Mit dem

[5]) Corresponding author; e-mail: klaus@bio.uni-frankfurt.de

[6]) e-mail: j.prieto@lrz.uni-muenchen.de

Nachweis der potamiden Süßwasserkrabbe *Potamon* cf. *speciosus* (Von Meyer, 1844) (Decapoda, Potamidae) in der Fundstelle Giggenhausen bei Freising (spätestes Mittleres Miozän) tragen wir zur Kenntnis dieser Familie am Ende des Mittleren Miozäns bei. Wir geben einen kurzen Überblick zur Biogeographie des fossilen *Potamon*, und diskutieren detaillierter die Verwendung des ersten Auftretens von *Potamon* zur Kalibrierung von molekularen Uhren. Wir nehmen an, dass der früheste *Potamon quenstedti* (Zittel, 1885) zur Stammgruppe von *Potamon* gehört, auch wenn er wahrscheinlich nicht in direkter Linie zum gegenwärtigen *Potamon* steht, der, basierend auf bekannten molekularen und biogeographischen Mustern, sehr wahrscheinlich im Nahen oder Mittleren Osten evolviert ist. Bei der Verwendung zur zeitlichen Kalibrierung von Phylogenien sollte diese taxonomische Unsicherheit in die Analyse einfließen, um gut begründete Ergebnisse zu erhalten.

INTRODUCTION

The fossil record of freshwater decapods is rather scarce compared to that of their marine relatives and this is particularly true for the Brachyura. Although extant primary freshwater crabs (the families Gecarcinucidae, Potamidae, Potamonautidae, Pseudothelphusidae and Trichodactylidae) comprise about 20% of the currently described brachyuran species (Ng et al., 2008), there are only eight exclusively fossil species described: six from Europe (excluding the dubious *Potamon hungaricum* Kőrössy, 1940), one from the Indian subcontinent (*Potamon sivalense* Glaessner, 1933), and one from East Africa (*Tanzanonautes tuerkayi* Feldmann et al., 2005) which is the earliest known fossil freshwater crab. Freshwater crab fragments, especially of the strongly calcified claws, are found more frequently and include additional Neogene sites in Africa (see references in Klaus & Gross, 2010), South America (Rodríguez & Diaz, 1977; Melchor et al., 2010), and Asia (Naruse et al., 2003; Klaus et al., 2011). However, most fossil freshwater crab sites known to date are in Europe and the Near East (fig. 1; Klaus & Gross, 2010).

The German part of the North Alpine Foreland Basin (NAFB, fig. 1) holds a special historical place in the study of fossil freshwater crabs because it is where the first fossils of any freshwater crabs were reported at the beginning of the XIXth century (Karg, 1805). Moreover, the Miocene deposits of South Germany have yielded the oldest known Eurasian fossil freshwater crabs (Klaus & Gross, 2010). The record of fossil freshwater crabs from the Upper Freshwater Molasse from Engelswies in Germany dates back to *Potamon quenstedti* (Zittel, 1885) around 17.0 ± 0.1 Myr (Böhme et al., 2011; early Upper Freshwater Molasse (OSM); OSM C+D in the local biostratigraphical scale; see Abdul Aziz et al., 2008, 2010). The same species is probably also a part of the faunas of two other localities in the OSM C+D (Langenenslingen and Oggenhof), thus extending the stratigraphic range of the species to ~16 Myr. Several

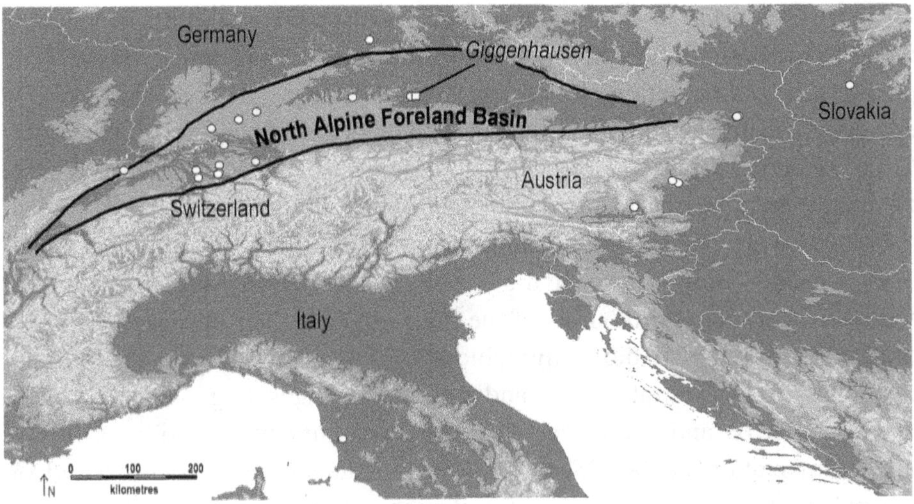

Fig. 1. Map of southern Central Europe and the adjacent Mediterranean area showing known fossil freshwater crab localities (white circles, after Klaus & Gross, 2010), including the site of Giggenhausen, Germany. Black lines indicate the approximate extent of the North Alpine Foreland basin (NAFB) based on Kuhlemann & Kempf (2002).

sites in the Suisse OSM that yielded freshwater crab fossils (including ichnofossils) are even younger (see Klaus & Gross, 2010 and references therein), ranging from ~16 Myr (Vermes) to 14.8-?14.5 (Schwamendingen). However, as complete carapaces are lacking, the exact taxonomic assignment is ambiguous, although geographic and stratigraphic proximities suggest perhaps a close relationship. *Potamon speciosus* (Von Meyer, 1844) was described at the end of the Badenian and the early Sarmatian s. str. from Öhningen and this species also occurs in Kleineisenbach (Klaus & Gross, 2010), a fossil vertebrate locality close to Munich that belongs to the younger deposits of the NAFB (Mayr, 1979; Prieto, 2007). The most recent evidence of Potamidae from the NAFB is *Potamon hegauense* Klaus & Gross, 2010, from the Late Miocene deposits (10.3 Myr) at Höwenegg (Giersch et al., 2010).

In the unpublished material of the locality Giggenhausen, a classic locality for micro-mammalian palaeontology (Fahlbusch, 1964), remains of freshwater crabs have been excavated and are presented herein.

LOCALITY AND STRATIGRAPHY

The NAFB is a classical foreland basin at the northern rim of the Alps spanning from the French Haute-Savoie in the west to Lower Austria (Vienna area) in the east (fig. 1). It is bounded to the North by the Swiss Jura Mountains, the Black forest, the Franconian Platform and the Bohemian Massif. The post-flysch sedimentation, following the early stages of the orogenesis, is characterized by an alternation of marine, brackish and continental Oligo-Miocene deposits (e.g., Kuhlemann & Kempf, 2002). The youngest deposits of interest here belong to the continental Upper Freshwater Molasse (OSM, abbreviation of the German terminology "Obere Süßwassermolasse") that begins in the German part of the NAFB at around 18 Myr; the youngest sediments correlate to the basis of the Late Miocene. The chronostratigraphy of the OSM is linked to biostratigraphic studies, most often based on terrestrial mammals, magnetostratigraphy and radiometric ages (e.g., Abdul Aziz et al., 2008, 2010 and references therein). Environmental conditions during the sedimentation of the OSM are mainly deduced from proxies based on the paleobotanic and vertebrate fossil record (e.g., Böhme, 2003; Böhme et al., 2006, 2007). They show that the beginning of the sedimentation of the OSM (Late Ottnangian) was completed under an especially warm period characterized by evergreen *Carapoxylon* (*Xylocarpus*) forests (Böhme et al., 2007) and the presence of ectothermic vertebrates adapted to dryer habitats (Böhme, 2003). Indeed this period was significantly warmer than the following Mid-Miocene Climate Optimum which ends by a climatic deterioration between 14 and 13 Myr. The massive and stepwise drop of temperature at that time in Germany led to the extinction of chameleons, alligators and giant turtles, and to drastic changes in the mammalian faunal composition.

The deposits in which the freshwater crab fragments were found were exposed along a small road in a forest near to the village of Giggenhausen (SE-Germany, Bavaria, 7 km SW of Freising, 48.3655°N 11.6682°W, for lithological details see Fahlbusch, 1964). Beside molluscs, reptiles, amphibians and fishes (Böhme, 2003), suids and proboscidians (Eronen & Rössner, 2007), a rich small mammal fauna characterises the locality (e.g., Fahlbusch, 1964, 1975; Black, 1966; Mayr, 1979; de Bruijn et al., 1993; Prieto, 2007, 2010). The fauna is biostratigraphically correlated to the Assemblage Zone of Anwil (*sensu* Kälin et al., 2001) or *Deperetomys hagni* taxon range zone (*sensu* Kälin & Kempf, 2009). The fauna lacks the cricetid rodent *Megacricetodon germanicus* Aguilar, 1980, which belongs to a lineage that has been used for relative age estimation (Kälin et al., 2001; Prieto & Rummel, 2009). The fauna is biochronologically close to that of the Swiss locality Anwil (Engesser, 1972; Prieto, 2007, age estimation 13.3 Myr after Kälin & Kempf, 2009).

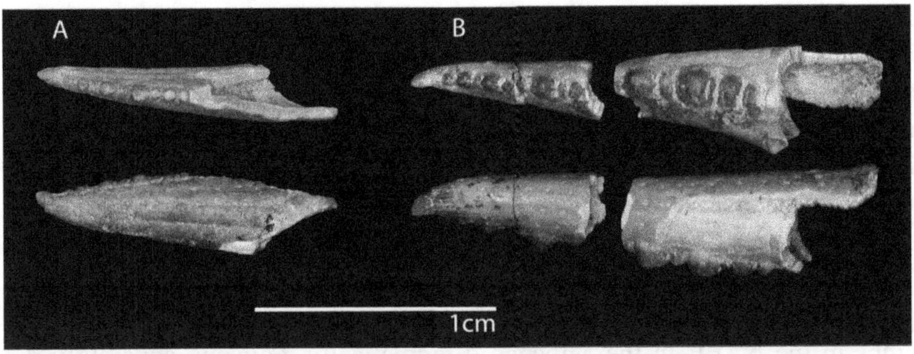

Fig. 2. Fragments of two claws of *Potamon* cf. *speciosus* (A, B) from Giggenhausen, Germany (A, BSPG 1952 XIV 414; B, BSPG 1952 XIV 415). Upper row: view showing teeth; lower row: lateral view.

MATERIAL

The material comprises carapace fragments and the damaged remains of the tips of the chelae (fig. 2), and their assignment to either the dactylus or pollex of the propodus is difficult. The larger claw shows several larger teeth at its base (fig. 2B). However, there is no prominent crushing cusp as in the adult male cheliped of *Potamon potamios* (Olivier, 1804). Larger teeth are separated by one to three smaller teeth. The lateral sides of the claw fragments have distinct grooves and small pits. These fossils are housed in the Bayerische Staatssammlung für Paläontologie und Geologie in Munich (BSPG) under the reference 1952 XIVx.

DISCUSSION

Taxonomic assignment and biogeography. — The fossil brachyuran claw fragments described here are assigned to the genus *Potamon* because this is the only freshwater crab genus known from Europe, for both fossil and extant species. The assignment of the fossil material to species is a little more difficult because the freshwater crabs remains from Giggenhausen are restricted to claw fragments. Despite this, we have assigned the fragments described here to *Potamon* cf. *speciosus* because of a strong geographical and biostratigraphical connection between Giggenhause and the nearby locality Kleineisenbach (Prieto, 2007), were the fossil freshwater crab *Potamon speciosus* is found.

The recent findings of freshwater crabs in the NAFB, usually fragments of the strongly calcified claws, confirm that these brachyuran crabs were a more widespread component of the Early and Middle Miocene environment in the NAFB than previously thought (Berger, 2010; Klaus & Gross, 2010 and references therein). Extant freshwater crabs can contribute a major part of the invertebrate biomass in freshwaters (Dobson et al., 2007) and play an important role as detritivores (Dobson et al., 2002), and the Miocene fossil freshwater crabs may have played a similar key ecological role. However, at the current state of knowledge, this is difficult to test.

The current understanding of the biogeography and phylogeny of the genus *Potamon* is challenged by the occurrence of the fossil freshwater crabs in the NAFB in the Miocene. Enigmatically, *Potamon* apparently did not disperse into present day France and to the Iberian Peninsula. Until now, no freshwater crab fossils are known from Western Europe; and also extant *Potamon* only occurs in Southern France with a recently introduced population of *P. ibericum* (Bieberstein, 1808). This will definitely need further palaeogeographic investigation; especially as for other groups (e.g., mammals) no such biogeographic break can be detected.

Studies on extant *Potamon* based on molecular phylogenetic methods argue for an initial diversification of all species in the Eastern Mediterranean/Near East during the Late Miocene (Shih et al., 2009; Jesse et al., 2011). However, the earliest species of this genus, *Potamon quenstedti* and *P. speciosus*, already existed at this time. Italian *Potamon fluviatile* (Herbst, 1785) dispersed to the Italian peninsula only after the last glacial maximum (Jesse et al., 2009), long after the Late Miocene *Potamon castellinense* (Capellini, 1874) from Italy became extinct. Therefore, the Middle Miocene freshwater crabs of the NAFB could represent a separate lineage of extant Mediterranean *Potamon* that diverged early on from their Near Eastern relatives. Alternatively, the Central European *Potamon* could be part of the direct ancestral group that gave rise to the extant Mediterranean species. However, the latter hypothesis would require an initial dispersal from the Near East to the NAFB before the Karpatian, a back dispersal to Asia Minor before the earliest phylogenetic split within extant Near Eastern *Potamon* (8.3-5.5 Myr according to Jesse et al., 2011), followed by subsequent diversification in the Mediterranean area and secondary dispersal of the extant species to the west. We consider this complex biogeographic scenario as highly unlikely and less parsimonious than the assignment to a separate lineage within the stem group of *Potamon*.

Fossil *Potamon* and molecular clocks. — Three available phylogenies for the Old World freshwater crab families give divergence times using fossil *Potamon* to calibrate molecular clocks (Potamonautidae: Daniels et al., 2006;

Potamidae: Shih et al., 2009; Gecarcinucidae: Klaus et al., 2010). As shown below, all three studies represent different approaches as each assigns the oldest known record of *Potamon* to the ancestry of a different clade, and therefore calibrates different nodes with the same fossil.

Daniels et al. (2006), in their phylogeny of the African Potamonautidae, used the palaeontological data available from Glaessner (1969) for calibration. Glaessner (1969) dated the first occurrence of freshwater crabs in Europe to the Lower Miocene, leading Daniels et al. (2006) to assume 24 Myr (Late Oligocene) as a maximum age of fossil *Potamon*. Note that in this case the "U. Miocene" of Glaessner (1969) refers to the Lower Miocene (23.03-16.00 Myr) and not to the Upper Miocene (probably confused with German "Unteres Miozän", i.e., Lower Miocene). The phylogeny of Daniels et al. (2006) included the potamids *Potamon fluviatile* and *Isolapotamon consobrinum* (De Man, 1899) as outgroup taxa and estimated the minimum age of their common ancestor to be 24 Myr (Daniels et al., 2006). The age of the African Potamonautidae is given with 75-73 Myr (Upper Cretaceous, Campanian). This implies that the Potamidae are considerably older than these dates because they diverged earlier in that phylogeny (Daniels et al., 2006).

Shih et al. (2009) presented a dated phylogeny of the Potamidae, basing their calibration explicitly on Daniels et al. (2006). However, they set the minimum age of the whole family Potamidae to 24 Myr, probably because, given their much broader sampling of the Potamidae, it is apparent that the most recent common ancestor (MRCA) of *Isolapotamon* and *Potamon* is placed at the root of the Potamidae. Applying a relaxed molecular clock their final estimation of the divergence time at the root of the Potamidae was 25.1 Myr (± 1.1 Myr standard deviation).

A dated phylogeny of the Gecarcinucidae used the age of fossil *P. quenstedti* (16.5 Myr) to approximate the split between *P. fluviatile* in the West and *Potamon persicum* Pretzmann, 1962, in the East, and assumed that *P. quenstedti* is more closely related to *P. fluviatile* than to other species of *Potamon* from the Near East (Klaus et al., 2010). Those authors set a gamma prior distribution for this split (95% quantile of 24-18 Myr) and as a result, the root of the Potamidae is placed within a 95% credibility interval of about 54-25 Myr (Eocene-Oligocene).

Summarizing the three different approaches, the first occurrence of *Potamon* is used to either (1) calibrate the divergence of the genus *Potamon* from other potamids, (2) set the age of the most recent common ancestor of the entire Potamidae, or (3) to calibrate a split within the genus *Potamon*. Molecular clock calibration methods include many assumptions and uncertainties that arise from the underlying taxon sampling, sequence data, method of phylogenetic inference, method and model parameters of the calibration, and the molecular clock estimation itself (Ho, 2007; Inoue et al., 2010; Warnock et al., 2011). Therefore, node dating can still be regarded only as a coarse approximation of the true divergence time. The estimated age of the Potamidae differs profoundly between the above three studies. We do not want to judge the accuracy and precision of these time estimates, as all three studies involve further calibration points, different phylogenies, and different software. Nevertheless, we want to point out that each of these three approaches has drawbacks concerning the calibrations that use fossil *Potamon*.

To calibrate the MRCA of a given clade, the phylogeny should include representatives of the two furthest related subclades (i.e., going back in time, their lineages are connected only by the MRCA of the clade). This prerequisite is violated both in the study of Daniels et al. (2006) and Klaus et al. (2010). Daniels et al. (2006) calibrated the MRCA of *Potamon* on the split between *Isolapotamon* and *Potamon*. However, the genera *Paratelphusula* and *Himalayapotamon* are more closely related to *Potamon* (see Shih et al., 2009) and divergence times would be underestimated if an older split was calibrated instead. The problem of underestimation is most probably neutralised by the other calibration points used, explaining the estimation of a Mesozoic origin for the Potamidae in this study.

Jesse et al. (2011) showed that the MRCA of *P. fluviatile* and *P. persicum*, used for calibration by Klaus et al. (2010) cannot be *P. quenstedti*, because this split appears to be much younger. However, calibrating the split between *P. fluviatile* and *P. algeriense* using the end of the Messinian salinity crisis (Jesse et al., 2011) can lead to an overestimation of divergence times. As proposed above, *P. quenstedti* possibly belongs to a separate lineage within the stem group of *Potamon*, and the phylogenetic placement of *P. gedrosianum*, the easternmost species of *Potamon*, has not yet been resolved (Jesse et al., 2011). Assuming that *Potamon* dispersed from east to west, *P. gedrosianum* could potentially represent the sister group to the western species of *Potamon* (unpublished data), and consequently, this split might be the more appropriate node to be calibrated with fossil *P. quenstedti*.

Shih et al. (2009) calibrated the MRCA of the whole Potamidae and assumed that the first occurrence of the Potamidae as represented by the fossil *Potamon quenstedti* represents the minimum age. This appears to be problematic, because the genus *Potamon* is nested within the subfamily Potaminae (see Shih et al., 2009), making an underestimation of the family's age likely.

Therefore, we recommend when considering a fossil to be used in molecular clock calibration to evaluate carefully its taxonomic assignment, its age based on the most recent geological literature, and its correct placement on the phylogeny. However, because first occurrence dates of fossils represent only minimum ages, we cannot escape a certain error when calibrating molecular phylogenies, even if taxonomic assignment of the fossil is unambiguous, and a split between species that truly represent sister groups is calibrated. Certainly, using fossil ages as simple point estimates should be avoided, as should applying calibration points as minimum ages with any time point older than the fossil's age given the same probability to be the actual divergence time, as this can distort the time estimates towards older ages (Ho, 2007). Instead, uncertainty because of node age can be modelled with the application of probability distributions ('calibration densities') in Bayesian methods of molecular clock calibration, although choosing appropriate parameter values still remains difficult (Warnock et al., 2011). A conservative approach will inevitably involve loss of precision (as in the study of Klaus et al., 2010), but large credibility intervals for divergence times can at least prevent incorrect conclusions based on erroneously 'precise' estimates.

ACKNOWLEDGEMENTS

Gertrud Rößner (Munich) is thanked for kindly making the Giggenhausen fossil material available for study. This research received the support of the DFG grant BO 1550/16. We also thank Tohru Naruse, Neil Cumberlidge and Shane Ahyong for critically reviewing and improving this manuscript.

REFERENCES

ABDUL AZIZ, H., M. BÖHME, A. ROCHOLL, J. PRIETO, J. R. WIJBRANS, V. BACHTADSE & A. ULBIG, 2010. Integrated stratigraphy and $^{40}Ar/^{39}Ar$ chonology of the early to middle Miocene Upper freshwater molasse in western Bavaria (Germany). International Journal of Earth Sciences (Geologische Rundschau), **99**: 1859-1886.

ABDUL AZIZ, H., M. BÖHME, A. ROCHOLL, A. ZWING, J. PRIETO, J. R. WIJBRANS, K. HEISSIG & V. BACHTADSE, 2008. Integrated stratigraphy and $^{39}Ar/^{40}Ar$ chronology of the Early to Middle Miocene Upper Freshwater Molasse in eastern Bavaria (Germany). International Journal of Earth Sciences (Geologische Rundschau), **97**: 115-134.

BERGER, G., 2010. Die miozäne Flora und Fauna (MN5) der historischen Fossil-Lagerstätte Georgensgmünd (Mfr.) unter Berücksichtigung der Ablagerung des Urmaintals zwischen Roth und Treuchtlingen. Abhandlungen der Naturhistorische Gesellschaft Nürnberg, **46**: 191.

BLACK, C. C., 1966. Tertiary Sciuridae (Mammalia, Rodentia) from Bavaria. Mitteilungen der bayerischen Staatsammlung für Paläontologie und historische Geologie, **6**: 51-63.

BÖHME, M., 2003. Miocene Climatic Optimum: evidence from lower vertebrates of Central Europe. Palaeogeography, Palaeoclimatology, Palaeoecology, **195**(3-4): 389-401.

BÖHME, M., H. ABDUL AZIZ, J. PRIETO, V. BACHTADSE & G. SCHWEIGERT, 2011. Biomagnetostratigraphy and environment of the oldest Eurasian hominoid from the Early Miocene of Engelswies (Germany). Journal of Human Evolution, **61**(3): 332-339.

BÖHME, M., A. A. BRUCH & A. SELMEIER, 2007. The reconstruction of Early and Middle Miocene climate and vegetation in southern Germany as determined from the fossil wood flora. Palaeogeography, Palaeoclimatology, Palaeoecology, **253**: 91-114.

BÖHME, M., A. ILG, A. OSSIG & H. KÜCHENHOFF, 2006. New method to estimate paleoprecipitation using fossil amphibians and reptiles and the middle and late Miocene precipitation gradients in Europe. Geology, **34**(6): 425-428.

BRUIJN, H. DE, V. FAHLBUSCH & E. ÜNAY, 1993. Early Miocene rodent faunas from the eastern Mediterranean area. Part III. The genera *Deperetomys* and *Cricetodon* with a discussion of the evolutionary history of the Cricetodontini. Proceedings of the Koninklijke Nederlandse Akademie van Wetenschappen, **96**(2): 151-216.

CAPELLINI, G., 1874. La Formazione Gessosa di Castellina Marittima e i suoi fossili. Memorie della Reale Accademia delle Scienze dell'Istituto di Bologna, **4**: 38.

DANIELS, S. R., N. CUMBERLIDGE, M. PÉREZ-LOSADA, S. A. E. MARIJNISSEN & K. A. CRANDALL, 2006. Evolution of Afrotropical freshwater crab lineages obscured by morphological convergence. Molecular Phylogenetics and Evolution, **40**: 227-235.

DOBSON, M. K., A. MAGANA, J. M. MATHOOKO & F. K. NDEGWA, 2002. Detritivores in Kenyan highland streams: more evidence for the paucity of shredders in the tropics? Freshwater Biology, **47**: 909-919.

— —, — —, — — & — —, 2007. Distribution and abundance of freshwater crabs (*Potamonautes* spp.) in rivers draining Mt Kenya, East Africa. Archiv für Hydrobiologie, **168**: 271-279.

ENGESSER, B., 1972. Die obermiozäne Säugertierefauna von Anwil (Baselland). Tätigkeitsberichte der Naturforschenden Gesellschaft Baselland, **28**: 37-363.

ERONEN, J. T. & G. E. RÖSSNER, 2007. Wetland paradise lost: Miocene community dynamics in large herbivore mammals from the German Molasse Basin. Ecology and Evolutionary Research, **9**: 471-494.

FAHLBUSCH, V., 1964. Die Cricetiden der Oberen Süßwassermolasse Bayerns. Bayerische Akademie der Wissenschaften; mathematisch-naturwissenschaftliche Klasse, Abhandlungen, neue Folge, **118**: 1-136.

— —, 1975. Die Eomyiden (Rodentia, Mammalia) der Oberen Süsswasser-Molasse Bayerns. Mitteilungen der bayerischen Staatsammlung für Paläontologie und historische Geologie, **15**: 63-90.

FELDMANN, R. M., P. M. O'CONNOR, N. J. STEVENS, M. D. GOTTFRIED, E. M. ROBERTS, S. NGASALA, E. L. RASMUSSON & S. KAPILIMA, 2005. A new freshwater crab (Decapoda: Brachyura: Potamonautidae) from the Paleogene of Tanzania, Africa. Neues Jahrbuch für Geologie und Paläontologie, Abhandlungen, **244**: 71-78.

GIERSCH, S., W. MUNK & R. ZIEGLER, 2010. The first record of a beaver — *Trogontherium* (*Euroxenomys*) *minutum* — in the Höwenegg fauna (Miocene, southern Germany). Palaeodiversity, **3**: 235-239.

GLAESSNER, M. F., 1933. Tertiary crabs in the collection of the British Museum. The Annals and Magazine of Natural History, Zoology, Botany and Geology, **12**: 12-13, 20.

— —, 1969. Decapoda. In: R. C. MOORE (ed.), Treatise on invertebrate paleontology, Part R. Arthropoda 4, vol. 2: R400-R533. (Geological Society of America & University of Kansas Press, Boulder & Lawrence).

HERBST, J. F. W., 1785. Versuch einer Naturgeschichte der Krabben und Krebse, nebst einer systematischen Beschereibung ihrer verschiedenen Arten: 1-226. (G.A. Lange, Berlin & Stralsund).

HO, S. Y. W., 2007. Calibrating molecular estimates of substitution rates and divergence times in birds. Journal of Avian Biology, **38**: 409-414.

INOUE, J., P. C. J. DONOGHUE & Z. YANG, 2010. The impact of the representation of fossil calibrations on Bayesian estimation of species divergence times. Systematic Biology, **59**: 74-89.

JESSE, R., M. GRUDINSKI, S. KLAUS, B. STREIT & M. PFENNINGER, 2011. Evolution of freshwater crab diversity in the Aegean region (Crustacea: Brachyura: Potamidae). Molecular Phylogenetics and Evolution, **59**: 23-33.

JESSE, R., M. PFENNINGER, S. FRATINI, M. SCALICI, B. STREIT & C. D. SCHUBART, 2009. Disjunct distribution of the Mediterranean freshwater crab *Potamon fluviatile* — natural expansion or human introduction? Biological Invasions, **11**: 2209-2221.

KÄLIN, D. & O. KEMPF, 2009. High-resolution stratigraphy from the continental record of the Middle Miocene northern Alpine Foreland Basin of Switzerland. Neues Jahrbuch für Geologie und Paläontologie, Abhandlungen, **254**(1): 177-235.

KÄLIN, D., M. WEIDMANN, B. ENGESSER & J.-P. BERGER, 2001. Paléontologie et âge de la Molasse d'eau douce supérieure (OSM) du Jura neuchâtelois. Schweizerische Paläontologische Abhandlungen, **121**: 63-99.

KARG, J. M., 1805. Über den Steinbruch zu Öningen, den Stein am Rheine und dessen Petrefakten. Denkschriften der vaterländischen Gesellschaft der Ärzte und Naturforscher Schwabens, Tübingen, **1**: 1-74.

KLAUS, S., M. BÖHME, S. SCHNEIDER, J. PRIETO & B. PHETSOMPHOU, 2011. Evidence of the earliest freshwater decapod fossil from Southeast Asia (Crustacea: Decapoda: Brachyura). Raffles Bulletin of Zoology, **59**: 47-51.

KLAUS, S. & M. GROSS, 2010. Synopsis of the fossil freshwater crabs of Europe (Brachyura: Potamoidea: Potamidae). Neues Jahrbuch für Geologie und Paläontologie, Abhandlungen, **256**: 107-119.

KLAUS, S., C. D. SCHUBART, B. STREIT & M. PFENNINGER, 2010. When Indian crabs were not yet Asian — evidence for Eocene proximity of India and Southeast Asia from freshwater crab biogeography. BMC Evolutionary Biology, **10**: 287.

KŐRÖSSY, L., 1940. Kurze geologische Beschreibung der Umgebung von Hernádzsadány. Földtani Közlöny (Geologische Mitteilungen), Budapest, **70**: 83-109, 143-144.

KUHLEMANN, J. & O. KEMPF, 2002. Post-Eocene evolution of the North Alpine Foreland basin and its response to Alpine tectonics. Sedimentary Geology, **152**: 45-78.

MAN, J. G. DE, 1899. Zoological results of the Dutch Scientific Expedition to Central Borneo. The Crustacea. Part II, Brachyura. Notes from the Leyden Museum, **21**: 53-144, pls. 5-12.

MAYR, H., 1979. Gebissmorphologische Untersuchungen an miozänen Gliriden (Mammalia, Rodentia) Süddeutschlands: 1-380. (Ph.D. thesis, Universität München).

MELCHOR, R. N., J. F. GENISE, J. L. FARINA, M. V. SÁNCHEZ, L. SARZETTI & G. VISCONTI, 2010. Large striated burrows from fluvial deposits of the Neogene Vinchina Formation, La Rioja, Argentina: a crab origin suggested by neoichnology and sedimentology. Palaeogeography, Palaeoclimatology, Palaeoecology, **291**: 400-418.

MEYER, H. VON, 1844. Zur Fauna der Vorwelt: fossile Säugethiere, Vögel und Reptilien von Öningen; über eine allgemeine Übersicht der fossilen Wirbelthiere; *Lagomys*, *Chelydra*, *Coluber oweni* u. a. Arten, *Grapsus speciosus*, *Homelys major* und *minor* und Vogel-Reste von Öningen. Neues Jahrbuch für Mineralogie, Geognosie, Geologie und Petrefakten-Kunde, **3**: 330-331.

NARUSE, T., H. KARASAWA, S. SHOKITA, T. TANAKA & M. MORIGUCHI, 2003. A first fossil record of the terrestrial crab, *Geothelphusa tenuimanus* (Miyake & Minei, 1965) (Decapoda, Brachyura, Potamidae) from Okinawa Island, central Ryukyus, Japan. Crustaceana, **76**: 1211-1218.

OLIVIER, G. A., 1804. Voyage dans l'Empire Ottoman, l'Égypte et la Perse, fait par ordre du gouvernment pendant les six premières années de la republique, **4**: 1-388. (Agasse, Paris).

PRETZMANN, G., 1962. Die mediterranen und vorderasiatischen Süßwasserkrabben (Potamoniden). Annalen des Naturhistorischen Museums Wien, **65**: 205-240.

PRIETO, J., 2007. Kleinsäuger-Biostratigraphie und Paläoökologie des höheren Mittelmiozäns (MN 8) Bayerns: Spaltenfüllungen der Fränkischen Alb und Lokalitäten der Oberen Süßwassermolasse im Vergleich: 1-213. (Ph.D. thesis, LMU München).

— —, 2010. The Middle Miocene mole *Desmanodon crocheti* sp. nov. (Talpidae, Mammalia): the last representative of the genus in the North Alpine foreland basin. Paläontologische Zeitschrift, **84**: 217-225.

PRIETO, J. & M. RUMMEL, 2009. Small and medium-sized Cricetidae (Mammalia, Rodentia) from the Middle Miocene fissure filling Petersbuch 68 (southern Germany). Zitteliana A, **48/49**: 89-102.

RODRÍGUEZ, G. & H. DIAZ, 1977. Note sur quelques restes des crabes d'eau douce (Pseudothelphusidae) provenant d'un "Kjoekken-Moedding" du Venézuéla. Contribution du Laboratoire de Carcinologie et d'Océanographie biologique, **79**: 107-108.

SHIH, H.-T., D. C. J. YEO & P. K. L. NG, 2009. Impact of the collision of the Indian plate with Asia on the evolution of the freshwater crab family Potamidae (Crustacea: Decapoda: Brachyura) revealed morphological and molecular evidence. Journal of Biogeography, **36**: 703-719.

WARNOCK, R. C. M., Z. YANG & P. C. J. DONOGHUE, 2011. Exploring uncertainty in the calibration of the molecular clock. Biology Letters, **8**(1): 156-159.

ZITTEL, K. A., 1885. Handbuch der Paläontologie. Band I, Paläozoologie, Abt. 2, Mollusca und Arthropoda: 1-893. (R. Oldenbourg, München & Leipzig).

First received 2 November 2011.
Final version accepted 29 July 2013.

DIFFERENTIATION WITHIN A RIVER SYSTEM: ECOLOGY OR GEOGRAPHY DRIVEN? EVOLUTIONARY SIGNIFICANT UNITS AND NEW SPECIES IN JAMAICAN FRESHWATER CRABS

CHRISTOPH D. SCHUBART[1]) and TOBIAS SANTL

Biologie 1, Institut für Zoologie, Universität Regensburg, 93040 Regensburg, Germany

ABSTRACT

Freshwater habitats of the Caribbean island Jamaica are unique, in so far that they are not inhabited by freshwater Decapoda Reptantia with a long evolutionary history in fresh water, like crayfish or old lineages of freshwater crabs. Instead, a relatively young invasion and radiation of originally coastal crabs from the family Sesarmidae took place, resulting in currently ten endemic sesarmid species that are recognized from the island. Six of those have been described from Jamaican brooks and streams (river crabs), whereas the other four species thrive in caves and more terrestrial habitats. After establishing and describing the diversity of Jamaican river crabs at the species level, ongoing studies are designed to highlight presumed intraspecific differentiation within the recognized species, as a means of understanding the diversification and rapid speciation processes of this adaptive radiation. Here we use mitochondrial DNA sequences of the ND1 gene and morphometrics to document diversity within river crabs from western Jamaica so far considered to belong to *Sesarma dolphinum*, complementing a recent population genetics study with nuclear DNA. Distinct evolutionary lineages can be recognized, of which two are so clearly separated that they do not share mitochondrial haplotypes nor do they show any overlap in morphometry. Interestingly, these lineages co-occur within the same river system, allowing first insights into the mechanisms of differentiation of these crabs. Ecological restriction to upper reaches of rivers isolates the crabs in different tributaries and thus genetic connectivity is apparently more likely to be maintained in overland dispersal between headwaters than within the river system. The distinct evolutionary lineage from the southeastern range of the distribution area of *S. dolphinum* is here described as a new species in order to highlight its uniqueness and to make it a management unit. A nested clade analysis reveals that the genetic relationship of populations of *S. dolphinum* is the result of restricted gene flow with isolation by distance. A literature review compiles other published reports for freshwater Crustacea with different evolutionary lineages in the same water catchment.

INTRODUCTION

Less than twenty years ago, secondary freshwater crabs from Jamaican brooks and streams (Grapsoidea: Sesarmidae) were considered to belong

[1]) Corresponding author; e-mail: christoph.schubart@biologie.uni-regensburg.de

to a single species, *Sesarma bidentatum* Benedict, 1892. A closer look at representatives of different regions of the island, however, allowed recognition and description of six distinct species, based on morphological characters alone (Türkay & Diesel, 1994; Schubart et al., 1997; Reimer et al., 1998) or a combination of morphological and molecular features (Schubart et al., 1998a, 1999; Schubart & Koller, 2005). The distinctness of the two closest species, *Sesarma windsor* Türkay & Diesel, 1994, and *Sesama meridies* Schubart & Koller, 2005 from central Jamaica, previously established by morphology and mitochondrial DNA (mtDNA) (Schubart & Koller, 2005), was later confirmed with the highly variable ITS1-5.8S-ITS2 region of the nuclear genome by Schubart et al. (2010). The latter study also revealed measurable intraspecific genetic structure within these two sister species as well as in the river crab species *Sesarma dolphinum* Reimer, Schubart & Diesel, 1998. Our own unpublished results confirm this for the other three river crab species and two of the more terrestrial crab species. The role of geographic structure and ecology in shaping intraspecific genetic variability, and the recognition of evolutionary significant units combined with their nomenclatural treatment, will be the topic of this and upcoming contributions concerning the radiation of endemic Sesarmidae in Jamaica, first documented in Schubart et al. (1998b).

The concept of evolutionary significant units (ESU) was originally introduced by Ryder (1986) and defined as "*Subsets of the more inclusive entity species, which possess genetic attributes significant for the present and future generations of the species in question*". Moritz (1994), Crandall et al. (2000) and many others adjusted and re-defined the concept for conservation biology purposes. This was reviewed by Fraser & Bernatchez (2001) resulting in a return to a more general definition of the concept of ESU as "*A lineage demonstrating highly restricted gene flow from other such lineages within the higher organizational level (lineage) of the species*". Used in this sense, ESUs can be applied to recognized intraspecific isolated lineages, regardless of the conservation status or the respective ecological exchangeability (cf. Crandall et al., 2001) or, as stated by Fraser & Bernatchez (2001: 2742): "*We emphasize that the strengths and weaknesses of various operational criteria should not encumber conservation efforts but rather aid managers in conducting sound conservation plans specific to the situation at hand*".

The comparison of 232 clones corresponding to the nuclear ITS regions of 32 specimens of *Sesarma dolphinum* throughout its distribution range in westernmost Jamaica revealed several genetic lineages (Schubart et al., 2010: table 3), of which three alone are encountered in the Cabarita River system. Even before studying the genetic population structure of this species, in the original description of the species Reimer et al. (1998) already noted some morphological differences of individuals inhabiting the Roaring River, a southeastern tributary of the Cabarita River. Those authors point out that "The three specimens of *S. dolphinum* from Roaring River (Westmoreland) differed from the other material of the species by having shorter legs (4th pereiopod length/carapace length: 1.91 ± 0.06). In these specimens, the dorsal row of granules on the palm of the chela is not always continuous and the number of tubercles on the dactylus of the chela varied between 0 and 4. These specimens might possibly be a distinct subspecies, but further material will be needed for confirmation" (Reimer et al., 1998: 194). Those authors also note that preliminary results based on the 16S rRNA gene of the mtDNA did not support a separation. This genetic marker, however, is relatively conserved within the mitochondrial genome and thus population studies including more populations and more variable markers would be necessary. The molecular data provided by Schubart et al. (2010) based on the ITS nDNA region confirm genetic separation of the Roaring River Deans Valley River populations. However, with one exception, this analysis also distinguished between all other sampled populations within this species (partly due to non-independence of the data by including several clones from single individuals) and suggested limited connectivity among all of the inhabited river systems.

The question remains of how to deal with genetically isolated and in part quite differentiated evolutionary units. In the present study, we provide new genetic (mtDNA) and morphometric evidence for the distinctness of the geographic population comprising crabs from the Deans Valley and Roaring rivers and compare these to other representatives of *Sesarma dolphinum*, including individuals from the type locality from the Hog-Davis Cove river system (near Kingsvale). In conclusion we consider the Deans Valley River and Roaring River populations to comprise a well-defined lineage of *Sesarma* that is described here as a new species.

MATERIAL AND METHODS

More than one hundred specimens of *Sesarma dolphinum* (Decapoda: Brachyura: Thoracotremata: Sesarmidae) were collected from 13 localities from throughout its distributional range in western Jamaica (fig. 1) during several sampling trips between 1997 and 2005. From those, 83 were used for morphometric comparisons and 42 for DNA isolation (see table I) and 40 for amplification and sequencing of mitochondrial DNA.

Genetics. — Tissue was extracted from the muscle of a walking leg and DNA isolation was performed using a modified Puregene method from Gentra Systems. Dried genomic DNA was resuspended in 20 μl TE buffer and the concentration was ascertained on agarose gels. From the corresponding dilutions of the DNA solution 1 μl was used for polymerase chain reactions. A >800 basepair (bp) fragment of the mitochondrial NADH1 dehydrogenase subunit 1 (ND1) was amplified with NDL4 (5′-AAAADKCTAATTRTTTTGTG-3′) (corrected from Schubart et al., 2011) and NDH2 (5′-GCTAAATATATWAGCTTATCATA-3′) (Schubart, 2009) and the internal primers NDL5 (5′-TTGCTGGWTGRTCTTCWAATTG-3′) (new) and NDH5 (5′-GCYAAYCTWACTTCATAWGAAAT-3′) (Schubart, 2009). For PCR, a standard 25 μl reaction was prepared containing 2.5 μl of 10× buffer, 2.5 μl of 1.25 mM dNTPs, 0.5 μl of both primer (20 mM), 2 μl of 25 mM $MgCl_2$, 1 μl of 0.5 U/μl TAQ and 15 μl of double-distilled water in addition of 1 μl gDNA. 40 cycles were run with 45 s at 94°C, 1 min at 48°C as an annealing temperature, and 1 min at 72°C. PCR products were cleaned using QuickClean (Bioline) and sequenced with an ABI-PRISM 310 (Applied Biosystems). Sequences were proofread for possible errors made by the computerized analysis provided with ChromasLite (http://www.technelysium.com). Due to the lack of indels, the corrected sequences were aligned by eye using BioEdit (Hall, 1999). Sequences of all detected haplotypes were submitted to the EMBL molecular database and archived as HF678402-HF678413. The nexus file of the ND1 dataset was then used to construct a statistical parsimony network using the algorithm outlined in Templeton et al. (1992) and implemented in the TCS software package version 1.21 (Clement et al., 2001). Based on the obtained haplotype network of the ND1 data a nested clade analysis (NCA) was performed (Templeton et al., 1995; Templeton, 2004) to test the null hypothesis of no association between the geographic distribution of haplotypes. The haplotype network was converted into a nested statistical design using the instruction given in Templeton and Sing (1993) and in Crandall & Templeton (1996). To test for an association between the genetic composition and the geographic distribution of the haplotypes, two distances were calculated. First, the clade distance Dc, which estimates how geographically widespread a clade is, and second, the nested clade distance Dn, which measures the relative distribution of a clade compared to the other clades in the same higher clade level. All calculations were

TABLE I

List of western Jamaican localities, from where *Sesarma abeokuta* n. sp. and *S. dolphinum* were collected for genetic and morphometric analyses, including the number of individuals used for both methods and the museum collection number

	Location	N_{DNA}	Haplotypes	N_{morpho}	Museum #
Sesarma abeokuta					
Deans Valley R.-Abeokuta, 2005	18°14′43.08″N/78°02′38.58″W	2	12	1	NHMW 25418
Deans Valley R.-Galloway, 1997	~18°15.5′N/78°02′W	3	12	2	SMF-34538
Roaring R.-Spring, 1997 & 2005	18°17′15.54″N/78°02′43.80″W	2	7	15	SMF-34539
Sesarma dolphinum					
Flint R.-Cascade, 2003	18°23′52.19″N/78°05′39.12″W	2	1	8	SMF-34540
Upper Cabarita-Buxton, 2003	18°22′24.00″N/78°06′1.38″W	4	1, 4	4	SMF-34541
Lucea East R.-Tom Spring, 2003	18°24′11.95″N/78°08′6.38″W	0	n.a.	12	SMF-34542
Lucea West R.-Harvey, 2003	18°24′6.81″N/78°09′15.24″W	3	1, 5	5	SMF-34543
Lucea West R.-Askenish, 2003	18°22′42.47″N/78°09′07.30″W	0	n.a.	5	SMF-34544
Green Island R.-Rock Spring, 2003	18°21′30.42″N/78°14′32.58″W	5	1, 6, 8	5	SMF-34545
New Savannah R.-Jerusalem, 2003	18°18′53.73″N/78°14′10.26″W	3	2, 3	12	SMF-34546
Lances R.-Dias, 2005	18°23′59.88″N/78°11′29.64″W	4	8	0	SMF-34547
Morgan R.-Flamstead, 2005	18°22′28.18″N/78°12′20.02″W	4	8, 11	8	SMF-34548
Upper Cabarita-Bath Mt., 2005	18°20′26.22″N/78°05′57.42″W	2	1	2	SMF-34549
Upper Cabarita-nr. Cash Hill, 2005	18°22.400′N/78°06.023′W	0	n.a.	3	SMF-34550
Green Island R.-Kendal, 2005	18°21′56.49″N/78°13′50.41″W	4	1, 8	0	SMF-34551
Davis Cove R.-Paradise, 1997	18°23′0.17″N/78°12′45.79″W	4	8, 9, 10	1	Not available

done with the software GEODIS 2.5 (Posada et al., 2000) using 1 000 000 permutations and direct distances. The direct distances option was favoured over river distances as all species in this study are freshwater species with no marine form, which would be necessary to connect certain rivers. The direct distances between the single sample locations were measured in GoogleEarth (http://earth.google.com). To infer the historical events that caused the observed genetic population structure we used the most recent inference key from Templeton (http://darwin.uvigo.es/software/geodis.html).

Morphometry. — Data on the relative proportions of morphological characters of specimens were collected to detect phenotypic differences among populations. A mechanical calliper gauge with a digital display was used to record the corresponding measurements. The following characters were recorded: carapace width, measured at two separate positions: at the widest anterior part including the exorbital tooth (CWf) and at the posterior broadest part of the carapace (CWb); carapace length (CL) and body height (BH), measured along the median line of the carapace; frontal width (FW) between the two orbits, and length of the exorbital tooth (ET). Three chelar measurements were taken, the height (PrH) and ventral length (PrL) of the propodus and the dorsal length of the dactylus (DaL). From pereiopods three and four, the length (3L, 4L) and width (3W, 4W) of the meri were recorded. Finally, the pleon (PlW) was measured at its widest part. To minimize possible errors due to allometric growth (see Reuschel & Schubart, 2006) only individuals considered to be adult (or very close to it: CL > 12 mm) were measured. All measurements were logarithmically transformed to further minimize the effect of possible allometric growth. Measurements were tested for normal distribution using the one-sample Kolmogorov-Smirnov test. Those which showed normal distribution were included in a canonical discriminant function analysis. The variable which had the greatest weight on the outcome of the discriminant function analysis was calculated. The discriminant function analysis was then re-done without this variable to assure that the observed differences are not the result of a single factor. All calculations were performed in SPSS version 16 (SPSS Inc., Chicago, IL).

RESULTS

Genetics. — A fragment of 834 bp mtDNA of the ND1 gene was successfully sequenced from 40 individuals of *Sesarma dolphinum* and aligned, resulting in 12 different haplotypes (ht). Base frequencies were A = 0.2870,

C = 0.0855, G = 0.1795 and T = 0.4480. Initial tests of pairwise distinctness between neighbouring localities with low representation failed to show differences. Therefore the following populations were pooled for subsequent analyses: two tributaries of Deans Valley River; the upper Cabarita River and its tributary near Bath Mountain; West and East Lucea rivers; neighbouring Davis Cove and Lances rivers (Davis/Lances rivers). The 13 collecting sites for which DNA sequences became available were pooled into 9 populations (see codes in map fig. 1). The statistical parsimony network (fig. 1) revealed the existence of two major clades (ht I-VI versus ht VII-XII) separated by a relatively high number of 12 (with ht VI) or 16 (without ht VI) mutations. The software considers ht-I as the ancestral haplotype. This haplotype can be found in the northernmost samples from the Flint River, both Lucea rivers, the upper Cabarita tributaries, and in two individuals from the Green Island River (n = 10). Four haplotypes (ht II-ht V) are derived from ancestral ht-1 in a star-like pattern and differ by only one or two substitutions. Haplotypes II (n = 1) and III (n = 2) are exclusively found in samples from the southwestern New Savannah River system, ht-V is present twice in the Lucea rivers, and ht-IV (n = 1) was found in one sample from the upper Cabarita system. Haplotype VI was encountered in one individual from the Green Island River and differs from the first major clade with ht-I by four substitutions and from the second major clade by at least 12. The second major clade consists of three common haplotypes, of which two are characteristic and exclusive for single populations. Haplotype VIII has the highest frequency (n = 10). It contains most samples from the Green Island River, the Morgan River (western tributary to Cabarita River) and the Davis/Lances rivers. Also three of the four haplotypes which are connected to ht VIII by one or two substitutions belong to individuals collected in these three river systems (ht IX and ht X from the Davis Cove River, and ht XI from the Morgan River, all n = 1). All these river systems are found in the northwestern part of the distribution area. Haplotype VII is found on the connecting line between ht-I and ht-VIII, only one mutational step away of the latter. This haplotype was exclusively found in the five sequenced samples from the Roaring River, which is another tributary to the Cabarita River, but in the southeastern distribution limit of *Sesarma dolphinum*. Haplotype XII departs from ht VII and is separated by four mutations. This haplotype was found in all sequenced animals from the Deans Valley River (n = 5). This river lies even further southeast than the Roaring River and marks the border between *Sesarma dolphinum* and *Sesarma fossarum* Schubart, Reimer, Türkay & Diesel, 1997.

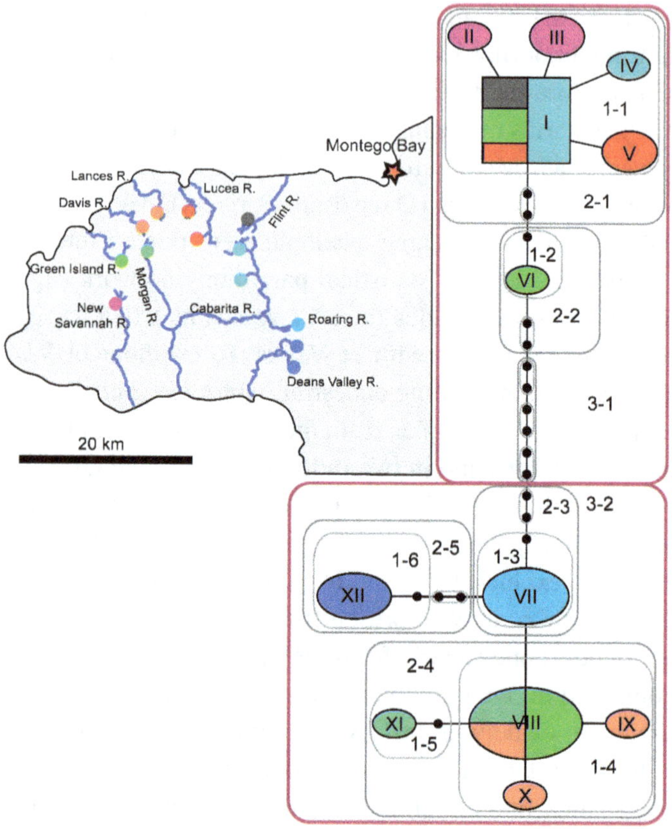

Fig. 1. Left: map of the western part of Jamaica (Greater Antilles) showing selected rivers and sampling sites where specimens of *Sesarma dolphinum* Reimer, Schubart & Diesel, 1998 and *S. abeokuta* n. sp. were collected. Right: Statistical parsimony network constructed with TCS based on a 834-basepair fragment of the ND1 gene (N = 37) and the corresponding nesting design for the Nested Clade Analysis. Each line represents one substitution; dots on the lines indicate additional substitutions separating two haplotypes. Coloration corresponds to sample sites on map. The size of the circle is representative for the frequency of the haplotypes (small: N = 1; medium: N = 2-3; large: N = 4-5; largest: N > 10; square: proposed ancestral haplotype). Light grey lines enclose the 1-step clades (1-..), dark grey lines enclose 2-step clades (2-..) and red lines enclose 3-step clades (3-1 and 3-2).

The resulting ND1 statistical parsimony network displays the genetic relationships between the different populations of *Sesarma dolphinum*. However, it does not tell us how these genetic relationships originated. To investigate such a question a nested clade analysis was performed. The twelve haplotypes constitute units at the smallest level, i.e. the 0-level clades. Based on these 0-level clades, seven 1-level clades, five 2-level clades and two 3-level clades were constructed, which formed the total cladogram. At the 1-level, two clades show geographic association, but only the clade 1-1 has significant values for the within- and nested-clade distances and the interior to tip within- and nested-clade distances. The analysis of clade 1-1 with the inference key resulted in an inconclusive outcome. From the 2-level clades, only one has geographic associations, but does not show any significant values. At the 3-level, again two clades have geographic associations, of which clade 3-2 displayed significant values. Analyzing these values with the inference key, gives three scenarios, how the present state could come into place: either by past fragmentation, or by long distance colonization, or by a combination of these two possibilities. The total cladogram also infers geographic associations and the analysis produced significant values for the within- and nested-clade distances. This time, the inference key suggested a single outcome: the present state of the *Sesarma dolphinum* populations under research is the result of restricted gene flow with isolation by distance, which seems to be the most important result to understand differentiation in this species. The nesting design is shown in fig. 1 and the analysis of the geographic associations in table II.

Morphometry. — The same nine pooled populations as above, and as depicted in fig. 1, were used for the statistical analyses of the morphometry, in which 83 individuals were included (see table I). From the 15 characters mea-

TABLE II

Nested clade distance analysis of ND1 haplotypes observed in *Sesarma dolphinum* and *S. abeokuta*. The nested design is given in fig. 1. Dc and Dn are clade and nested clade distances, respectively (for details see Templeton et al., 1995). Interior vs. tip contrasts for Dc and Dn are indicated with 'I-T' in the corresponding clade. Interior clades are identified by shading. The S and L superscripts refer to significantly small and large values at the 0.05 level, respectively. Significance of values is based on permutation analysis using 1 000 000 resamples

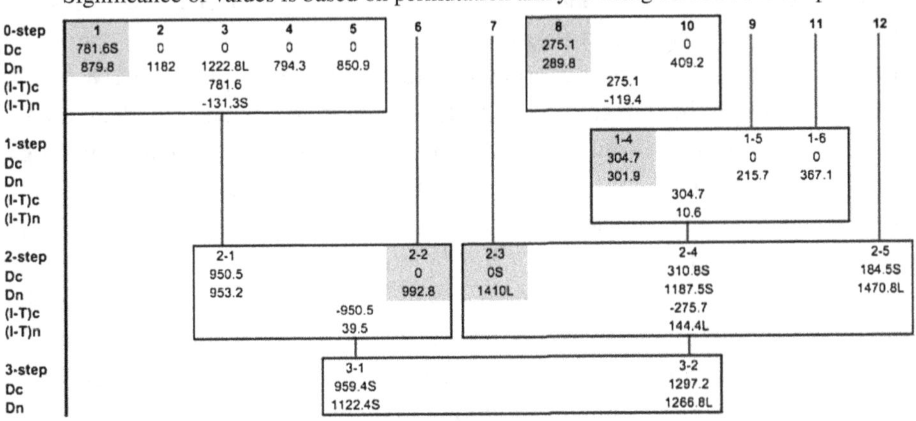

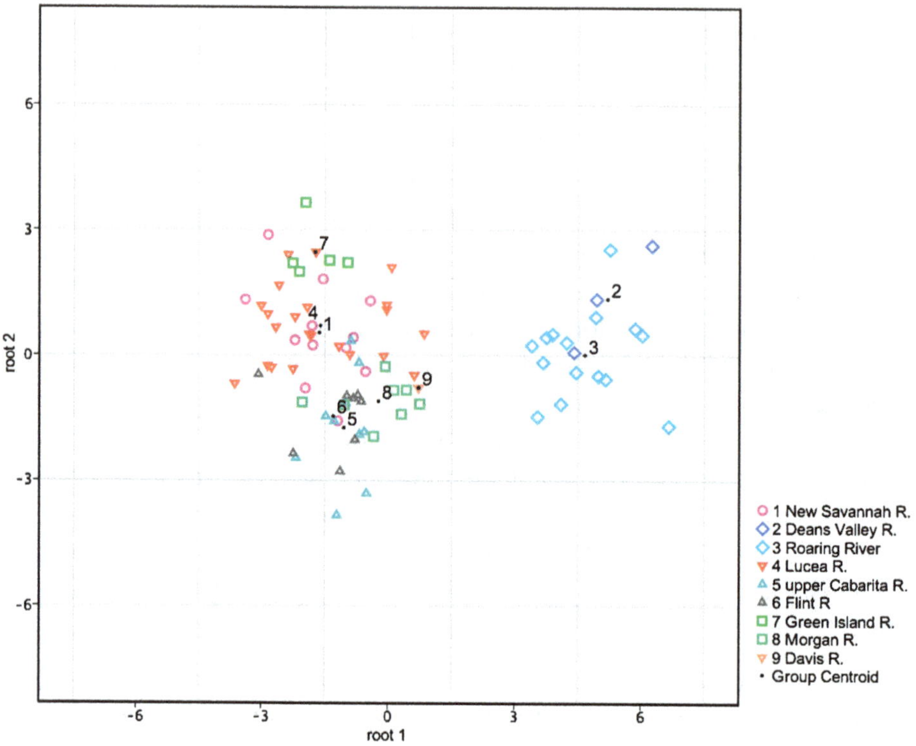

Fig. 2. Canonical analysis showing discrimination by morphometric measurements between nine populations of *Sesarma dolphinum* Reimer, Schubart & Diesel, 1998 and *S. abeokuta* n. sp. from western Jamaica; plot of the first discriminant function (root 1) against the second (root 2).

sured, all showed normal distribution in the Kolmogorov-Smirnov test. The Wilk's Lambda for the overall model is 0.015 with $p \leqslant 0.001$, which indicates a very good discrimination, also shown in the overall correct classification of 71.4%. In fig. 2, a two-dimensional plot of the first two canonical values is shown. These two canonical values together explain 85.1% of the variables found in the dataset. In the plot, two separate clusters are clearly visible. One cluster contains only samples from the southeastern distribution range of *Sesarma dolphinum*, namely from the Deans Valley River system and the Roaring River. In the second cluster, the remaining populations do not appear much differentiated, but certain clusters can nevertheless be recognized. This is also reflected in the classification matrix (table III). This matrix indicates the percentage of animals from each site which are correctly grouped according to the discriminant analysis. The two southeastern sites have 100% correct placement, if pooled. Wrongly placed individuals can be found in the corresponding

TABLE III

Percentage of correct classification based of the morphometric classification function for nine western Jamaican populations of *Sesarma dolphinum* and *S. abeokuta* n. sp. (overall correct classification: 71.4%)

Population	1	2	3	4	5	6	7	8	9
1 New Savannah R.	75	0	0	8.3	0	0	8.3	8.3	0
2 Deans Valley R.	0	66.7	33.3	0	0	0	0	0	0
3 Roaring R.	0	26.7	73.3	0	0	0	0	0	0
4 Lucea rivers	9.1	0	0	68.2	0	0	18.2	4.5	0
5 Upper Cabarita R.	10	0	0	10	40	40	0	0	0
6 Flint R.	0	0	0	0	37.5	62.5	0	0	0
7 Green Island R.	0	0	0	0	0	0	100	0	0
8 Morgan R.	0	0	0	0	0	0	0	100	0
9 Davis Cove R.	0	0	0	0	0	0	0	0	100

other population. Similarly, the Flint River and upper Cabarita River taken together almost reach 100% correct classification, with 62.5% correct placement of Flint River animals, whereas the remaining 37.5% are all attributed to the nearby upper Cabarita River population. Vice versa, the upper Cabarita population has 40% correct placements and another 40% are assigned to the Flint River population. The three northwestern sites together, Green Island River, Davis Cove River and Morgan River, have all correct placement at 100% respectively. They also belong to different drainage systems and are geographically quite close. The remaining New Savannah River and pooled Lucea rivers also have remarkably high correct individual placements with 75% and 68.2% respectively. Overall, the populations show good to very good classification by means of their morphometric characters.

Taxonomy. — According to the above results, in combination with those of Schubart et al. (2010), we consider to have gathered sufficient evidence to recognize the southeastern population from *Sesarma dolphinum*, colonizing the Deans Valley and the Roaring rivers, as an evolutionary significant unit. The mtDNA network from fig. 1 gives evidence for genetic isolation even between these two rivers and suggests possible founder effects resulting in genetic bottlenecks (both rivers having unique ND1 haplotype, separated by four mutation steps), but the nuclear DNA (Schubart et al., 2010) and the morphometry (see fig. 2) suggest a close relationship and common origin of these two populations. We therefore describe individuals from both populations jointly as a new species.

Sesarma abeokuta new species
fig. 3A-D

Sesarma dolphinum Reimer, Schubart & Diesel, 1998, specimens from Roaring River (Reimer et al., 1998); specimens from Roaring & Deans Valley rivers (Schubart et al., 2010).

Material examined. — Holotype: 1 male (20.9 × 17.6 (carapace width × carapace length in mm)) (SMF-34537), Jamaica, Westmoreland, Galloway, Deans Valley River system, wet gravel at foot of hillside, 17 Mar. 1997, leg. R. Diesel, J. Reimer, C. D. Schubart (DNA extr. 7 Jul. 2004 ×4). Paratypes: 3 males (19.0 × 16.4, 19.06 × 16.53, 17.96 × 15.32), 2 females (16.7 × 14.14, 16.96 × 14.1) (SMF-34538) same data as holotype (DNA extr. 10 Oct. 1997; 7 Jul. 2004 ×3); 2 males (15.73 × 13.12, 13.33 × 10.94), 2 females (19.2 × 15.8, 16.6 × 14.0) (NHMW 25418), Jamaica, Westmoreland, Abeokuta Nature Park, near Galloway, tributary to Deans Valley River, 128 m, 18°14.718′N-78°02.643′W, 16 Oct. 2005, leg. C. D. Schubart, S. Reuschel, T. Santl; 2 males (19.70 × 16.50, 15.16 × 12.91), 2 females (18.02 × 15.11, 16.62 × 14.19) (ZRC 2013.0451), same data as previous; 2 males (19.08 × 16.09, 17.19 × 14.21), 2 females (18.97 × 15.73, 16.88 × 14.54) (RMNH.CRUS.D.55074), same data as previous; 9 males (21.35 × 18.72, 18.0 × 15.89, 17.23 × 14.55, 15.66 × 13.71, 15.6 × 13.65, 15.5 × 13.08, 14.72 × 12.78, 12.91 × 11.02, 12.63 × 10.78), 1 female (18.75 × 16.2) (SMF-34539), Jamaica, Westmoreland, spring of Roaring River nr. Shrewsbury, tributary to Cabarita River, 98 m, 18°17′16″N-78°02′44″W, 16.3.1997, leg. C. D. Schubart, J. Reimer, R. Diesel (DNA extr. 26 Jul. 2005 ×10); 2 males (14.98 × 12.83, 13.27 × 11.58), 4 females (21.2 × 18.43, 16.85 ×

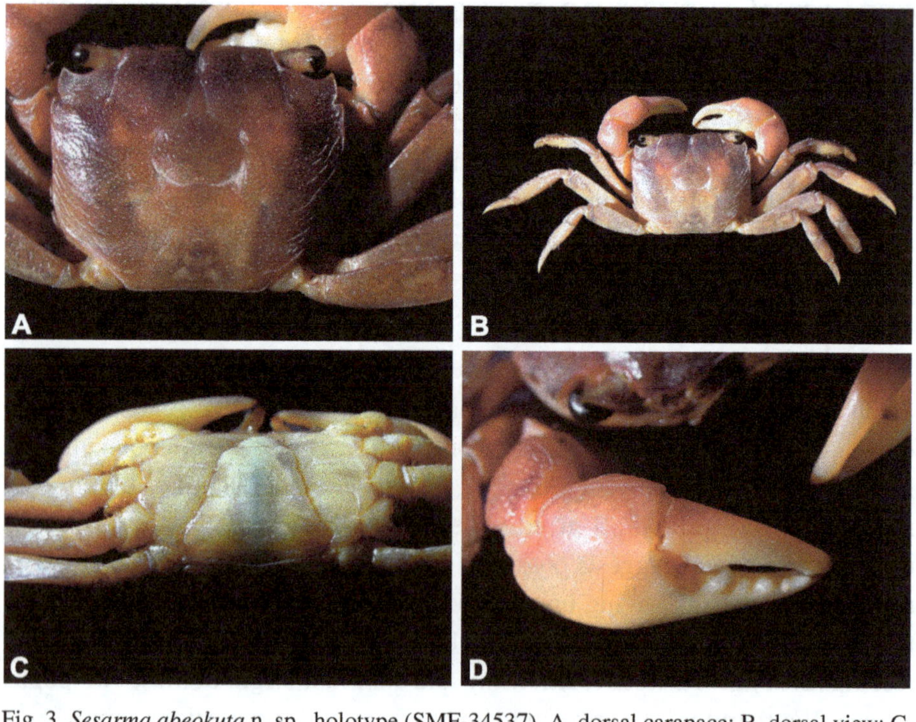

Fig. 3. *Sesarma abeokuta* n. sp., holotype (SMF-34537). A, dorsal carapace; B, dorsal view; C, sternum and male pleon; D, dorsal and outer face of chela.

14.65, 13.82 × 11.88, 11.96 × 10.28) (ZSMA20130013), same locality as previous, 16 Oct. 2005, leg. C. D. Schubart, S. Reuschel, T. Santl (DNA extr. 14 Nov. 2005 ×6).

Other material.— 1 male (16.36 × 14.19), 2 females (19.23 × 16.19, 17.23 × 14.61) (Collection R. Diesel), Jamaica, Westmoreland, Roaring River Spring, tributary to Cabarita River, 18°17′16″N-78°02′44″W, 19 Mar. 1995, leg. R. Diesel, C. D. Schubart; 21 juveniles (collection C. D. Schubart), Jamaica, Westmoreland, Abeokuta Nature Park, near Galloway, tributary to Deans Valley River, 18°14.718′N-78°02.643′W, 16 Oct. 2005, leg. C. D. Schubart, S. Reuschel, T. Santl.

Comparative material. — *Sesarma dolphinum* Reimer, Schubart & Diesel, 1998: Holotype: 1 male (25.3 × 21.44) (SMF-23304), Jamaica, Hanover, tributary to Hog River-Davis Cove River system, road between Paradise Great House and Kingsvale (∼18°23.1′N/78°12.75′W), 19 Mar. 1995, leg. R. Diesel, C. D. Schubart; 4 specimens (CDS private), same locality as holotype, Mar. 1997, leg. R. Diesel, J. Reimer, C. D. Schubart (DNA extr. ×4); 7 males, 5 females (SMF-34540), Jamaica, Hanover, Flint River between Cascade and Pondside (18°23′52.19″N/78°05′39.12″W), 17 Mar. 2003, leg. C. D. Schubart, T. Weil, T. Santl (DNA extr. ×2); 8 males, 7 females, 2 juv. (SMF-34541), Jamaica, Hanover, upper Cabarita River near Buxton (18°22′24.00″N/78°06′1.38″W), 17 Mar. 2003, leg. C. D. Schubart, T. Weil, T. Santl (DNA extr. ×4); 7 males, 3 females, 1 juv. (SMF-34542), Jamaica, Hanover, Lucea East River near Tom Spring (18°24′11.95″N/78°08′6.38″W), 17 Mar. 2003, leg. C. D. Schubart, T. Weil, T. Santl (DNA extr. ×2); 2 males, 3 females (SMF-34543), Jamaica, Hanover, nr. Harvey River, right tributary to Lucea West River (18°24′6.81″N/78°09′15.24″W), 17 Mar. 2003, leg. C. D. Schubart, T. Weil, T. Santl (DNA extr. ×1); 4 males, 5 females (SMF-34544), Jamaica, Hanover, Askenish, tributary to Lucea West River (18°22′42.47″N/78°09′07.30″W), 17 Mar. 2003, leg. C. D. Schubart, T. Weil, T. Santl (DNA extr. 1 male 8.5.2003); 2 males, 3 females (SMF-34545), Jamaica, Hanover, Rock Spring, Green Island River tributary (18°21′30.42″N/78°14′32.58″W), 18 Mar. 2003, leg. C. D. Schubart, T. Weil, T. Santl (DNA extr. ×5); 9 males, 5 females (SMF-34546), Jamaica, Westmoreland, 3 km E of Jerusalem Mountain, New Savannah River tributary (18°18′53.73″N/78°14′10.26″W), 18 Mar. 2003, leg. C. D. Schubart, T. Weil, T. Santl (DNA extr. ×3); 2 males, 2 females, 1 juvenile, 1 exuvia (SMF-34547), Jamaica, Hanover, Dias, upper Lances River (198 m, 18°23.920′N/78°11.542′W), 15 Oct. 2005, leg. C. D. Schubart, S. Reuschel, T. Santl (DNA extr. ×4); 2 males, 5 females (SMF-34548), Jamaica, Hanover, Flamstead, South of Kingsvale, Morgan River (90 m, 18°22′28.18″N/78°12′20.02″W), 15 Oct. 2005, leg. C. D. Schubart, S. Reuschel, T. Santl (DNA extr. ×4); 4 males, 6 females (SMF-34549), Jamaica, Hanover, tributary to Jackass River and Cabarita River near Bath Mountain (94 m, 18°20′26.22″N/78°05′57.42″W), 22 Oct. 2005, leg. C. D. Schubart, S. Reuschel, T. Santl (DNA extr. ×2); 7 males, 8 females, 3 juv. (SMF-34550), Jamaica, Hanover, upper Cabarita River, 2.8 km south of Cash Hill (219 m, 18°22.400′N/78°06.023′W), 22 Oct. 2005, leg. C. D. Schubart, S. Reuschel, T. Santl (DNA extr. ×2); 2 males, 3 females, 1 juv. (SMF-34551), Jamaica, Hanover, Green Island River near Kendal (62 m, 18°21′56.49″N/78°13′50.41″W), 15 Oct. 2005, leg. C. D. Schubart, R. Brodie, S. Reuschel, T. Santl (DNA extr. ×4) (see table I).

Diagnosis. — See species diagnosis of *S. dolphinum* (Reimer et al., 1998) with exception of: Legs relatively short and stouter, with ratio 4th pereiopod length/carapace length smaller than 2. Carapace broadest anteriorly, next to epibranchial teeth. MtDNA sequences of the ND1 gene rendering two unique haplotypes, corresponding to haplotype VII (Roaring River, GenBank

HF678408) and haplotype XII (Deans Valley River, GenBank HF678413) in fig. 1.

Description. — See species description of *S. dolphinum*, with exception of walking legs which are relatively short and stout with 4th pereiopod length/carapace length = 1.91 ± 0.06 (see Reimer et al., 1998); merus of fourth pereiopod/anterior carapace width in the range of 0.58-0.66 (as opposed to 0.63-0.75 in *S. dolphinum*, see Gaß, 2012: fig. 48). Carapace broadest anteriorly, next to epibranchial teeth: CWb/CWf in the range of 0.96-1.005 (as opposed to CWb/CWf in the range of 0.99-1.06 in *S. dolphinum*, see Gaß, 2012: fig. 47).

Etymology. — This species is named after the locality Abeokuta Hills in the Galloway area in southwestern Jamaica (Province Westmoreland), in the vicinity of which this species was encountered in three different localities, belonging to two river systems. The name is used as a noun in apposition.

Remarks. — While documenting and mapping the distribution of *Sesarma dolphinum* for the present study, we noticed that coordinates given in the original description of the species by Reimer et al. (1998) are imprecise, or mistaken by the magnitude of 10 min. Therefore we here provide correct coordinates (see also table I): *Sesarma dolphinum*: Holotype SMF-23304 from upper Hog-Davis Cove river system, ~18°23.1′N-78°12.75′W. Paratypes: SMF-19576-77, USNM 284155 from Askenish, Lucea West River system, ~18°22.75′N-78°09.12′W; SMF-23305 from Flamstead, tributary to Morgan River, 18°22′28.18″N-78°12′20.02″W. Other material: Collection R. Diesel from Roaring River Spring (now *Sesarma abeokuta* n. sp.): 18°17′15.54″N-78°02′43.80″W. The latter is the material that Reimer et al. (1998) referred to when describing morphological differences and considering the establishment of a new subspecies (not material from Galloway as wrongly quoted in Schubart et al., 2010: 346).

DISCUSSION

Entirely freshwater organisms that do not exit their native waters via terrestrial migrations or via planktonic dispersal in coastal waters are genetically isolated in specific drainage systems (watersheds). These watersheds thus represent units and interesting case studies in island biogeography in the sense of MacArthur and Wilson (1967) and Diamond (1975). Depending on its ecology, a species may inhabit the entire drainage area, which most often consists of a main river and varying number of tributaries, or only thrive in specific sections of the drainage that meet certain ecological demands. This habitat specificity (for example in higher reaches of rivers) may cause isolation of freshwater organisms even within a river system. Examples of such isolation are summarized in table IV and in Decapoda include among others the Japanese freshwater palaemonid shrimp *Macrobrachium nipponense* (see Mashiko & Numachi, 2000); the Australian atyid shrimp *Paratya australiensis* (see Cook et al., 2006, 2007); the European astacid crayfish *Austropotamobius torrentium* in the Danube system (Trontelj et al., 2005; Schubart & Martin, 2006); North American cambarid crayfish (Fetzner & Crandall, 2003; Buhay et al., 2007), and Taiwanese potamid freshwater crabs of the genus *Geothelphusa* (see Shih et al., 2004, 2007, 2010).

Better known are the cases of diversification and speciation of Crustacea in ancient lakes, representing a similar case of within-catchment genetic isolation and speciation, but with less evident allopatric mechanisms: Amphipoda Gammaridea in Lake Baikal (Sherbakov et al., 1998), Atyidae and Potamonautidae in East African Lake Tanganyika (Fryer, 2006; Marijnissen et al., 2006, 2009) and Southeast Asian Atyidae and Gecarcinucidae in ancient lakes of Sulawesi (von Rintelen et al., 2007, 2010; Schubart et al., 2008). There are thus two important scenarios of isolation and genetic diversification in freshwater Crustacea, one is more obvious and results from allopatric separation in unconnected fresh waters, whereas the other case appears less obvious, as it takes place in the same catchment area and may be driven by the ecology and/or behaviour of the species, and thus can be parapatric or sympatric. This is in agreement with an ongoing paradigm shift away from allopatric speciation towards an ecological adaptive one, as recently advocated and summarised by Dieckmann et al. (2004).

Also for Jamaican river crabs of the family Sesarmidae, two cases of different genetic lineages within single catchment systems are mentioned by Schubart et al. (2010). In one case, *Sesarma bidentatum* and *S. meridies* inhabit different tributaries of the Rio Cobre. In another case, three different genetic lineages of the westernmost species *Sesarma dolphinum* are found in the Cabarita River system (Schubart et al., 2010: 343). This second case has been more thoroughly studied and results are confirmed within the present paper. Interestingly, the populations inhabiting the different tributaries of the Cabarita system are closer related to other populations from nearby rivers, corresponding to different catchment areas (Upper Cabarita R.-Flint R.; Morgan R.-Davis/Lances and Green Island rivers; Roaring R.-Deans Valley R.)

TABLE IV

Compilation of the known cases in which differentiation and speciation has been recorded within freshwater stream for Decapoda and other Crustacea

Species name	Taxonomy	Locality	Species status	References
Kinnecaris solitaria complex	Copepoda: Parastenocarididae	Australia	Species	Karanovic & Cooper, 2011
Asellus aquaticus	Isopoda: Asellidae	Danube catchment	Intraspecific	Verovnik et al., 2004
Eophreatoicus species	Isopoda: Phreatoicidae	Danube catchment	Intraspecific	Wilson et al., 2009
Paratya australiensis	Atyidae	Eastern Australia	Intraspecific	Hurwood et al., 2003; Cook et al., 2006, 2007; etc.
Troglocaris anophthalmus	Atyidae	Danube catchment	Intraspecific	Zakšek et al., 2009
Macrobrachium nipponense	Palaemonidae	Japan	Intraspecific	Mashiko & Numachi, 2000
Austropotamobius pallipes	Astacidae	Western Europe	Intraspecific	Trontelj et al., 1995
Austropotamobius torrentium	Astacidae	Danube catchment	Intraspecific	Trontelj et al., 1995
Orconectes luteus	Cambaridae	Southern USA	Intaspecific	Fetzner & Crandall, 2003
Cambarus hamulatus	Cambaridae	Southern USA	Intaspecific	Buhay et al., 2007
Euastacus species	Parastacidae	Eastern Australia	Species	Baker et al., 2004
Geothelphusa species	Potamidae	Taiwan	Species	Shih et al., 2004, 2007, 2011
Potamonautes perlatus	Potamonautidae	South Africa: Olifants R.	Intraspecific	Daniels et al., 1999
Sesarma dolphinum-abeokuta	Sesarmidae	Jamaica: Cabarita R.	Species	Current study
Sesarma meridies-bidentatum	Sesarmidae	Jamaica: Cobre T.	Species	Schubart & Koller, 2005

than they are among each other. This suggests that overland dispersal between nearby headwaters in forested areas is more likely than genetic mixing in the lowland part of the rivers, despite the fact that all river crabs from Jamaica with known development have two larval stages (Hartnoll, 1964; Anger & Schubart, 2005; González-Gordillo et al., 2010) and thus relatively high distribution potential compared to other freshwater crabs with direct development. However, in both these cases documented in Jamaica, i.e. the Cabarita and the Cobre rivers, the lower sections of the corresponding rivers are broad and fast-flowing lowland streams that lack rocky structures and are thus not suitable for adult nor larval crabs. This explains, how different crab species or populations can become ecologically and geographically isolated and evolve independently, even within single river systems, providing important insights for the reconstruction of the rapid adaptive radiation of Jamaican crabs that took place during the past 4.5 million years (Schubart et al., 1998b). Future analyses with larger sample sizes and more variable markers are planned to allow statistical quantification of gene flow and its direction among all sites, and thus provide a better understanding of the underlying patterns of diversification and speciation.

ACKNOWLEDGEMENTS

We are grateful to all the students and colleagues who helped to collect specimens of *Sesarma dolphinum* and *S. abeokuta* n. sp. over the years, i.e., Rudolf Diesel, Jens Reimer, Rene Brodie, Silke Reuschel, Luise Heine, Peter Koller as well as to the Discovery Bay Marine Laboratory and the University of the West Indies for support. Wolfgang Gass allowed access to unpublished morphometric data and Andrea Sailer-Muth helped with formatting. Two anonymous reviewers provided useful comments.

REFERENCES

ANGER, K. & C. D. SCHUBART, 2005. Experimental evidence of food-independent larval development in endemic Jamaican freshwater-breeding crabs. Physiological and Biochemical Zoology, **78**(2): 246-258.

BAKER, A. M., J. M. HUGHES, J. C. DEAN & S. E. BUNN, 2004b. Mitochondrial DNA reveals phylogenetic structuring and cryptic diversity in Australian freshwater macroinvertebrate assemblages. Marine and Freshwater Research, **55**: 629-640.

BENEDICT, J. E., 1892. Decapod Crustacea of Kingston Harbour. Johns Hopkins University Circular, **11**: 77.

BUHAY, J. E., G. MONI, N. MANN & K. A. CRANDALL, 2007. Molecular taxonomy in the dark: evolutionary history, phylogeography, and diversity of cave crayfish in the subgenus *Aviticambarus*, genus *Cambarus*. Molecular Phylogenetics and Evolution, **42**: 435-448.

CLEMENT, M., D. POSADA & K. A. CRANDALL, 2000. TCS: a computer program to estimate gene genealogies. Molecular Ecology, **9**: 1657-1660.

COOK, B. D., A. M. BAKER, T. J. PAGE, S. C. GRANT, J. H. FAWCETT, D. A. HURWOOD & J. M. HUGHES, 2006. Biogeographic history of an Australian freshwater shrimp, *Paratya australiensis* (Atyidae): the role life history transition in phylogeographic diversification. Molecular Ecology, **15**: 1083-1093.

COOK, B. D., S. E. BUNN & J. M. HUGHES, 2007. A comparative analysis of population structuring and genetic diversity in sympatric lineages of freshwater shrimp (Atyidae: *Paratya*): concerted or independent responses to hydrographic factors? Freshwater Biology, **52**: 2156-2171.

CRANDALL, K. A., O. R. P. BININDA-EMONDS, G. M. MACE & R. K. WAYNE, 2000. Considering evolutionary processes in conservation biology. Trends in Ecology and Evolution, **15**(7): 290-295.

CRANDALL, K. A. & A. R. TEMPLETON, 1996. Applications of intraspecific phylogenetics. In: P. H. HARVEY, A. J. L. BROWN, J. M. SMITH & S. NEE (eds.), New uses for new phylogenies: 81-99. (Oxford University Press, New York, Oxford).

DANIELS, S. R., B. A. STEWART & M. J. GIBBONS, 1999. Genetic structure among populations of *Potamonautes perlatus* (Decapoda: Potamonautidae) from the Olifants River system in the Western Cape, South Africa. Journal of Zoology, London, **249**: 137-142.

DIAMOND, J. M., 1975. The island dilemma: lessons of modern biogeographic studies for the design of natural reserves. Biological Conservation, **7**(2): 129-146.

DIECKMANN, U., M. DOEBELI, J. A. J. METZ & D. TAUTZ, 2004. Adaptive speciation. Cambridge Studies in Adaptive Dynamic: 488 pp. (Cambridge University Press).

FETZNER, J. W. JR. & K. A. CRANDALL, 2003. Linear habitats and the nested clade analysis: an empirical evaluation of geographic versus river distances using an Ozark Crayfish (Decapoda: Cambaridae). Evolution, **57**(9): 2101-2118.

FRASER, D. J. & L. BERNATCHEZ, 2001. Adaptive evolutionary conservation: towards a unified concept for defining conservation units. Molecular Ecology, **10**: 2741-2752.

FRYER, G., 2006. Evolution in ancient lakes: radiation of Tanganyikan atyid prawns. Hydrobiologia, **568**: 131-142.

GASS, W., 2012. Morphometrische Differenzierung und allometrisches Wachstum bei brachyuren Krabben: 70 pp. (Unpublished Master thesis, Universität Regensburg).

GONZÁLEZ-GORDILLO, J. I., K. ANGER & C. D. SCHUBART, 2010. Morphology of the larval and first juvenile stages of two Jamaican endemic crab species with abbreviated development, *Sesarma windsor* and *Metopaulias depressus* (Decapoda, Brachyura, Sesarmidae). Journal of Crustacean Biology, **30**(1): 101-121.

HALL, T. A., 1999. BioEdit: a user-friendly biological sequence alignment editor and analysis program for Windows 95/98/NT. Nucleic Acids Symposium Series, **41**: 95-98.

HARTNOLL, R. G., 1964. The freshwater grapsid crabs of Jamaica. Proceedings of the Linnean Society London, **175**: 145-169

HURWOOD, D. A., J. M. HUGHES, S. E. BUNN & C. CLEARY, 2003. Population structure in the freshwater shrimp (*Paratya australiensis*) inferred from allozymes and mitochondrial DNA. Heredity, **90**: 64-70.

KARANOVIC, T. & S. J. B. COOPER, 2011. Molecular and morphological evidence for short range endemism in the *Kinnecaris solitaria* complex (Copepoda: Parastenocarididae), with descriptions of seven new species. Zootaxa, **3026**: 1-64.

MACARTHUR, R. H. & E. O. WILSON, 1967. The theory of island biogeography: 203 pp. (Princeton University Press, Princeton, New Jersey).

MARIJNISSEN, S. A. E., E. MICHEL, D. F. R. CLEARY & P. B. MCINTYRE, 2009. Ecology and conservation status of endemic freshwater crabs in Lake Tanganyika, Africa. Biodiversity and Conservation, **18**: 1555-1573.

MARIJNISSEN, S. A. E., E. MICHEL, S. R. DANIELS, D. ERPENBECK, S. B. J. MENKEN & F. R. SCHRAM, 2006. Molecular evidence for recent divergence of Lake Tanganyika endemic crabs (Decapoda: Platythelphusidae). Molecular Phylogenetics and Evolution, **40**: 628-634.

MASHIKO, K. & K. I. NUMACHI, 2000. Derivation of populations with different-sized eggs in the palaemonid prawn *Macrobrachium nipponense*. Journal of Crustacean Biology, **20**: 118-127.

MORITZ, C., 1994. Defining 'Evolutionary Significant Units' for conservation. Trends in Ecology and Evolution, **9**: 373-375.

POSADA, D., K. A. CRANDALL & A. R. TEMPLETON, 2000. GeoDis: a program for the cladistic nested analysis of the geographical distribution of genetic haplotypes. Molecular Ecology, **9**: 487-488.

REIMER, J., C. D. SCHUBART & R. DIESEL, 1998. Description of a new freshwater crab of the genus *Sesarma* Say, 1817 (Brachyura: Grapsidae: Sesarminae) from western Jamaica. Crustaceana, **71**: 186-196.

RINTELEN, K. VON, M. GLAUBRECHT, C. D. SCHUBART, A. WESSEL & T. VON RINTELEN, 2010. Adaptive radiation and ecological diversification of Sulawesi's ancient lake shrimps. Evolution, **64**(11): 3287-3299.

RINTELEN, K. VON, T. VON RINTELEN & M. GLAUBRECHT, 2007. Molecular phylogeny and diversification of freshwater shrimps (Decapoda, Atyidae, Caridina) from ancient Lake Poso (Sulawesi, Indonesia) — the importance of being colourful. Molecular Phylogenetics and Evolution, **45**: 1033-1041.

RYDER, O. A., 1986. Species conservation and systematics: the dilemma of subspecies. Trends in Ecology and Evolution, **1**: 9-10.

SCHUBART, C. D., 2009. Mitochondrial DNA and decapod phylogenies; the importance of pseudogenes and primer optimization. In: J. W. MARTIN, K. A. CRANDALL & D. L. FELDER (eds.), Decapod crustacean phylogenetics. Crustacean Issues, **18**: 47-65 (Taylor & Francis/CRC Press, Boca Raton, Florida).

SCHUBART, C. D., R. DIESEL & S. B. HEDGES, 1998b. Rapid evolution to terrestrial life in Jamaican crabs. Nature, **393**: 363-365.

SCHUBART, C. D. & M. G. J. HUBER, 2006. Genetic comparisons of German populations of the stone crayfish, *Austropotamobius torrentium* (Crustacea: Astacidae). Bulletin Français de la Pêche et de la Pisciculture, **380-381**: 1019-1028.

SCHUBART, C. D. & P. KOLLER, 2005. Genetic diversity of freshwater crabs (Brachyura: Sesarmidae) from central Jamaica with description of a new species. Journal of Natural History, **39**(6): 469-481.

SCHUBART, C. D., J. REIMER & R. DIESEL, 1998a. Morphological and molecular evidence for a new endemic freshwater crab, *Sesarma ayatum* sp. n., (Grapsidae, Sesarminae) from eastern Jamaica. Zoologica Scripta, **27**: 373-380

— —, — — & — —, 1999. Gonopod morphology and 16S mtDNA sequence alignment of the freshwater crabs *Sesarma bidentatum* and *S. ayatum* (Grapsidae: Sesarminae) from Jamaica. Zoologica Scripta, **28**(3-4): 367-369.

SCHUBART, C. D., J. REIMER, R. DIESEL & M. TÜRKAY, 1997. Taxonomy and ecology of two endemic freshwater crabs from western Jamaica with the description of a new *Sesarma* species (Brachyura: Grapsidae: Sesarminae). Journal of Natural History, **31**: 403-419.

SCHUBART, C. D., T. SANTL & P. KOLLER, 2008. Mitochondrial patterns of intra- and interspecific differentiation among endemic freshwater crabs of ancient lakes in Sulawesi. Contributions to Zoology, **77**: 83-90.

SCHUBART, C. D., T. WEIL, J. T. STENDERUP, K. A. CRANDALL & T. SANTL, 2010. Ongoing phenotypic and genotypic diversification in adaptively radiated freshwater crabs from Jamaica. In: M. GLAUBRECHT (ed.), Evolution in action — adaptive radiations and the origins of biodiversity: 323-349. (Springer-Verlag, Berlin Heidelberg).

SHERBAKOV, D. YU., R. M. KAMALTYNOV, O. B. OGARKOV & E. VERHEYEN, 1998. Patterns of evolutionary change in Baikalian gammarids inferred from DNA sequences (Crustacea, Amphipoda). Molecular Phylogenetics and Evolution, **10**: 160-167.

SHIH, H.-T., P. K. L. NG & H.-W. CHANG, 2004. Systematics of the genus *Geothelphusa* (Crustacea: Decapoda, Brachyura, Potamidae) from southern Taiwan: a molecular appraisal. Zoological Studies, **43**(3): 561-570.

SHIH, H.-T., P. K. L. NG, T. NARUSE, S. SHOKITA & M.-Y. LIU, 2011. Pleistocene speciation of freshwater crabs (Crustacea: Potamidae: *Geothelphusa*) from northern Taiwan and southern Ryukyus, as revealed by phylogenetic relationships. Zoologischer Anzeiger, **250**: 457-471.

SHIH, H.-T., P. K. L. NG, C. D. SCHUBART & H.-W. CHANG, 2007. Phylogeny and phylogeography of the genus *Geothelphusa* (Crustacea: Decapoda, Brachyura, Potamidae) in southwestern Taiwan based on two mitochondrial genes. Zoological Science, **24**: 57-66.

TEMPLETON, A. R., 2004. Statistical phylogeography: methods of evaluating and minimizing inference errors. Molecular Ecology, **13**: 789-809.

TEMPLETON, A. R., K. A. CRANDALL & C. F. SING, 1992. A cladistic analysis of phenotypic associations with haplotypes inferred from restriction endonuclease mapping and DNA sequence data. III. Cladogram estimation. Genetics, **132**: 619-633.

TEMPLETON, A. R., E. ROUTMAN, & C. A. PHILLIPS, 1995. Separating population structure from population history: a cladistic analysis of the geographical distribution of mitochondrial DNA haplotypes in the tiger salamander, *Ambystoma tigrinum*. Genetics, **140**: 767-782.

TRONTELJ, P., Y. MACHINO & B. SKET, 2005. Phylogenetic and phylogeographic relationships in the crayfish genus *Austropotamobius* inferred from mitochondrial COI gene sequences. Molecular Phylogenetics and Evolution, **34**: 212-226.

TÜRKAY, M. & R. DIESEL, 1994. Description of a new species of *Sesarma* from Jamaica with notes on its occurrence and biology (Crustacea: Decapoda: Brachyura). Senckenbergiana biologica, **74**: 157-161.

VEROVNIK, R., B. SKET & P. TRONTELJ, 2004. Phylogeography of subterranean and surface populations of water lice *Asellus aquaticus* (Crustacea: Isopoda). Molecular Ecology, **13**: 1519-1532.

WILSON, G. D. F., C. L. HUMPHREY, D. J. COLGAN, K. GRAY & R. JOHNSON, 2009. Monsoon-influenced speciation patterns in a species flock of *Eophreatoicus* Nicholls (Isopoda; Crustacea). Molecular Phylogenetics and Evolution, **51**: 349-364.

ZAKŠEK, V., B. SKET, S. GOTTSTEIN, D. FRANJEVIĆ & P. TRONTELJ, 2009. The limits of cryptic diversity in groundwater: phylogeography of the cave shrimp *Troglocaris anophthalmus* (Crustacea: Decapoda: Atyidae). Molecular Ecology, **18**: 931-946.

First received 1 October 2011.
Final version accepted 18 November 2013.

THE AEGLIDAE OF URUGUAY (DECAPODA, ANOMURA), WITH THE DESCRIPTION OF A NEW SPECIES OF *AEGLA*

BY

SANDRO SANTOS[1,5]), GEORGINA BOND-BUCKUP[2]), LUDWIG BUCKUP[2]), TAINÃ G. LOUREIRO[2]), ALBERTO S. GONÇALVES[1]), ANA VERDI[3]), FABRIZIO SCARABINO[4]) and CHRISTIAN CLAVIJO[4])

[1]) Programa de Pós-Graduação em Biodiversidade Animal, Universidade Federal de Santa Maria, Av. Roraima, 1000, 97105-900, Santa Maria, RS, Brazil

[2]) Programa de Pós-Graduação em Biologia Animal, Universidade Federal do Rio Grande do Sul, Av. Bento Gonçalves, 9500, 90501-970, Porto Alegre, RS, Brazil

[3]) Universidad de la República do Uruguay, Facultad de Ciencias, Iguá 4225, C.P. 11400, Montevideo, Uruguay

[4]) Museo Nacional de Historia Natural y Antropología del Uruguay, Avda. de las Instrucciones 948, C.P. 12900, Montevideo, Uruguay

ABSTRACT

Specimens of aeglid anomuran freshwater crabs (Anomura: Aeglidae) from Uruguay deposited in Uruguayan and Brazilian scientific collections included a new species of *Aegla* from the Pampa biome. This discovery is the first new species of this genus to be described from Uruguay since 1942, and raises the number of species of *Aegla* found in that country to four. The new species is described and illustrated, and a map of all Uruguayan species of this genus is provided.

RESUMO

Aproximadamente 70 anos após a descrição de três espécies de aeglídeos registrados para a República do Uruguai (*Aegla platensis* Schmitt, 1942; *Aegla prado* Schmitt, 1942 y *Aegla uruguayana* Schmitt, 1942), animais deste país, depositados em coleções científicas uruguaias e brasileiras, foram examinados. Este trabalho aprofundou nosso conhecimento sobre a distribuição dos eglídeos e resultou na descrição de uma nova espécie da família Aeglidae, a qual ocorre no bioma Pampa.

[5]) Corresponding author; e-mail: sandro.santos30@gmail.com

INTRODUCTION

The uplift of the Andean Cordillera in South America began more than 90 Mya, and has shaped the hydrographic basins of the inner part of the continent, especially those near the Atlantic coast, including Uruguay (Ribeiro, 2006). These movements have dramatically altered the drainage areas of many parts of the continent and this has either restricted the dispersal of many aquatic organisms, or has allowed an extension of the distributional range of others (Alexander & Lamp, 2008).

Freshwater decapod anomuran crustaceans of the genus *Aegla* Leach (family Aeglidae) are restricted to the temperate parts of southern South America (Bond-Buckup & Buckup, 1994). These animals first colonized continental fresh waters during a marine transgression of the Pacific Ocean coast some 70 Mya (Pérez-Losada et al., 2004) and (with one exception) are the only anomurans that complete their life cycle entirely in fresh water habitats.

There are currently 74 species of aeglids (McLaughlin et al., 2010; Santos et al., 2013) of which 44 are found in Brazil, 21 in Chile, 14 in Argentina, one in Paraguay, and one in Bolivia. Three species were previously known to occur in Uruguay (*Aegla platensis* Schmitt, 1942, *Aegla prado* Schmitt, 1942, and *Aegla uruguayana* Schmitt, 1942) (Bond-Buckup & Buckup, 1994), were all described in 1942. Since then (a period of over 70 years) no new species have been described from this country (despite the addition of several aeglids to the scientific collections of the Museo Nacional de Historia Natural y Antropología del Uruguay and the Facultad de Ciencias de Montevideo). This makes the present description of a fourth species of *Aegla* an important contribution to Uruguay's biodiversity.

The Republic of Uruguay is the only South American country that lies completely within the temperate zone. The absence of important orographic systems in Uruguay contributes to its low spatial variation and most of this country's area consists of the meadows that form the Pampa Biome. This ecosystem is characterized by low hills (up to 514 m a.s.l.), and its main hydrographic basins are those of the Uruguay and Negro Rivers, the Mirim Lagoon, and the sub-basins of the Plate River.

We present here the first species of *Aegla* from Uruguay to be described for over 70 years based on material previously deposited in scientific collections. The distributions of the three described species from Uruguay are updated here and compared to that of the new species (table I and fig. 1). The new

TABLE I
Numbers of lots and specimens of the genus *Aegla* analysed in each scientific institution

Institution	Species	Number of lots	Number of specimens
FC-UDELAR	*A. platensis*	15	131
	A. prado	09	172
	A. uruguayana	31	139
MNHN	*A. platensis*	12	112
	A. prado	03	21
	A. uruguayana	19	176
UFRGS	*A. platensis*	04	13
	A. prado	10	43
	A. uruguayana	11	33
Total		114	840

species is described by G. Bond-Buckup and T. Gonçalves Loureiro who are the taxonomic authorities for *A. carinata* sp. nov.

MATERIAL AND METHODS

Specimens of Aeglidae from the following collections were examined: Facultad de Ciencias, Universidad de La República, Montevideo, Uruguay (FC-UDELAR); Museo Nacional de Historia Natural, Montevideo, Uruguay (MNHN); Departamento de Zoologia, Instituto de Biociências, Universidade Federal do Rio Grande do Sul, Porto Alegre, RS, Brazil (UFRGS).

The description of the new species was based on examination of the characters of the holotype and of the type-series. Measurements of the specimens were taken according to the methodology described by Bond-Buckup & Buckup (1994), using the following abbreviations: CL — total cephalothorax length: between the tip of the rostrum and the midpoint of the posterior margin of the carapace; AL — areola length: length of the longitudinal median line of the areola; AW — areola width: distance between the lateral margins of the areola, taken on their anterior curvature; FW — frontal width: between the tips of the spines of the anterolateral angles of the carapace; PCW — pre-cervical width: carapace width measured at the height of the third hepatic lobes; m = male, f = female, f ov = ovigerous female, and j = juvenile.

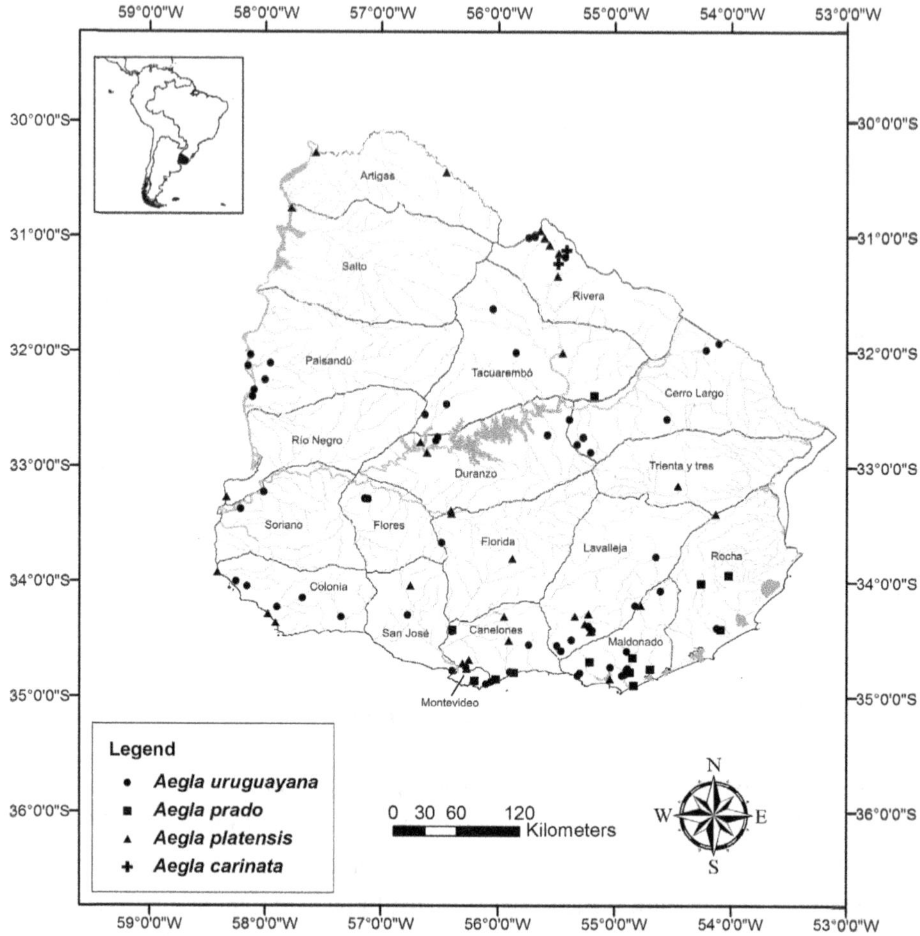

Fig. 1. The distributions of the four known species of *Aegla* in Uruguay.

TAXONOMY

Family AEGLIDAE Dana
Genus **Aegla** Leach
Aegla carinata sp. nov. Bond-Buckup & Loureiro
(fig. 2)

Material examined. — Holotype, male MZUSP 24432, Uruguay, Department of Rivera, Negro River Drainage, Cuñapiru Creek, km 12.3 Ruta (Route) 27, 31°02′21″S 55°29′31″W, coll. L. R. Malabarba, 8 Dec. 2001. Paratypes, male UFRGS 4440, same data as holotype; 7 m, UFRGS 4439, Uruguay, Department of Rivera, Negro River Drainage, Cuñapiru Creek, km 12.3 Ruta (Route) 27, 31°02′21″S 55°29′31″W, coll. L. R. Malabarba, 27 May 2005; 2 m,

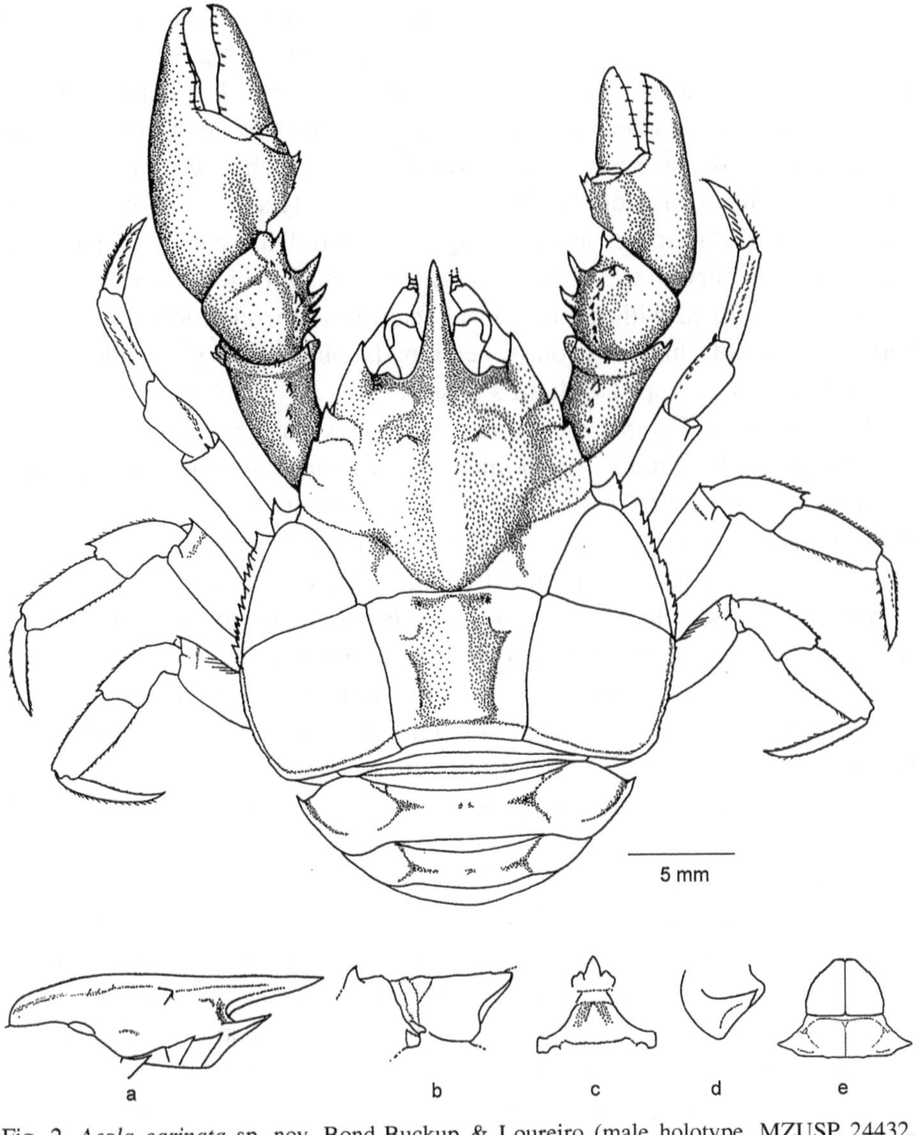

Fig. 2. *Aegla carinata* sp. nov. Bond-Buckup & Loureiro (male holotype, MZUSP 24432, scale: 5 mm). Below: a, anterior portion of carapace (lateral view); b, basis-ischium of cheliped (ventral view); c, third and fourth thoracic sternites (ventral view); d, epimeron 2 (lateral view); e, sixth abdominal segment and telson (dorsal view).

1 f UFRGS 4238, Uruguay, Department of Rivera, Negro River Drainage, Cuñapiru Creek, 31°02′13″S 55°29′31″W, 172 m a.s.l., coll. G. Bond-Buckup & L. Buckup, 9 Dec. 2006.

Diagnosis. — Antero-lateral spine of carapace extending beyond half of cornea; protogastric lobes very elevated and with scales; rostrum styliform, carinate along its entire length; cephalothorax with longitudinal dorsal carina on median line, ornamented with scales up to anterior region, meeting the posterior areola; lateral margins of branchial anterior and posterior areas of carapace arched, expanded as a lamina and with tubercles; extra-orbital sinus wide; outer proximal margin of movable finger of cheliped lacking lobe; fingers of cheliped with lobular denticle; palmar crest modest, sub-rectangular, projected in distal spine; anterior angle of ventral margin of epimeron 2 projected in recurved and robust spine; inner margin of ventral surface of ischium of cheliped with modest distal scaliform tubercle; dorsal margin of carpus of second, third and fourth pereiopods with distal spine followed by scaliform tubercles tipped with tufts of setae.

Description. — Carapace sub-oval, moderately convex; area of epigastric region elevated longitudinally, forming carina on median line; dorsal surface scabrous with small scales, punctations, setae. Longitudinal elevation extending from rostral carina to end of gastric region, tipped by rows of scales, tufts of setae; deep depression present near transverse dorsal line, areola. Front wide; PCW/FW ratio of holotype male = 2.13. Rostrum styliform, long, carinate to apex. Sub-rostral process absent. Rostral carina elevated, with two to three parallel, very close rows of scales, tufts of short setae in distal third. Rostral carina margins strongly excavated at height of protogastric lobes; oblique in distal third. Lateral margins of rostrum with tufts of long setae. Orbits wide, deep. Orbital margin with tufts of long setae. Extra-orbital sinus U-shaped, shallow, moderately wide. Antero-lateral angle of carapace projecting anteriorly in spine, extending past half of cornea. Outer margin of antero-lateral lobe with scaliform tubercles, tufts of setae; inner margin with tufts of short setae. Hepatic lobes arched, detached from margin. First hepatic lobe U-shaped, delimited anteriorly by prominent spine distinct from margin; second hepatic lobe V-shaped, projecting anteriorly in tubercle; third hepatic lobe slightly delimited, with modest incision extending to inner portion of carapace; lateral margins of three lobes ornamented with sub-equal scaliform tubercle, tufts of setae. Epigastric prominences moderate, with undefined shape, elongated toward base of first hepatic lobe, with punctations, sparse setae. Protogastric lobes prominent, elevated, tipped by scales, tufts of setae. Transverse line slightly sinuous. Areola quadrate, elevated especially in median region constituting prolongation of longitudinal dorsal carina tipped by scales, tufts of setae; margins sub-parallel along entire length, elevated in region of longitudinal dorsal carina. AL/AW ratio of holotype male = 1.57. Epibranchial area well developed, projecting in prominent anterior spine, followed by tubercles, setae. Lateral margins of anterior branchial area arched, laminate, with tubercles of varied size, scales, tufts of setae; posterior lateral margin arched, laminate, with scaliform tubercle, tufts of setae. Dorsal median area of abdominal segments 2 to 4 elevated, suggesting continuity of dorsal carina of cephalothorax. Sixth dorsal segment of pleon split by longitudinal suture. Anterior angle of ventral margin of epimeron 2 recurved, projecting in carina, robust apical spine; ventro-lateral margin slightly convex; posterior angle of ventral margin without projection, with scales, tufts of long setae. Epimera of third to sixth segments projecting in several setae; third, fourth epimera each with lateral projection ornamented with small apical tubercle, long setae. Telson divided by longitudinal suture. Anterior extremity of third sternite projected between coxae of exopods of third maxillipeds. Fourth thoracic sternite elevated, with scales, long setae, lateral margins not recurved. Fifth thoracic sternite slightly elevated in anterior portion. Chelipeds unequal, hand subrectangular. Larger cheliped with propodus with scales, tufts of short setae on dorsal surface, with slight depression in antero-medial region and slightly inflated in posterolateral region. Palmar crest scarcely projected, margin ornamented with prominent distal spine. Pre-dactylar lobe ornamented with scales and tufts of setae, forming small step with anterior margin of propodus. Fingers thickened, covered by scales, tufts of setae. Proximal outer margin of movable finger without lobe. Prehensile margins of fingers with scaliform denticles, tufts of setae along entire length, fitted opposed lobular teeth, with space between fingers. Dorsal surface of carpus rugose, with scales, tubercles, setae; inner margin with prominent distal spine on antero-lateral angle, followed by four robust spines that decrease in size proximally; spines bearing scales on lateral margins, tufts of setae; antero-dorsal margin with scales depression beginning close to first carpal crest becoming shallower toward outer margin of carpus. Carpal crest very prominent along entire length, formed by elevations with variable heights, tipped by scales, tufts of setae. Outer ventral angle of carpus projecting in tubercle. Dorsal margin of merus of cheliped with very prominent median spine, followed by three to four spines that decrease in size proximally. Lateral faces scabrous, with scales. Inner ventral margin of merus with robust distal spine, outer ventral margin with smaller spine. Dorsal margin of ischium ornamented with scales; inner margin of ventral face without ornamentation, only one distal tubercle present. Inner margin of ventral surface of coxae ornamented with distal tubercle smaller, more robust proximal tubercle projecting toward fourth thoracic sternite. Second, third, fourth pereiopods, dorsal margin of dactylus, propodus, carpus with rows of scales arranged in longitudinal rows, tufts of setae; dorsal margin of carpus with prominent distal spine, scaliform tubercles, tufts of setae distributed along segment. Coxae of third, fourth, fifth pereiopods with proximal laminar projection.

Variations. — In some paratypes there is a second, less prominent median palmar crest parallel to the main one. In small specimens the dorsal division of the sixth abdominal segment is indistinct.

Measurements. — CL of male holotype: 23.77 mm. Mean CL of paratypes (N = 14) 14.51 mm. PCW/FW mean ratio of paratypes: 1.80, ranging from 1.69 to 1.97. AL/AW mean ratio of paratypes: 1.57, ranging from 1.40 to 1.98.

Distribution. — Uruguay, Department of Rivera, Rio Negro basin.

Biology. — These crustaceans burrow in the sandy substratum of the river bed.

Etymology. — Specific name carinata: from Latin carina (keel or hull of a boat), referring to the keel-shaped prominent carina that extends along the dorsal portion of the carapace.

DISCUSSION

Although the biodiversity of the temperate Pampa biome cannot be compared with tropical rain forests in terms of species richness and ecological complexity, this biome represents a unique part of global biodiversity (TGCI, 2008). For some groups such as birds and mammals there is extensive information about species richness, which allows comparisons with other biomes; however, estimates of invertebrate species richness are few. The diurnal butterflies of the subfamily Satyrinae (Santos et al., 2008) and the bees Andrenidae and Colletidae (Blochtein & Harter-Marques, 2003) are the best-studied invertebrate groups. There are no reliable data about freshwater crustaceans from the Pampa biome.

The new species, *Aegla carinata* is recognized by its unique combination of morphological characters. No other aeglid, either in Uruguay, or anywhere else on the Atlantic side of South America is known to have a sub-oval carapace with a pronounced dorsal carina on the cephalothorax, and well ornamented margins of the anterior and posterior branchial areas.

On the other side of the Andes in Chile there is a species of *Aegla* (*A. denticulata* (Nicolet, 1849)) that also has a dorsal carina on the cephalothorax and ornamentation on the margins of the anterior and posterior branchial areas. However, *A. carinata* from Uruguay differs from the Chilean species in other characters: the dorsal carina of cephalothorax is very prominent and elevated, the rostrum is more elevated and styliform, a rostral carina is present, the margins of the branchial areas are more strongly arched and have more tubercles, protogastric lobes are present and pronounced, and the palmar crest has a different shape.

Aegla prado from Uruguay and the southern part of the Brazilian state of Rio Grande do Sul, has a modest longitudinal elevation on the cephalothorax, which does not form a carina. *Aegla prado* also has a styliform rostrum, although it is deflected, and the carapace is less sub-oval than that of the new species, lacking the ornamentation on the margins of the branchial areas.

The new species is known only from northern Uruguay, occurring in areas where *A. platensis* and *A. uruguayana* are also recorded, in Rivera Department, sub-basin of the Rio Negro (fig. 1). The distinction between these two species and *A. carinata* is evident in the presence of dorsal carinae in the latter.

Most lots deposited in collections of Uruguayan institutions are more than 40 years old (~70%). The best investigated regions are near Montevideo, in the departments of Maldonado, Canelones and Lavalleja. The three species already recorded from Uruguay are present in all of these localities.

According to Bond-Buckup & Buckup (1994), *A. prado* is distributed along the sub-basins of the Plate River in southern Uruguay, and small basins along the Atlantic Ocean in eastern Uruguay. In this zone its distribution also extends to the tributaries of the Mirim Lagoon and reaches south eastern Rio Grande do Sul, a region dominated by plains, with fields and marshes that border the coastal lagoons. Our data extend the known distribution of this species to the basin of the Negro River and to the boundary between the departments of Canelones and San José.

Aegla uruguayana has a broad geographical distribution in several hydrographical basins of Argentina, Uruguay, and Brazil. Its known localities are concentrated in tributaries of the Uruguay River, near the boundary with Brazil, and extend through the sub-basins of the Plate River (Bond-Buckup & Buckup, 1994). Beyond this main axis, there are also records of *A. uruguayana* in sub-basins of the Paraná River (Argentina) and near the Andean Cordillera (Mendoza, Argentina). In Uruguay *A. uruguayana* is undoubtedly the species with the broadest distribution, and has been recorded in most departments either in tributaries of the Negro River sub-basin or of the Plate River. The present study extends the known distribution of this species inland to the tributaries of the Negro River that drain waters of the Departments of Tacuarembó, Durazno, Cerro Largo, and Flores.

Aegla platensis has a wide distribution in the sub-basins of the Uruguay River, northern Rio Grande do Sul in Brazil, sub-basins of the Paraná River in

Paraguay and Argentina, and near the mouth of the Plate River (Bond-Buckup & Buckup, 1994). In Uruguay this species is found in a small tributary of the Plate River (Bond-Buckup & Buckup, 1994). The new records reported on here place *A. platensis* in the northern region of Uruguay at border with Brazil, where it is present in tributaries of the Tacuarembó River (Negro River basin: Rivera) and the Quaraí River (Uruguay basin: Artigas). Further inland *A. platensis* is present in small streams feeding the Passo de los Toros Reservoir (Negro River basin: Durazno) and in the Cebollati River (basin of Mirim Lagoon: Trienta y Tres).

Despite the high number of specimens examined in this study many parts of Uruguay have either only a few records or no records of aeglids (fig. 1). Besides these still-unstudied regions, areas that were surveyed more than 40 years ago may have undergone severe alterations since then that may have significantly altered the structure and composition of the fauna. This situation underlines the need for further surveys on the Uruguayan aeglids to reveal the true diversity and distribution patterns of aeglids in this geologically and geographically complex region of South America.

ACKNOWLEDGEMENTS

We are grateful to the curators of the crustacean collections of the Facultad de Ciencias (Universidad de La República, Montevideo, Uruguay), Museo Nacional de Historia Natural (Montevideo, Uruguay) and Universidade Federal do Rio Grande do Sul (Porto Alegre, RS, Brazil). We also thank Ana Rossi for the illustrations of the new species, and CNPq for productivity grants to GBB (306490/2007-2) and SS (308598/2011-3).

REFERENCES

ALEXANDER, L. C. & W. O. LAMP, 2008. Mayfly population density, persistence and genetic structure in fragmented headwater habitats. In: F. R. HAUER, J. A. STANFORD & R. L. NEWELL (eds.), International advances in the ecology, zoogeography, and systematics of mayflies and stoneflies: 39-50. (University of California Press, Berkeley).

BLOCHTEIN, B. & B. HARTER-MARQUES, 2003. Himenópteros. In: C. S. FONTANA, G. A. BENCKE & R. E. REIS (eds.), Livro vermelho da fauna ameaçada de extinção no Rio Grande do Sul: 95-109 (Ed. PUCRS, Porto Alegre).

BOND-BUCKUP, G. & L. BUCKUP, 1994. A família Aeglidae (Crustacea, Decapoda, Anomura). Arquivos de Zoologia, **32**: 159-347.

McLaughlin, P. A., R. Lemaitre & K. A. Crandall, 2010. Annotated checklist of anomuran decapod crustaceans of the world (exclusive of the Kiwaoidea and families Chirostylidae and Galatheidae of the Galatheoidea) Part III — Aegloidea. Raffles Bulletin of Zoology, (Supplement) **23**: 131-137.

Pérez-Losada, M., G. Bond-Buckup, C. G. Jara & K. A. Crandall, 2004. Molecular systematics and biogeography of the southern South American fresh-water "crabs" *Aegla* (Decapoda: Anomura: Aeglidae) using multiple heuristic tree search approaches. Systematic Biology, **53**: 767-780.

Ribeiro, A. C., 2006. Tectonic history and the biogeography of the freshwater fishes from the coastal drainages of eastern Brazil: an example of faunal evolution associated with a divergent continental margin. Neotropical Ichthyology, **4**: 225-246.

Santos, E. C., O. H. H. Mielke & M. M. Casagrande, 2008. Inventários de borboletas no Brasil: estado da arte e modelo de áreas prioritárias para pesquisa com vistas à conservação. Natureza & Conservação, **6**: 68-90.

Santos, S., C. G. Jara, M. L. Bartholomei-Santos, M. Pérez-Losada & K. A. Crandall, 2013. New species and records of the genus *Aegla* Leach, 1820 (Crustacea, Anomura, Aeglidae) from the West-Central region of Rio Grande do Sul, Brazil. Nauplius, **21**(2): 211-223.

Schmitt, W. L., 1942. The species of *Aegla*, endemic South American fresh-water crustaceans. Proceedings of the United States National Museum, **91**(3132): 431-520.

TGCI (Temperate Grasslands Conservation Initiative), 2008. Life in a working landscape: towards a conservation strategy for the world's temperate grasslands. A record of the World Temperate Grasslands Conservation Initiative Workshop Hohhot, China — June 28 and 29, 2008. (TGCI/WCPA/IUCN, Vancouver).

First received 1 July 2011.
Final version accepted 29 July 2013.

ATYID SHRIMPS OF HAINAN ISLAND, SOUTHERN CHINA, WITH THE DESCRIPTION OF A NEW SPECIES OF *CARIDINA* (CRUSTACEA, DECAPODA, ATYIDAE)

BY

YIXIONG CAI[1])
National Biodiversity Centre, National Parks Board, 1 Cluny Road, Singapore 259569

ABSTRACT

The atyid shrimps of Hainan Island, China, are reviewed. A taxonomic synopsis is given for the seven species of *Caridina* so far known from Hainan, including two new records, *C. clinata* and *C. elongapoda*, and one new species, *Caridina baoting* sp. nov. A detailed morphological description and illustrations are provided for the new species. *Caridina baoting* sp. nov. is morphologically similar to *C. braviata* Ng & Cai, 2000, from Yangjiang, Guangdong Province, but can be distinguished by its longer antennular peduncle, slender endopod of the male first pleopod, and much larger egg size. Taxonomic notes are provided for the remaining six species of *Caridina* from Hainan.

INTRODUCTION

To date, only four species of atyid shrimps, all in the genus *Caridina* H. Milne Edwards, 1837, have been recorded from Hainan Island, southern China. Yu (1936) reported *C. nilotica gracilipes* De Man, 1892, and described *C. lanceifrons* Yu, 1936, from Hai-Kiu-sche (Hai Kou). Liang & Yan (1983) described another species, *C. hainanensis* Liang & Yan, 1983, from Wenchang County of Hainan, and added *C. macrophora* Kemp, 1918, which was also found in Guangxi, mainland China. Cai & Shokita (2006) subsequently synonymised *C. hainanensis* with *C. propinqua* De Man, 1908.

The aim of the present study is to review and update the status of the *Caridina* species found on Hainan Island based on all available specimens. A detailed description and illustrations are provided for one new species and several

[1]) e-mail: caiyixiong@yahoo.com

rarely reported species. Specimens examined are deposited in the Zoological Reference Collection of the Raffles Museum of Biodiversity Research, National University of Singapore, Singapore (ZRC); Muséum national d'Historie naturelle, Paris, France (MNHN); and Shanghai Fisheries University, Shanghai, People's Republic of China (SFU). The abbreviation cl is used for carapace length measured in mm from the post-orbital margin to the posterior margin of the carapace. All specimens examined were collected by Y. Cai & N. K. Ng from the Hainan Island unless stated otherwise.

SYSTEMATIC ACCOUNT

Family Atyidae De Haan, 1849

Genus **Caridina** H. Milne Edwards, 1837

Caridina gracilipes De Man, 1892

(figs. 1-3)

Caridina Wyckii var. *gracilipes* De Man, 1892: 387, pl. 24 figs. 29-29e [type localities: Sulawesi (Celebes), and Selajar, Indonesia].
Caridina nilotica gracilipes — De Man, 1908: 270, fig. 7a, b.
Caridina acuticaudata Dang, 1975: 70, fig. 4 [type locality: Hoa Binh, North Vietnam].
Caridina gracilipes — Wowor et al., 2004: 341, fig. 6C, D; Cai & Shokita, 2006a: 250.
Non *Caridina nilotica gracilipes* — Kemp, 1918: 275; Liu, 1955: 28, figs. 11-17.

Material examined. — 6 males, cl 2.5-4.1 mm, 9 ov. females, cl 5.0-7.1 mm, ZRC 2012.1164, river in Dayuan village, Wenchang County, Hainan Province, China, 19°35.83′N 110°46.01′E, pH 6.5, coll. 30 Nov. 1998; 2 males, cl 2.6-3.5 mm, 4 females, cl 2.8-5.2 mm, 1 ov. female, cl 5.3 mm, ZRC 2012.1039, HN23, Salong Bridge, stream across Changpo-Huiwen road, Wenchang County, 19°21.55′N 110°39.34′E, pH 7.3, coll. 1 Dec. 1998; 13 males, cl 3.6-4.4 mm, 5 females, cl 4.0-5.6 mm, 11 ov. females, cl 5.0-6.3 mm, ZRC 2012.1144, HN24, stream across Changpo-Huiwen road, Wenchang County, 19°23.03′N 110°40.38′E, pH 7.2, coll. 1 Dec. 1998; 2 males, cl 3.8-3.9 mm, 2 ov. females, cl 5.5-6.4 mm, ZRC 2012.1140, HN25, seashore along Changpo-Huiwen road, Wenchang County, coll. 1 Dec. 1998; 15 males, cl 3.2-4.2, 3 females, cl 4.5-6.5 mm, 40 ov. females, cl 4.5-6.5 mm, ZRC 2012.1141, HN26, stream at Wenchang-Qionghai border, 19°23.79′N 110°40.76′E, pH 7.3, coll. 1 Dec. 1998; 3 males, cl 3.8-3.9 mm, 6 females, cl 5.6-6.2 mm, 15 ov. females, cl 5.4-7.1 mm, ZRC 2012.1142, HN27, stream across Changpo-Huiwen road, station 3, Wenchang County, 19°23.83′N 110°40.93′E, pH 7.2, coll. 1 Dec. 1998; 14 males, cl 2.1-3.2 mm, 30 ov. females, cl 2.6-4.2 mm, ZRC 2012.1143, HN28, stream across Yandun village, Wenchang County, 19°25.40′N 110°42.50′E, pH 7.2, coll. 1 Dec. 1998; 11 males, cl 3.2-3.8 mm, 24 females, cl 3.0-5.7 mm, 27 ov. females, cl 4.9-6.7 mm, eggs 0.53 × 0.33 mm, ZRC 2012.1145, HN29a, stream outside Yandun village, Wenchang County, pH 6.6, coll. 1 Dec. 1998; 1 male, cl 3.8 mm, 3 females, cl 2.8-4.8 mm, 8 ov. females, cl 5.0-6.3 mm, ZRC 2012.1146, HN29bcd, stream across Huiwen Bridge, Huiwen Town, Wenchang County, pH 6.5, coll. 1 Dec. 1998; 5 males, cl 2.3-3.0 mm, 9 females, cl 2.2-3.6 mm, 9 ov. females, cl 3.6-4.5 mm, ZRC 2012.1147, HN32, Huishan Reservoir, 19°05.08′N 110°17.45′E, pH 6.9, coll. 2 Dec. 1998; 3 males, cl 2.7-2.9 mm, 6 females, cl 3.0-4.0 mm,

19 ov. females, cl 3.7-3.8 mm, ZRC 2012.1165, HN59, river across Tongzha-Shanya road, 18°25.98′N 109°40.59′E, pH 6.8, coll. 6 Dec. 1998; 6 males, cl 3.8-4.5 mm, 29 females, cl 2.5-5.5 mm, 12 ov. females, cl 3.9-6.7 mm, ZRC 2012.1148, stream in Yacheng Town, Sanya city, pH 6.8, 18°22.15′N 109°09.74′E, coll. 6 Dec. 1998; 1 male, cl 2.4 mm, ZRC 2012.1149, HN 11, Jinjiang, 27 Nov. 1998; 1 male, cl 33 mm, 1 female, cl 3.7 mm, ZRC 2012.1150, HN 6, stream along road from Jinjiang to Yongfa, Chengjiang County, 26 Nov. 1998; 2 males, cl 3.9-4.0 mm, 3 females, cl 3.9-5.3 mm, 9 ov. females, cl 5.0-6.8 mm, ZRC 2012.1151, HN 30, stream at Huiwen Bridge, Huiwen Town, Wenchang County, 1 Dec. 1998; 7 males, cl 2.7-4.3 mm, 4 females, cl 2.7-3.7 mm, ZRC 2012.1152, HN 4, stream at Tingchao bridge, Luonui Mountain near Qiongshan city, 25 Nov. 1998.

Habitat. — Lowland freshwater rivers with seawater influence.

Remarks. — Wowor et al. (2004) recorded *Caridina gracilipes* from Sarawak, northern Borneo, without any discussion. Cai & Shokita (2006a) recorded this species from the Philippines, provided an informative diagnosis and compared it with *C. elongapoda* Liang & Yang, 1977, and *C. longirostris* H. Milne Edwards, 1837. These authors confirmed *C. gracilipes* as a valid species. Two specimens of *C. gracilipes* from Hainan, a male and an ovigerous female, are illustrated in figs. 1-3.

Distribution. — Sulawesi, Taiwan, southern China, Peninsular Malaysia, Philippines, and Borneo.

Caridina macrophora Kemp, 1918
(figs. 4, 5)

Caridina nilotica macrophora Kemp, 1918a: 277, fig. 9a-f [type locality: Tale Sap Lake, Peninsular Siam (Thailand)].

Caridina nilotica macrophora — Liang & Yan, 1983: 211; Li & Liang, 2002: 708.

Caridina subnilotica Dang, 1975: 69, fig. 3 [type locality: Hanoi, North Vietnam].

Material examined. — 1 male, cl 2.9 mm, 1 female, cl 3.8 mm, 1 ov. female, cl 4.3 mm, eggs 0.95 × 0.55 mm, MNHN Na 769, syntype of *Caridina nilotica* var. *macrophora* Kemp, 1918a, exchanged from Indian Museum in 1921, Tale Sap, Siam (Thailand). 8 males, cl 2.4-2.7 mm, 22 females, cl 2.1-3.5 mm, 3 ov. females, cl 3.3-3.6 mm, eggs 0.80-0.90 × 0.50-0.55 mm, ZRC 2012.1153, pond near Qiongshan city, 19°58.09′N 110°22.93′E, coll. 25 Nov. 1998; 2 males, cl 2.5-2.8 mm, 4 females, cl 2.4-2.9 mm, ZRC 2012.1154, Nandujiang River, Jing Jiang Town, coll. 27 Nov. 1998; 3 males, cl 3.0-3.1 mm, 7 females, cl 2.3-4.7 mm, 6 ov. females, cl 3.9-4.4 mm, ZRC 2012.1155, stream across Dongtai Bridge, Wanquan River, Huishan Town, 19°04.23′N 110°15.41′E, pH 6.5, coll. 2 Dec. 1998.

Description. — Rostrum long, upturned anteriorly, reaching beyond end of scaphocerite; rostral formula: 2-3 + 11-19 + 1-2/6-20. Antennal spine lower than inferior orbital angle. Pterygostomian angle broadly rounded.

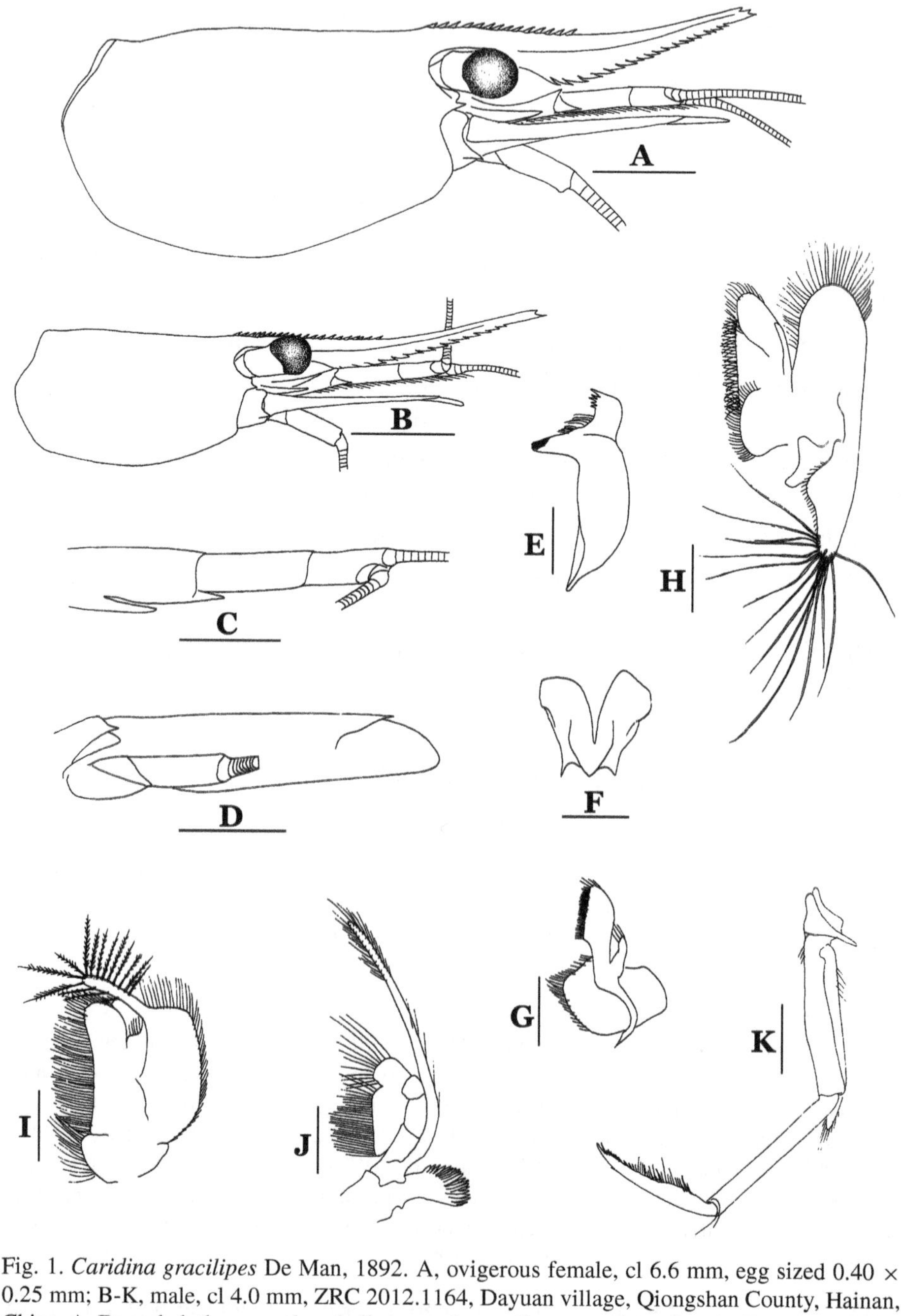

Fig. 1. *Caridina gracilipes* De Man, 1892. A, ovigerous female, cl 6.6 mm, egg sized 0.40 × 0.25 mm; B-K, male, cl 4.0 mm, ZRC 2012.1164, Dayuan village, Qiongshan County, Hainan, China. A, B, cephalothorax and cephalic appendages; C, antennular peduncle; D, scaphocerite; E, mandible; F, paragnathus; G, maxillula; H, maxilla; I, first maxilliped; J, second maxilliped; K, third maxilliped. Scales: A-D = 2 mm; E-K = 0.5 mm.

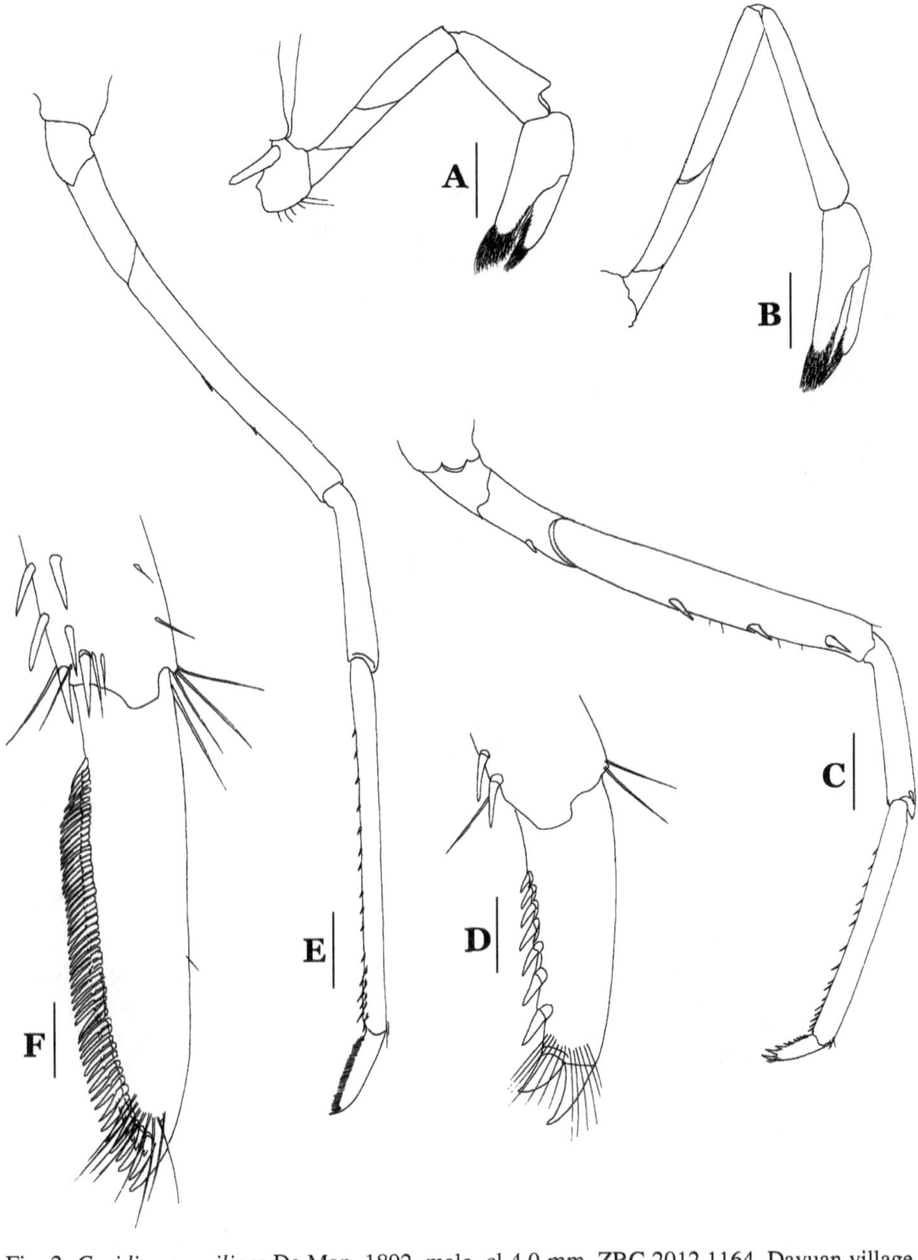

Fig. 2. *Caridina gracilipes* De Man, 1892, male, cl 4.0 mm, ZRC 2012.1164, Dayuan village, Qiongshan County, Hainan, China. A, first pereiopod; B, second pereiopod; C, third pereiopod; D, dactylus of third pereiopod; E, fifth pereiopod; F, dactylus of fifth pereiopod. Scales: A-C, E = 0.5 mm; D, F = 0.1 mm.

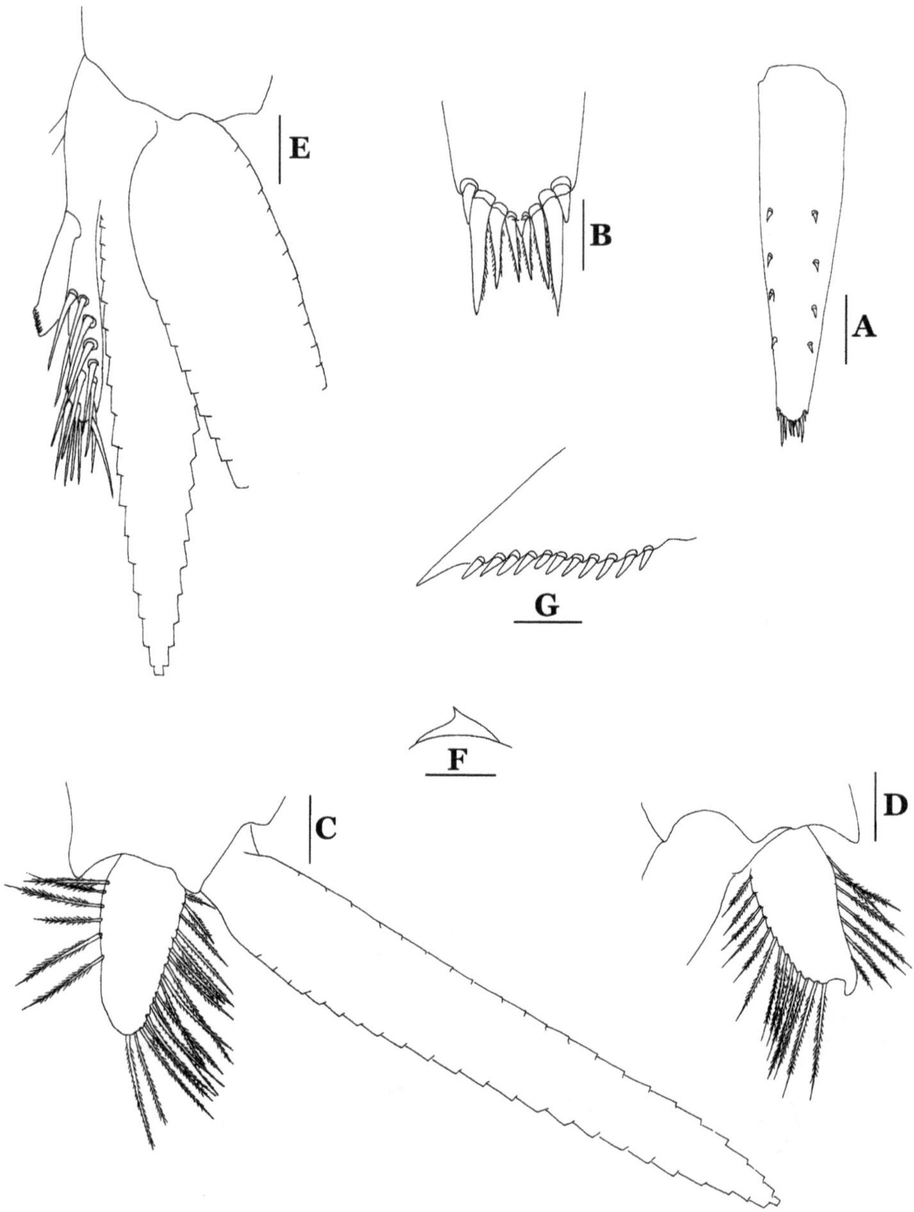

Fig. 3. *Caridina gracilipes* De Man, 1892, male, cl 4.0 mm, ZRC 2012.1164, Dayuan village, Qiongshan County, Hainan, China. A, telson; B, distal portion of telson; C, male first pleopod; D, endopod of male first pleopod, show variation of another specimen; E, male second pleopod; F, preanal carina; G, uropodal diaeresis. Scales: A, F = 0.5 mm; B-E, G = 0.2 mm.

Sixth abdominal somite 0.72 times length of carapace, 2.2 times as long as fifth somite, as long as telson. Telson 3.5 times as long as wide, not terminating in projection, with 3-5 pairs dorsal spinules, one pair dorsolateral spinules; distal margin with 4 pairs distal spines, lateral pair distinctly longer than intermediate pairs, sublateral pair shortest. Preanal carina small, rounded, without spine.

Eyes well developed, anterior end reaching 0.7 times length of basal segment of antennular peduncle. Antennular peduncle 0.86 times as long as carapace length; basal segment of antennular peduncle longer than sum of second and third segment length, anterolateral angle pointed, reaching 0.25 length of the second segment, second segment distinctly longer than third segment. Stylocerite reaching to 0.8 length of basal segment of antennular peduncle. Scaphocerite 4.5 times as long as wide.

Incisor process of mandible ending in irregular teeth, molar process truncated. Lower lacinia of maxillule broadly rounded, upper lacinia elongated, with several distinct teeth on inner margin, palp slender. Upper endites of maxilla subdivided, palp short, scaphognathite tapering posteriorly with long, curved setae at posterior end. Palp of first maxilliped ending in broad triangular structure. Second maxilliped typical of genus. Third maxilliped reaching to end of second segment of antennular peduncle, with ultimate segment shorter than penultimate segment.

Epipods on first 4 pereiopods. First pereiopod reaching to end of eyestalk; merus 2.2 times as long as broad, as long as carpus; carpus excavated anteriorly, shorter than chela, 1.6 times as long as high; chela 1.8 times as long as broad; fingers as long as, or slightly longer than, palm. Second pereiopod reaching to middle of second segment of antennular peduncle; merus shorter than carpus, 3.8 times as long as broad; carpus 1.2 times as long as chela, 4.3 times as long as high; chela 2.2 times as long as broad; fingers slightly longer than palm. Third pereiopod reaching near end of antennular peduncle, propodus 12 times as long as broad, 3.8 times as long as dactylus; dactylus 3.5 times as long as wide (spines included), with 6-10 accessory spines on flexor margin. Fifth pereiopod reaching to end of second segment of antennular peduncle, propodus 13 times as long as broad, 3.1 times as long as dactylus, dactylus 3.7 times as long as wide (spinules included), with 34-46 spinules on flexor margin. Endopod of male first pleopod subtriangular, 0.25 times length of exopod, no appendix interna. Appendix masculina of male second pleopod half length of endopod.

Uropodal diaeresis with 7-9 movable spinules.

Eggs 0.80-0.95 × 0.50-0.60 mm in diameter.

Habitat. — Lowland freshwater habitats: ponds, larger rivers, and streams.

Remarks. — Within the *Caridina nilotica* species group, *C. macrophora* most closely resembles *C. aruensis* J. Roux, 1911, e.g., in the form of the rostrum, the absence of an appendix interna on the endopod of the male first pleopod, and the relatively large eggs. However, it can be separated from *C. aruensis* by the telson not terminating in a posteromedian projection.

Dang (1975) described *Caridina subnilotica* Dang, 1975, from Hanoi, North Vietnam; however, he did not compare *C. subnilotica* with *C. macrophora*. Although Dang's types of *C. subnilotica* are untraceable and may have been lost (see Cai & Ng, 1999), the original description and illustrations, as well as the examination of topotypical specimens from Hanoi, all suggest that *C. subnilotica* is a junior synonym of *C. macrophora*, as already pointed out by Li & Liang (2002).

Distribution. — Thailand, Cambodia, Vietnam and South China (Guangxi and Hainan).

Caridina elongapoda Liang & Yan, 1977

Caridina nilotica elongapoda Liang & Yan, 1977: 220, figs. 5-8 [type locality: Xinzai, Gulei village, Zhangpu County, Fujian, southern China].
Caridina longirostris — Liang, 2004: 195, fig. 94.
Caridina elongapoda — Wowor et al., 2004: 341, fig. 6A, B; Cai & Shokita, 2006: 249.

Material examined. — 1 male, cl 4.3 mm, ZRC 2012.1166, HN23, Salong Bridge, stream across Changpo-Huiwen road, Wenchang County, 19°21.55′N 110°39.34′E, pH 7.3, coll. 1 Dec. 1998; 7 males, cl 2.8-3.4 mm, 24 females, cl 2.8-4.6 mm, ZRC 2012.1167, HN33, Wanquan River, Shibi Town, Baoting County, 19°09.86′N 110°18.51′E, pH 6.9, coll. 2 Dec. 1998.

Habitats. — Lowland streams and rivers.

Remarks. — The present specimens represent the first record of *Caridina elongapoda* for Hainan Island. Liang (2004) synonymised *C. nilotica elongapoda* Liang & Yan, 1973 (lap. cal. for 1977) with *C. longirostris* H. Milne Edwards, 1837, without any discussion. Wowor et al. (2004) listed *C. elongapoda* as a valid species distributed in Peninsular Malaysia and Borneo (Sabah), however, without justifying the status change from subspecies to species. Subsequently, Cai & Shokita (2006) provided a detailed description and discussion on the taxonomy of this species.

Distribution. — Southern China (including Hainan), Peninsular Malaysia, Borneo (Sabah) and the Philippines.

Fig. 4. *Caridina macrophora* Kemp, 1918, ovigerous female, cl 3.6 mm, egg sized 0.8 × 0.5 mm, ZRC 2012.1153, Qiongshan, Hainan Island, China. A, cephalothorax and cephalic appendages; B, telson; C, distal protion of telson; D, antennular peduncle; E, scaphocerite; F, mandible; G, maxillula; H, maxilla; I, first maxilliped; J, second maxilliped; K, third maxilliped; L, uropodal diaeresis; M, preanal carina. Scales: A, D, E = 1 mm; B, F-K, M = 0.5 mm; C, L = 0.2 mm.

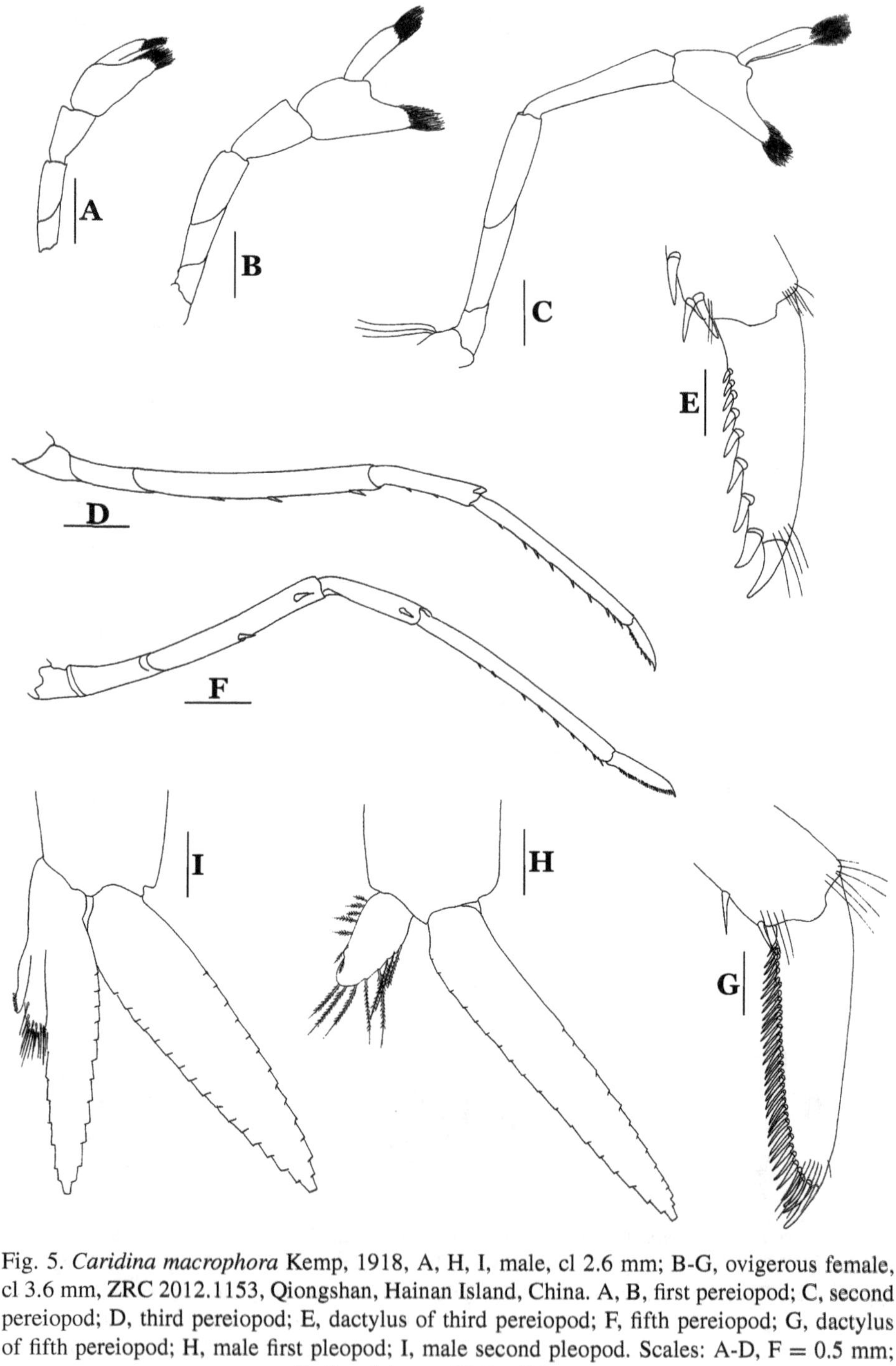

Fig. 5. *Caridina macrophora* Kemp, 1918, A, H, I, male, cl 2.6 mm; B-G, ovigerous female, cl 3.6 mm, ZRC 2012.1153, Qiongshan, Hainan Island, China. A, B, first pereiopod; C, second pereiopod; D, third pereiopod; E, dactylus of third pereiopod; F, fifth pereiopod; G, dactylus of fifth pereiopod; H, male first pleopod; I, male second pleopod. Scales: A-D, F = 0.5 mm; E, G = 0.1 mm; H, I = 0.2 mm.

Caridina lanceifrons Yu, 1936
(figs. 6, 7)

Caridina lanceifrons Yu, 1936: 89, figs. 5-7 [type locality: Haikou, Hainan Province, China].
Caridina lanceifrons — Dudgeon, 1985: 142; Liang & Yan, 1988: 15; Zheng, 1989: 6; Liang & Zhou, 1993: 231; Liang et al., 1993: 41; Li & Liang, 2002: 709.
Caridina flavilineata Dang, 1975: 71, fig. 5 [type locality: Nam Ha, North Vietnam].

Material examined. — Neotype: male, cl 3.0 mm, ZRC 2012.1168, pond near Qiongshan city, 19°58.09′N 110°22.93′E, 25 Nov. 1998. 6 males, cl 2.4-2.5 mm, 13 females, cl 2.5-3.8 mm, 1 ov. female, cl 3.2 mm, egg size 0.85 × 0.50 mm, ZRC 2012.1169, HN3, pond near Qiongshan city, 19°58.09′N 110°22.93′E, coll. 25 Nov. 1998; 3 males, 2.4-2.7 mm, 15 females, cl 2.1-4.3 mm, 3 ov. females, cl 3.7-5.0 mm, ZRC 2012.1170, HN 4, stream at Tingchao bridge, Luonui Mountain near Qiongshan city, coll. 25 Nov. 1998; 15 males, cl 2.3 to 2.9 mm, 23 females, cl 2.7-4.3 mm, ZRC 2012.1171, HN5, pond at Luonui Mountain, 25 Nov. 1998; 32 males, cl 2.8-3.4 mm, 35 females, cl 2.7-4.3 mm, 8 ov. females, 4.1-4.5 mm, ZRC 2012.1172, HN 6, stream along road from Jinjiang to Yongfa, Chengjiang County, 26 Nov. 1998; 21 males, cl 2.9-3.2 mm, 35 females, cl 2.4-5.3 mm, 22 ov. females, cl 4.2-4.7 mm, ZRC 2012.1173, HN 8, stream near the road from Chengmai to Yongfa, 26 Nov. 1998; 33 males, cl 2.8-3.7 mm, 56 females, cl 3.9-4.6 mm, 12 ov. females, cl 2.2-5.6 mm, ZRC 2012.1174, HN 9, tributary of Nando River at Yongfa Town, Chengjiang County, coll. 27 Nov. 1998; 11 males, cl 2.2-3.0 mm, 43 females, cl 2.3-4.2 mm, ZRC 2012.1176, HN 11, Jinjiang, 27 Nov. 1998; 1 male, cl 3.0 mm, 1 female, cl 3.3 mm, 2 ov. females, cl 4.7-5.1 mm, ZRC 2012.1177, HN16, river near Dayuan village, Wenchang County, 30 Nov. 1998; 3 males, cl 2.9-3.9 mm, 34 females, cl 2.1-4.3 mm, 17 ov. females, cl 4.5-5.3 mm, ZRC 2012.1178, HN29, stream outside Yandun village, Wenchang County, pH 6.6, coll. 1 Dec. 1998; 7 females, cl 3.3-4.3 mm, 2 ov. females, cl 4.2-4.5 mm, ZRC 2012.1179, stream near Huiwen Town, 1 Dec. 1998; 3 males, cl 2.6-3.1 mm, 25 females, cl 2.6-4.0 mm, 3 ov. females, cl 3.9-4.3 mm, egg size 0.82 × 0.52 mm, ZRC 2012.1180, HN 31, stream across Dongtai Bridge, Wanquan River, Huishan Town, 19°04.23′N 110°15.41′E, pH 6.5, coll. 2 Dec. 1998; 1 female, cl 3.2 mm, ZRC 2012.1181, HN33, Wanquan River, Shibi Town, 19°09.86′N 110°18.51′E, pH 6.9, coll. 2 Dec. 1998; 4 males, cl 2.9-3.2 mm, 2 females, cl 2.4-2.7 mm, ZRC 2012.1182, HN34, mountain stream across Shibi-Wanquan road, 19°10.49′N 110°20.56′E, pH 6.8, coll. 2 Dec. 1998; 3 males, cl 3.7-4.5 mm, 7 females, cl 3.3-5.8 mm, ZRC 2012.1183, HN 36, purchased from a market at Tongzha city, coll. 3 Dec. 2008; 2 males, cl 4.4-4.5 mm, 5 females, cl 4.7-6.0 mm, ZRC 2012.1184, mountain stream near the road from Maoan to Tongzha, 3 Dec. 1998; 2 males, 4.3-4.8 mm, 16 females, cl. 3.8-5.7 mm, 1 ov. female, cl 5.6 mm, eggs 0.92 × 0.6 mm, ZRC 2012.1185, HN 38, purchased from a market at Tongzha city, coll. 4 Dec. 2008; 11 males, cl 2.9-4.4 mm, 17 females, cl 2.5-5.3 mm, ZRC 2012.1186, HN39, stream across Mao'an-Boting road, pH 5.7, 18°38.24′N 109°35.48′E, coll. 4 Dec. 1998; 4 males, cl 3.3-3.4 mm, 3 females, cl 3.0-4.6 mm, ZRC 2012.1187, HN 42, stream near Shiluo village, Baoting County, 4 Dec. 1998; 1 male, cl 2.9 mm ZRC 2012.1188, HN 43, stream along road from Baoting to Lingshui, 4 Dec. 1998; 10 males, cl 3.4-4.1 mm, 7 females, cl 2.7-5.5 mm, 3 ov. females, cl 5.2-5.3 mm, ZRC 2012.1189, HN44, stream along road from Baoting to Lingshui, 4 Dec. 1998; 82 males, cl 3.2-3.8 mm, 185 females, cl 3.4-4.7 mm, 24 ov. females, 4.5-4.9 mm, ZRC 2012.1190, HN 45, stream across road from Baoting to Linsui, pH 6.7, 18°38.74′N 109°45.20′E, coll. 4 Dec. 1998; 10 males, cl 3.0-3.7 mm, 18 females, cl 2.5-4.5 mm, 4 ov. females, cl 4.4-6.0 mm, 5 juv., ZRC 2012.1191, HN46, Benhao village, stream across Baoting-Lingshui road, pH 6.4, 18°40.38′N 109°49.63′E, coll. 4 Dec. 1998; 10 males, cl 3.4-3.7 mm, 20 females, cl 2.7-4.6 mm, 4 ov. females, cl 4.3-4.9 mm, ZRC 2012.1192, stream across road from Baoting to Linsui, pH 6.7, 18°38′N 109°49.63′E, coll. 4 Dec. 1998; 29 males, cl 2.1-3.9 mm, 12 females, cl 3.0-5.0 mm, 2 ov. females, cl 4.5 mm, ZRC 2012.1193, HN51, stream across Tongzha-Wuzhi Mountain road, Maodan village, 18°55.43′N 109°32.38′E, pH 6.1, coll. 5 Dec. 1998; 8 females, cl 3.8-6.3 mm, ZRC 2012.1194, mountain stream across road from Tongzha to Wuzhishan Mountain, 5 Dec. 1998.

Comparative materials. — 1 male, cl 3.4 mm, 1 female, cl 5.0 mm, IZAS, Shangsi County, Guangxi Province, China, May 1978. 1 female, cl 3.8 mm, ZRC, Tin Liu, Sai Kung, New Territories, Hong Kong, coll. R. Yam, 16 Feb. 2001.

Description. — Rostrum straight, reaching near to or slightly beyond end of antennular peduncle; rostral formula: 3-5 + 12-18/1-6 (mode 2-3). Antennal spine fused with inferior orbital angle. Pterygostomian angle broadly rounded.

Sixth abdominal somite 0.51 times of carapace length, 1.6 times as long as fifth somite, shorter than telson. Telson 3.8 times as long as wide, terminating in projection, with 5 pairs dorsal spinules, one pair dorsolateral spinules; lateral pair of spines slightly longer than intermediate pairs of spiniform setae. Preanal carina high, lacking spine.

Eyes well developed, anterior end reaching to 0.8 times length of basal segment of antennular peduncle. Antennular peduncle 0.7 times as long as carapace; basal segment of antennular peduncle longer than sum of length of second and third segments, anterolateral angle reaching 0.30 length of second segment, second segment distinctly longer than third segment. Stylocerite reaching 0.85 length of basal segment of antennular peduncle. Scaphocerite 3.8 times as long as wide.

Incisor process of mandible ending in irregular teeth, molar process truncated. Lower lacinia of maxillule broadly rounded, upper lacinia elongate, several distinct teeth on inner margin, palp slender. Upper endites of maxilla subdivided, palp short, scaphognathite tapering posteriorly with some long, curved setae at posterior end. Palp of first maxilliped broadly triangular. Second maxilliped typical of genus. Third maxilliped reaching to end of antennular peduncle, with ultimate segment as long as penultimate segment.

Epipods on first four pereiopods. First pereiopod reaching to distal end of eyestalk; merus 2.0 times as long as broad, as long as carpus; carpus excavated anteriorly, shorter than chela, 1.1 times as long as high; chela 2.0 times as long as broad; fingers shorter than palm. Second pereiopod reaching to middle of second segment of antennular peduncle; merus as long as carpus, 4.0 times as long as broad; carpus 1.1 times as long as chela, 4.0 times as long as high; chela 2.6 times as long as broad; fingers 1.2 times as long as palm. Third pereiopod reaching to end of antennular peduncle, propodus 10 times as long as broad, 4.5 times as long as dactylus; dactylus 2.8 times as long as wide (spines included), terminating in one claw, with 5-6 accessory spines on its flexor margin. Fifth pereiopod reaching to end of second segment of antennular peduncle, propodus 13 times as long as broad, 4.0 times as long as dactylus, dactylus 3.6 times as long as wide (spinules included), terminating in one claw, with 33-35 spinules on its flexor margin. Endopod of male first pleopod subtriangular, 2.3 times as long as wide, 2/5 length of exopod, appendix interna exceed end of endopod by half of its length. Appendix masculina of male second pleopod about half length of endopod, with appendix interna 2/5 length of appendix masculina.

Uropodal diaeresis with 14-17 (mode 15) movable spinules.

Ovigerous females with egg sized 0.85-1.05 × 055-0.70 mm in diameter.

Habitat. — In the original description of *Caridina lanceifrons*, Yu (1936) mentioned that the type specimens were collected from "near the light house at Hai-kiu-sche in the salt water". Recent collections show that the species is very common in various lowland freshwater habitats of Hainan Island and North Vietnam.

Remarks. — *Caridina lanceifrons* is similar to *C. cantonensis* Yu, 1938, from Guangdong, Hong Kong and Vietnam, but can be distinguished from the latter species by the relatively shorter stylocerite, which never reaches the end of the basal segment of the antennular peduncle (vs. reaches distinctly beyond it in *C. cantonensis*).

Dudgeon (1985) recorded *C. lanceifrons* from Lam Tsuen River, New Territories, Hong Kong, and stated that "*N. serrata* and *C. lanceifrons* were most abundant in the summer and winter respectively...". A recent survey on the distribution of *C. lanceifrons* in Hong Kong shows that it is no longer found in the Lam Tsuen River, but is present in Tin Liu, Sai Kung, New Territories. Liang & Zheng (1988) recorded it in their list of *Caridina* species of Fujian Province, without mentioning a specific location. Zheng (1989) in a faunistic survey of freshwater shrimps of Jianxi River and its tributary in Fujian listed *C. lanceifrons* as one of the most abundant species, occurring in Nanping, Jian'ou, Pucheng, Jianyang, Chong'an, Zhenhe and Songxi Counties. Liang et al. (1993) reported *C. lanceifrons* from Fengfang County of Hunan Province; however, these records could not be ascertained by the present author as no specimens are available. Liang & Zhou (1993) also listed *C. lanceifrons* for Guangxi, China and this record was confirmed by re-examination of two specimens from Shangsi County.

Dang (1975) did not mention *C. lanceifrons* when he described *Caridina flavilineata* from Nam Ha, North Vietnam. Based on the description and illustrations he provided, *C. flavilineata* is most probably identical with *C. lanceifrons* as there are no obvious characters separating it from *C. lanceifrons*. As the type specimens of *C. flavilineata* and other species reported by Dang (1975) appear to be no longer extant (see Cai & Ng, 1999), the identity of *C. flavilineata* remains uncertain. Recent collections from many parts of northern Vietnam show that *C. lanceifrons* is very common in that area. Therefore, the author tentatively follows Li & Liang (2002), who synonymised *C. flavilineata* with *C. lanceifrons*

Distribution. — Known with certainty from China (Hainan, Guangxi, Hong Kong, possibly also Hunan and Fujian) and northern Vietnam.

Caridina propinqua De Man, 1908

Caridina propinqua De Man, 1908a: 227, pl. 19, fig. 6 [type locality: Dhappa, near Calcutta, India].

Caridina propinqua — Kemp, 1915: 309; 1918: 274; Bouvier, 1925: 181, figs. 375, 381; Johnson, 1961: 131, figs. 12-15; De Silva, 1982: 127, fig. 5; Ng & Choy, 1990: 17; Cai & Shokita, 2006a: 247; 2006b: 2150, figs. 10, 11.

Caridina blancoi Chace, 1997: 7, fig. 2 [type locality: Tayabus River, Luzon, Philippines].

Caridina hainanensis Liang & Yan, 1983: 211, fig. 1 [Wenchang County, Hainan Island, China].

Caridina hainanensis — Liang, 2004: 302, fig. 148.

Material examined. — 1 male, cl 2.2 mm, SFU 79-310-1, holotype of *C. hainanensis*, Wenchang County, Hainan Province, China, coll. X. Liang & S. Yan, 14 Apr. 1979; 1 female, cl 4.0 mm, SFU 79-310-2, paratype of *C. hainanensis*, same collection data as for holotype.

Remarks. — Liang & Yan (1983) described *Caridina hainanensis*, on the basis of specimens from Wenchang County, Hainan Island, which was subsequently synonomised with *C. propinqua* by Cai & Shokita (2006b).

Caridina baoting sp. nov.
(figs. 8, 9)

Material examined. — Holotype: female, cl 4.6 mm, diameter of eggs with eye spot 1.4 × 0.8 mm, ZRC 2012.1198, HN 39, stream across Mao'an-Boting road, pH 5.7, 18°38.24′N 109°35.48′E, coll. 4 Dec. 1998; paratypes: 32 males, cl 3.8-4.3 mm, 25 females, cl 3.0-4.9 mm, ZRC 2012.1199, same collection data as for holotype; 17 males, cl 3.9-4.4 mm, 21 females, cl 3.5-4.8 mm, IZAS, same collection data as for holotype.

Description. — Rostrum short, acute, reaching hardly to end of basal segment of antennular peduncle; usually unarmed, rarely armed dorsally with 1-4, ventrally with 1-2 teeth. Antennal spine fused with inferior orbital angle. Pterygostomian margin broadly rounded.

Sixth abdominal somite 0.46 times of carapace length, 1.5 times as long as fifth somite, slightly shorter than telson. Telson 2.7 times as long as wide,

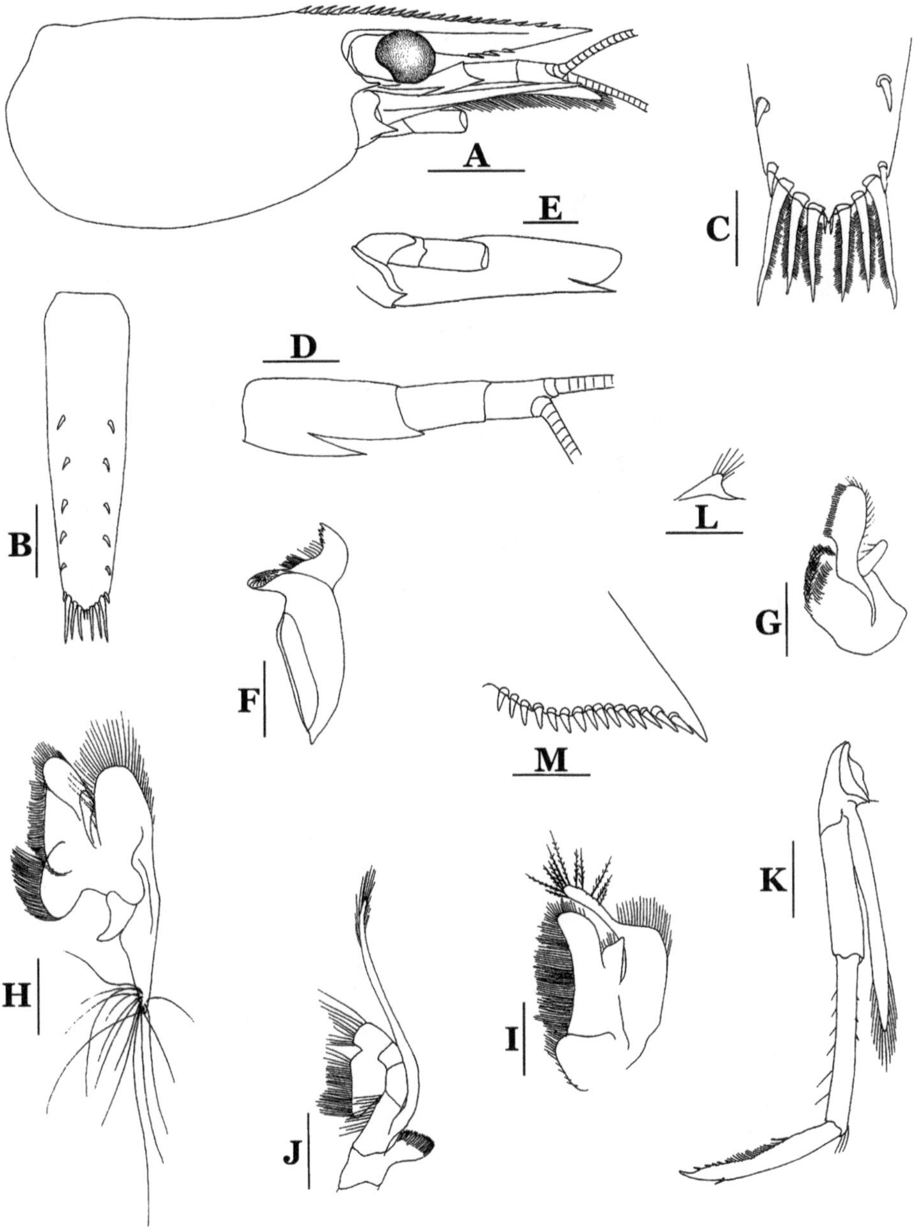

Fig. 6. *Caridina lanceifrons* Yu, 1936, male, cl 3.4 mm, ZRC 2012.1169, Hainan, China. A, cephalothorax and cephalic appendages; B, telson; C, distal portion of telson; D, antennular peduncle; E, scaphocerite; F, mandible; G, maxillula; H, maxilla; I, first maxilliped; J, second maxilliped; K, third maxilliped; L, preanal carina; M, uropodal diaeresis. Scales: A = 1 mm; B, D-L = 0.5 mm, C, M = 0.2 mm.

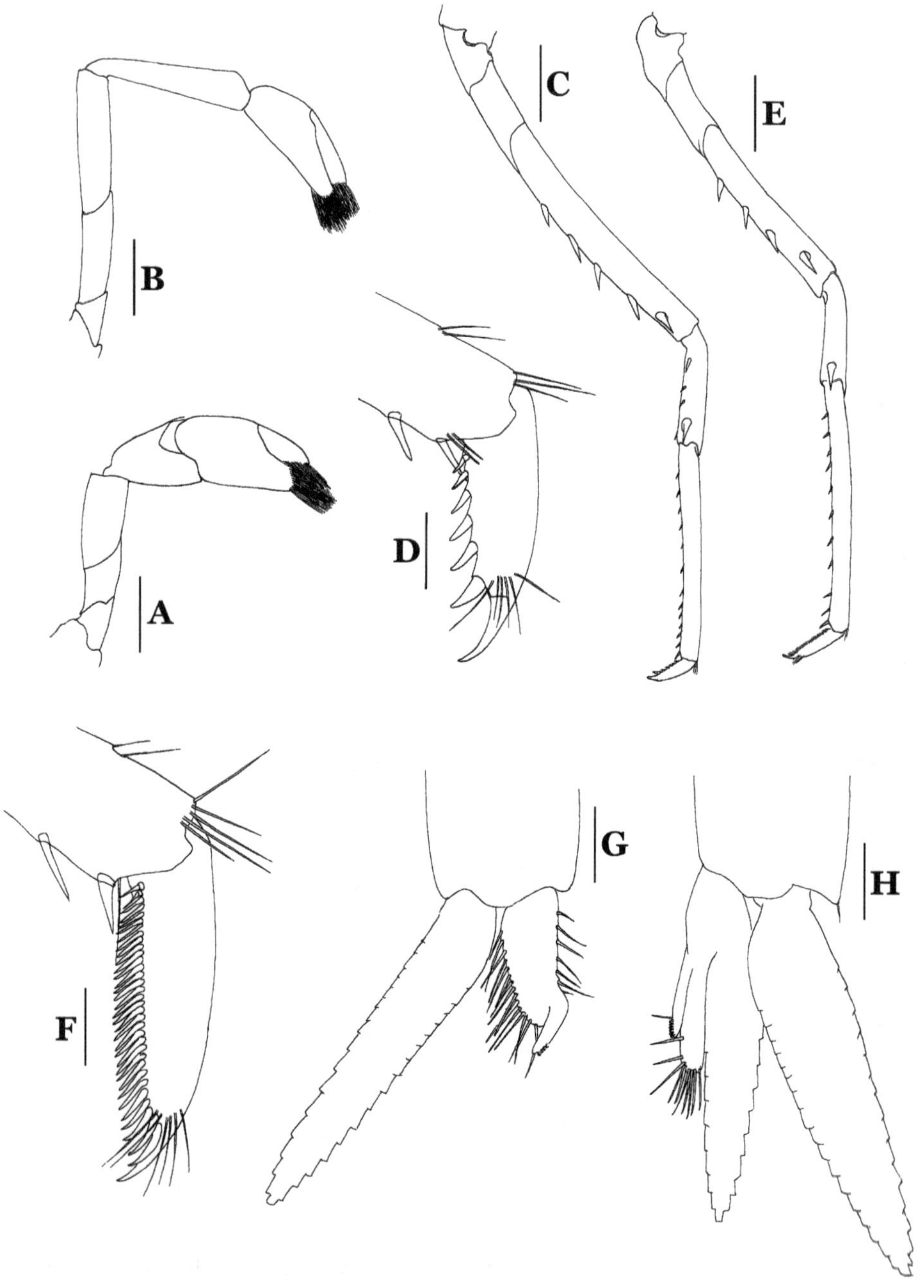

Fig. 7. *Caridina lanceifrons* Yu, 1936, male, cl 3.4 mm, ZRC 2012.1169, Hainan, China. A, first pereiopod; B, second pereiopod; C, third pereiopod; D, dactylus of third pereiopod; E, fifth pereiopod; F, dactylus of fifth pereiopod; G, male first pleopod; H, male second pleopod. Scales: A-C, E = 0.5 mm; D, F = 0.1 mm; G, H = 0.2 mm.

terminating in projection, with 5 pairs dorsal spinules, one pair dorsolateral spinules; lateral pair of spines distinctly longer than intermediate spines. Preanal carina high, lacking spine.

Eyes well developed, anterior end reaching to 0.8 times length of basal segment of antennular peduncle. Antennular peduncle 0.6 times as long as carapace; basal segment of antennular peduncle longer than sum of length of second and third segments, anterolateral angle reaching 0.2 length of the second segment, second segment distinctly longer than third segment. Stylocerite reaching 0.9-1.0 length of basal segment of antennular peduncle. Scaphocerite 3.4 times as long as wide.

Incisor process of mandible ending in irregular teeth, molar process truncated. Lower lacinia of maxillule broadly rounded, upper lacinia elongate, with several distinct teeth on inner margin, palp slender. Upper endites of maxilla subdivided, palp short, scaphognathite tapering posteriorly with some long, curved setae at posterior end. Palp of first maxilliped broadly triangular. Second maxilliped typical of genus. Third maxilliped reaching to end of antennular peduncle, with ultimate segment slightly long as penultimate segment.

Epipods on first four pereiopods. First pereiopod reaching to distal end of eyestalk; merus 2.4 times as long as broad, as long as carpus; carpus excavated anteriorly, shorter than chela, 1.7 times as long as high; chela 2.2 times as long as broad; fingers as long as palm. Second pereiopod reaching to end of scaphocerite; merus shorter than carpus, 4.8 times as long as broad; carpus 1.1 times as long as chela, 4.9 times as long as high; chela 2.4 times as long as broad; fingers 1.3 times as long as palm. Third pereiopod reaching beyond end of scaphocerite by entire dactylus, propodus 10 times as long as broad, 4.3 times as long as dactylus; dactylus 2.8 times as long as wide (spines included), terminating in one claw, with 5 accessory spines on flexor margin. Fifth pereiopod reaching to end of second segment of antennular peduncle, propodus 13 times as long as broad, 3.8 times as long as dactylus, dactylus 3.9 times as long as wide (spinules included), terminating in one claw, with 40-43 spinules on flexor margin.

Endopod of male first pleopod sub-rectangular, 2.8 times as long as wide, half length of exopod, appendix interna slightly exceed distal end of endopod. Appendix masculina of male second pleopod about half length of endopod, with appendix interna 2/5 length of appendix masculina.

Uropodal diaeresis with 20-23 movable spinules.

Ovigerous females with eggs sized 1.40 × 0.80 mm in diameter.

Habitat. — Mountain streams.

Etymology. — The species is named after the type locality, Baoting County, Hainan Island, southern China; used as a noun in apposition.

Remarks. — With regard to the short rostrum and the long stylocerite, *Caridina baoting* sp. nov. shows a great resemblance with *C. breviata* Ng & Cai, 2000, a species described from Yangjiang County of Guangdong Province, mainland China. However, *C. baoting* sp. nov. differs from *C. breviata* by the longer antennular peduncle (0.6 times as long as carapace vs. 0.4 times in *C. breviata*); the more slender endopod of the male first pleopod (2.8 times as long as wide vs. 2.1 times in *C. breviata*), and the larger egg size (diameter 1.40 × 0.80 mm vs. 0.80-0.90 × 0.53-0.56 mm).

Distribution. — Presently known only from the type locality, Baoting County, Hainan Island, southern China.

Caridina clinata Cai, Nguyen & Ng, 1999
(figs. 10, 11)

Caridina clinata Cai, Nguyen & Ng, 1999: 531-535, figs. 1, 2 [type locality: Cuc Phuong National Park, Ninh Binh Province, northern Vietnam].

Material examined. — 7 males, cl 3.5-4.5 mm, 11 females, cl 3.9-4.9 mm, ZRC 2012.1195, HN50, stream across Tongzha-Wuzhi Mountain road, pH 6.7, 5 Dec. 1998; 9 males, cl 3.8-4.5 mm, 16 females, cl 4.3-5.5 mm, ZRC 2012.1196, HN55, stream in Wuzhi Mountain, Zhacaoxi Bridge, 18°54.15′N 109°37.36′E, pH 6.5, coll. 5 Dec. 1998; 6 males, cl 3.8-5.0 mm, 14 females, cl 3.4-5.3 mm, ZRC 2012.1197, HN56, stream in Wuzhi Mountain, Zhashun Bridge, 18°54.09′N 109°37.09′E, pH 6.7, coll. 5 Dec. 1998.

Description. — Rostrum reaching slightly beyond end of basal segment of antennular peduncle, or near end of second segment; sloping ventral anteriorly, rostral formula: 3-5 + 11-14/1-3 (mode 2). Antennal spine fused with inferior orbital angle. Pterygostomian angle broadly rounded.

Sixth abdominal somite 0.52 times of carapace length, 1.7 times as long as fifth somite, shorter than telson. Telson 3.0 times as long as wide, not terminating in projection, with 5 pairs dorsal spinules, one pair dorsolateral spinules; length of lateral pair of spines subequal to intermediate pairs of spiniform setae. Preanal carina high, lacking spine.

Eyes well developed, anterior end reaching to 0.7 times length of basal segment of antennular peduncle. Antennular peduncle 0.63 times as long as carapace; basal segment of antennular peduncle longer than sum of length of second and third segments, anterolateral angle reaching 0.25 length of second segment, second segment distinctly longer than third segment. Stylocerite

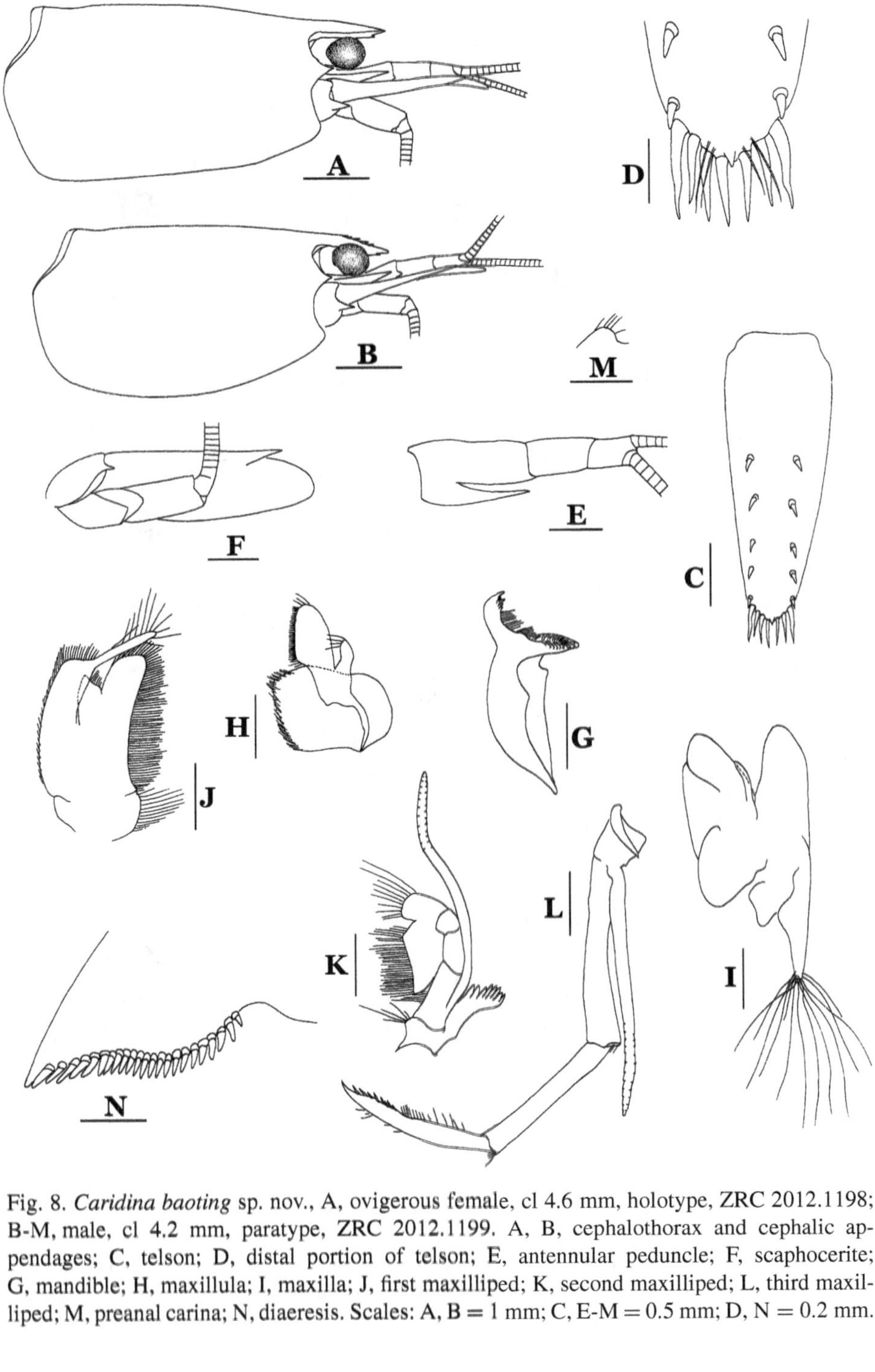

Fig. 8. *Caridina baoting* sp. nov., A, ovigerous female, cl 4.6 mm, holotype, ZRC 2012.1198; B-M, male, cl 4.2 mm, paratype, ZRC 2012.1199. A, B, cephalothorax and cephalic appendages; C, telson; D, distal portion of telson; E, antennular peduncle; F, scaphocerite; G, mandible; H, maxillula; I, maxilla; J, first maxilliped; K, second maxilliped; L, third maxilliped; M, preanal carina; N, diaeresis. Scales: A, B = 1 mm; C, E-M = 0.5 mm; D, N = 0.2 mm.

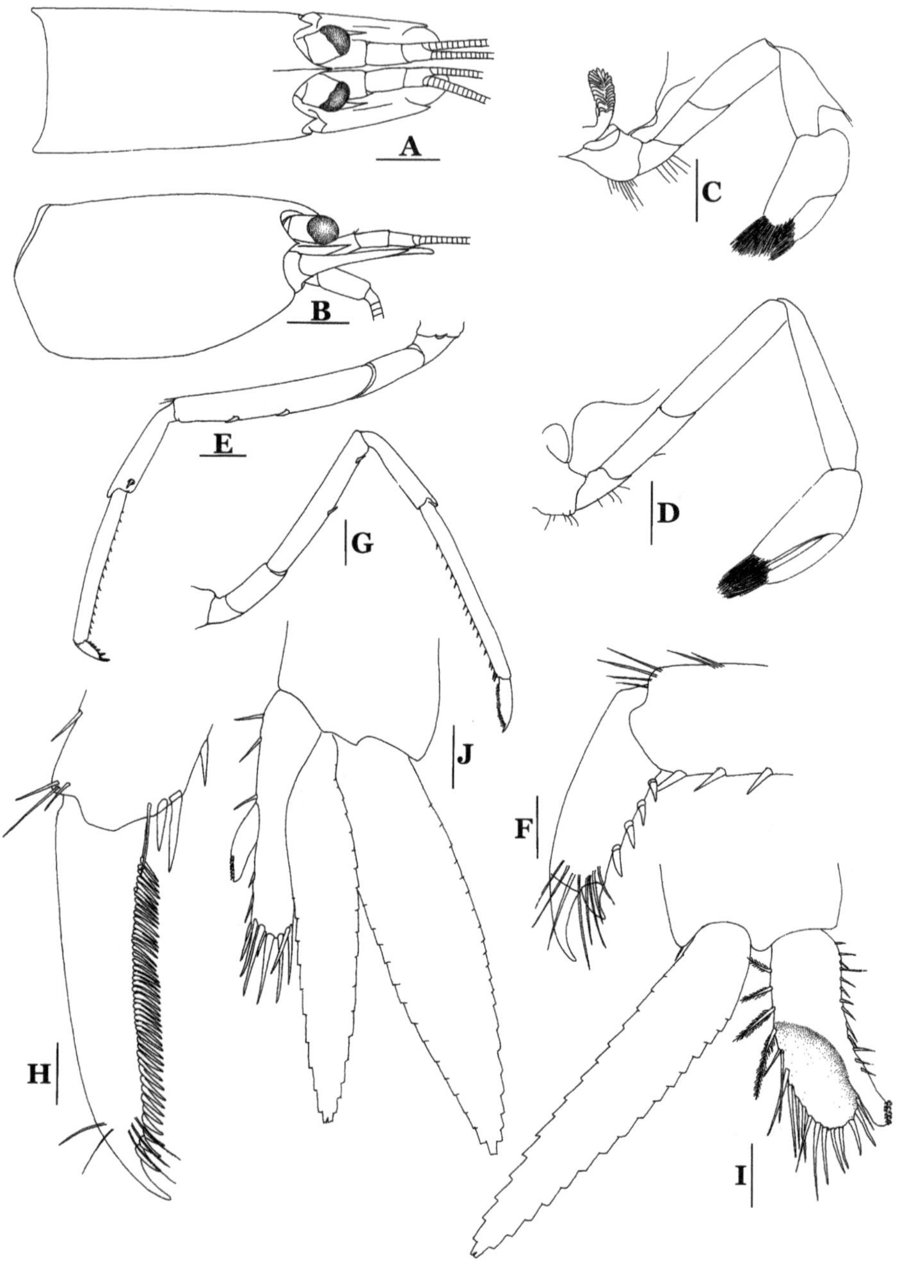

Fig. 9. *Caridina baoting* sp. nov., A, B, male, cl 4.0 mm; C-J, male, 4.2 mm, paratypes, ZRC 2012.1199. A, B, cephalothorax and cephalic appendages; C, first pereiopod; D, second pereiopod; E, third pereiopod; F, dactylus of third pereiopod; G, fifth pereiopod; H, dactylus of fifth pereiopod; I, male first pleopod; J, male second pleopod. Scales: A, B = 1 mm; C-E, G = 0.5 mm; F, H = 0.1 mm; I, J = 0.2 mm.

reaching 0.8 length of basal segment of antennular peduncle. Scaphocerite 2.8 times as long as wide.

Incisor process of mandible ending in irregular teeth, molar process truncated. Lower lacinia of maxillule broadly rounded, upper lacinia elongate, with several distinct teeth on inner margin, palp slender. Upper endites of maxilla subdivided, palp short, scaphognathite tapering posteriorly with some long, curved setae at posterior end. Palp of first maxilliped broadly triangular, ending in finger-like projection. Second maxilliped typical of genus. Third maxilliped reaching to end of antennular peduncle, with ultimate segment shorter than penultimate segment.

Epipods on first four pereiopods. First pereiopod reaching to distal end of eyestalk; merus 2.0 times as long as broad, as long as carpus; carpus excavated anteriorly, as long as chela, 1.5 times as long as high; chela 2.3 time as long as broad; fingers as long as than palm. Second pereiopod reaching to end of second segment of antennular peduncle; merus as long as carpus, 3.8 times as long as broad; carpus as long as chela, 3.5 times as long as high; chela 2.6 times as long as broad; fingers 1.3 times as long as palm. Third pereiopod reaching to end of scaphocerite, propodus 8-9 times as long as broad, 4.5 times as long as dactylus; dactylus 2.5 times as long as wide (spines included), terminating in one claw, with 5 accessory spines on flexor margin. Fifth pereiopod reaching to end of second segment of antennular peduncle, propodus 11 times as long as broad, 4.6 times as long as dactylus, dactylus 4.1 times as long as wide (spinules included), terminating in one claw, with 28-31 spinules on its flexor margin. Endopod of male first pleopod subtriangular, 2.1 times as long as wide, 1/3 length of exopod, appendix interna exceed end of endopod by half of its length. Appendix masculina of male second pleopod about half length of endopod, with appendix interna 1/4 length of appendix masculina.

Uropodal diaeresis with 14-17 movable spinules.

Habitat. — Mountain streams on the way from Tongzha to Wuzhishan Mountain, the summit of Hainan Island.

Remarks. — This is the first record of *Caridina clinata* for China. In Vietnam, the species is known only from the type series. In Hainan, *C. clinata* is morphologically most similar to *C. lanceifrons*, but can be distinguished by the shorter rostrum, which reaches from slightly beyond the end of the basal segment of the antennular peduncle almost to the end of the second segment (vs. reaches beyond the end of the antennular peduncle in *C. lanceifrons*); the palp of the first maxilliped ending in a finger-like process (this process is lacking in *C. lanceifrons*), the broader scaphocerite (2.8 times as long as

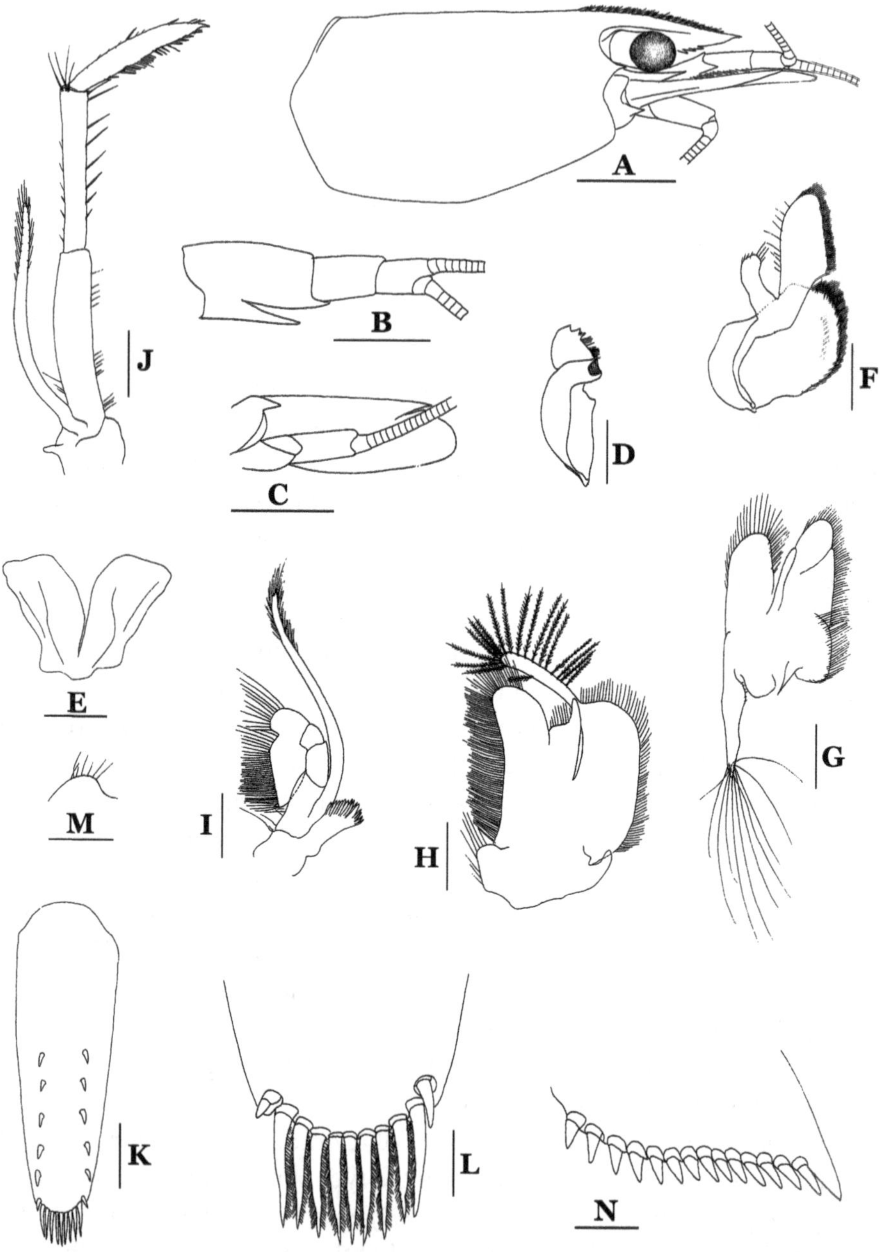

Fig. 10. *Caridina clinata* Cai et al., 1999, female, cl 5.3 mm, ZRC 2012.1197. A, cephalothorax and cephalic appendages; B, antennular peduncle; C, scaphocerite; D, mandible; E, paragnathus; F, maxillula; G, maxilla; H, first maxilliped; I, second maxilliped; J, third maxilliped; K, telson; L, distal portion of telson; M, preanal carina; N, uropodal diaeresis. Scales: A-C = 2 mm; D-K, M = 0.5 mm; L, N = 0.2 mm.

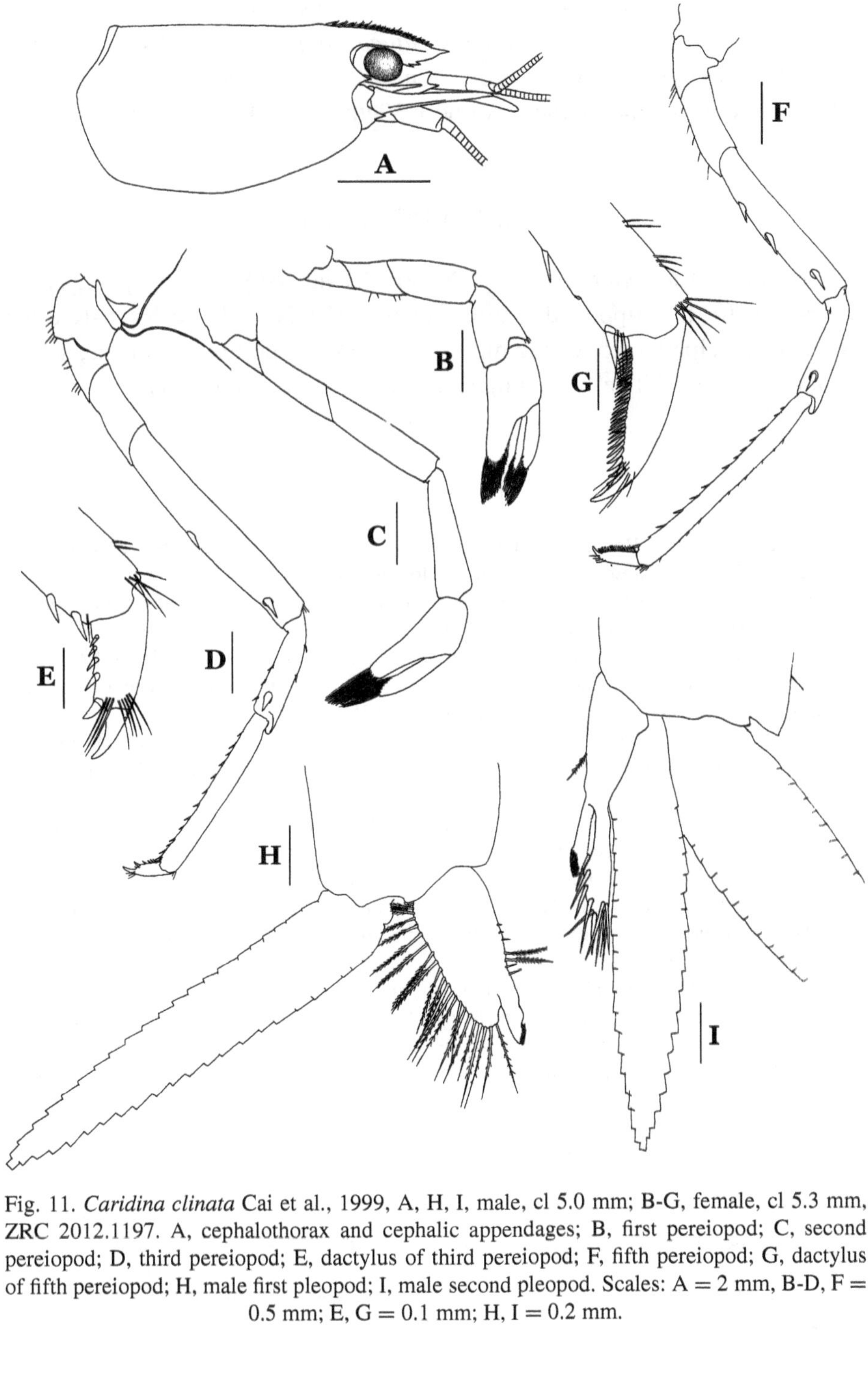

Fig. 11. *Caridina clinata* Cai et al., 1999, A, H, I, male, cl 5.0 mm; B-G, female, cl 5.3 mm, ZRC 2012.1197. A, cephalothorax and cephalic appendages; B, first pereiopod; C, second pereiopod; D, third pereiopod; E, dactylus of third pereiopod; F, fifth pereiopod; G, dactylus of fifth pereiopod; H, male first pleopod; I, male second pleopod. Scales: A = 2 mm, B-D, F = 0.5 mm; E, G = 0.1 mm; H, I = 0.2 mm.

wide vs. 3.8 times as long as wide in *C. lanceifrons*); and the wider telson (3.0 times as long as wide vs. 3.8 times in *C. lanceifrons*), which lacks a projection on the distal margin (vs. with a projection in *C. lanceifrons*).

Distribution. — Cuc Phuong, northern Vietnam and Hainan Island, China.

ACKNOWLEDGEMENTS

I am grateful to Peter K. L. Ng (National University of Singapore, NUS, Singapore) for his support during the course of this study, to Ngan-Kee Ng (NUS) who helped in collecting most of the herein reported specimens, and to Arthur Anker for critically reading an early version of the manuscript.

REFERENCES

BOUVIER, E. L., 1925. Recherches sur la morphologie, la geographique des crevettes des la famille des Atyides. Encyclopédie Entomologique, **4**(A): 1-370, figs. 1-761.

CAI, Y. & N. K. NG, 1999. A revision of *Caridina serrata* species group, with descriptions of five new species. Journal of Natural History, **33**: 1603-1638.

CAI, Y., X. Q. NGUYEN & P. K. L. NG, 1999. *Caridina clinata*, a new species of freshwater shrimp (Crustacea: Decapoda: Atyidae) from Northern Vietnam. Proceedings of the Biological Society of Washington, **112**(3): 531-535.

CAI, Y. & S. SHOKITA, 2006a. Report on a freshwater shrimp collection (Crustacea: Decapoda: Caridea) from Philippines, with descriptions of four new species. Raffles Bulletin of Zoology, **54**(2): 245-270.

— — & — —, 2006b. Atyid shrimps of the Ryukyu Islands, Southern Japan (Crustacea: Decapoda: Atyidae). Journal of Natural History, **40**(38-40): 2123-2172.

CHACE, F. A. JR., 1997. The caridean shrimps (Crustacea: Decapoda) of the Albatross Philippine expedition 1907-1910. Part 7: families Atyidae, Eugonatonotidae, Rhynchocinetidae, Bathypalaemonellidae, Processidae and Hippolytidae. Smithsonian Contributions to Zoology, **58**: 1-106.

DANG, N. T., 1975. Phan loai tom cua nuoc ngot mien bac vietnam (The identities of North Vietnamese freshweater shrimps and crabs). Tap san Sinh Vat-Dia Hoc (Journal of Biology and Geology), **13**(3): 65-78, figs. 1-8. [In Vietnamese with French summary.]

— —, 1980. Dinh loai dong khong xuong song nuoc noot bac Viet Nam (The identities of freshwater invertebrates of North Vietnam). Decapoda (Ha Noi, Vietnam): 380-418, figs. 219-236. [In Vietnamese.]

DE MAN, J. G., 1892. Decapoden des Indischen Archipels. In: M. WEBER (ed.), Zoologische Ergebnisse einer Reise in Niederlandisch Ost-Indien, **2**: 265-527, pls. 15-29.

— —, 1908a. The fauna of brackish ponds at port Canning, Lower Bengal. Part 10: Decapod Crustacea, with an account of a small collection from brackish water near Calcutta and in the Dacca District, Eastern Bengal. Records of the Indian Museum, **2**: 211-231, figs. 1-7.

— —, 1908b. On the *Caridina nilotica* (Roux) and its varieties. Records of the Indian Museum, **2**: 255-283, pls. 1-20.

DUDGEON, D., 1985. The population dynamics of some freshwater carideans (Crustacea: Decapoda) in Hong Kong, with special reference to *Neocaridina serrata* (Atyidae). Hydrobiologia, **120**: 141-149.

KEMP, S., 1915. Crustacea Decapoda. Fauna of the Chilka Lake. Memoirs of the Indian Museum, **5**: 199-325.

— —, 1918a. Zoological results of a tour in the Far East, Crustacea Decapoda Stomatopoda. Memoirs of the Asiatic Society of Bengal, **6**: 219-297.

— —, 1918b. Crustacea Decapoda of the Inle Lake Basin. Records of the Indian Museum, **14**: 81-102, figs. 1-3, pls. 24-25.

LI, S.-Q. & X.-Q. LIANG, 2002. Caridean prawns of northern Vietnam (Decapoda: Atyidae, Palaemonidae). Acta Zootaxonomica Sinica, **27**(4): 707-716.

LIANG, X., Z. L. GUO & J. GAO, 1993. Study on *Caridina* (Crustacea, Decapoda) from Hunan, China. Journal of Shanghai Fisheries University, **2**(1): 41-47, figs. 1-3. [In Chinese.]

LIANG, X. & J. ZHOU, 1993. Study on new atyid shrimps (Decapoda, Caridea) from Guangxi, China. Acta Hydrobiologica Sinica, **17**(3): 231-239, figs. 1-4. [In Chinese with English abstract.]

LIANG, X.-Q., 2004. Fauna Sinica: Invertebrata 36. Crustacea: Decapoda: Atyidae: 1-375. [In Chinese.] (Science Press, Beijing).

LIANG, X.-Q. & S.-L. YAN, 1977. New species and subspecies of *Caridina* (Decapoda, Caridea) from Fukien, China. Acta Hydrobiologica Sinica, **6**(2): 219-225, figs. 1-13.

— — & — —, 1983. New species and new records of freshwater shrimps (Crustacea Decapoda) from Hainan Island, China. Oceanologia et Limnologia Sinica, **14**(3): 211-216, figs. 1-3.

LIANG, X.-Q. & M.-Q., ZHENG, 1988. Notes on *Caridina* from Fujian, China. Acta Zootaxonomica Sinica, **13**(1): 15-19, figs. 1-9.

MILNE EDWARDS, H., 1837. Histoire naturalle des Crustacés, Comprenant l'Anatomie, la Physiologie et la Classification de ces Animaux. Volume 2: 1-532. (Libraire Encyclopedique de Roret, Paris).

NG, N. K. & Y. CAI, 2000. Two new species of atyid shrimps from southern China. Raffles Bulletin of Zoology, **48**(1): 167-175.

WOWOR, D., Y. CAI & P. K. L. NG, 2004. Crustacea: Decapoda: Caridea. In: C. YULE & H. S. YONG (eds.), The freshwater invertebrates of Malaysia and Singapore: 337-357. (Malaysian Academy of Sciences, Kuala Lumpur).

YU, S. C., 1936. Report on the macrurous Crustacea collected during the "Hainan Biological Expedition" in 1934. Chinese Journal of Zoology, **2**: 85-99, figs. 1-7.

— —, 1938. Study on Chinese *Caridina* with descriptions of five species. Bulletin Fan Memorial Institute of Biology, Zoology Series, **8**(3): 271-310, figs. 1-16.

ZHENG, M.-Q., 1989. Freshwater shrimps of the Jianxi River and its tributary. Chinese Journal of Zoology, **24**(6): 7-11, fig. 1.

First received 20 January 2012.
Final version accepted 29 July 2013.

ON THE PRESUMED PHYLOGENETIC POSITION OF THE XIPHOCARIDIDAE (DECAPODA, CARIDEA) BASED ON THE LARVAL MORPHOLOGY OF *XIPHOCARIS ELONGATA*

GUILLERMO GUERAO[1]), SILKE REUSCHEL[2]), KLAUS ANGER[3]) and CHRISTOPH D. SCHUBART[2,4])

[1]) IRTA, Carretera Poble Nou, km 5.5, 43540 Sant Carles de la Ràpita, Catalunya, Spain

[2]) Biologie I, Universität Regensburg, 93040 Regensburg, Germany

[3]) Alfred-Wegener-Institut für Polar- und Meeresforschung; Biologische Anstalt Helgoland, Meeresstation, 27498 Helgoland, Germany

ABSTRACT

The first zoeal stage of the freshwater shrimp *Xiphocaris elongata* from Jamaica is described and illustrated in detail from laboratory-hatched material. Newly hatched zoeae of *X. elongata* possess morphological characteristics (e.g., presence of maxillular exopod, with 3 long terminal plumose setae; a maxillular endopod with 2, 2 setae; an exopod of the maxillipeds with 4 long terminal plumose setae; a bilobed telson with a strong median indentation) that are shared by the larvae of Atyidae that have extended planktonic development. Comparisons with other caridean families support earlier classifications and recent phylogenetic insights that the Xiphocarididae is the sister group of the Atyidae.

INTRODUCTION

Freshwater shrimps of the Greater Antilles are taxonomically and ecologically diverse, comprising eight genera in three families: Atyidae (*Atya*, *Jonga*, *Micratya*, *Potimirim*, *Typhlatya*), Palaemonidae (*Macrobrachium*, *Troglocubanus*) and Xiphocarididae (*Xiphocaris*) (Chace & Hobbs, 1969). *Typhlatya*, *Jonga* and *Micratya* are all endemic to the Caribbean region, whereas species of *Potimirim* and *Atya* are also found in South America (Hunte, 1978; Almeida et al., 2008), and other species of *Atya* occur in West Africa and the Cape Verde Islands (Hobbs & Hart, 1982).

[4]) Corresponding author; e-mail: christoph.schubart@biologie.uni-regensburg.de

For a long time, *Xiphocaris* was considered to be a basal member of the caridean family Atyidae, because of its ancestral morphological characters (Bouvier, 1925; Christoffersen, 1990) and its preference for freshwater habitats. Chace (1992) grouped the Xiphocarididae Ortmann, 1895, within the Nematocarcinoidea, a classification that has been adopted by a number of authors (Holthuis, 1993; Martin & Davis, 2001; De Grave et al., 2009; De Grave & Fransen, 2011). However, molecular studies by Bracken et al. (2009) and Li et al. (2011) indicate that the superfamily Nematocarcinoidea may not be a natural (monophyletic) group which casts doubt on the continued inclusion of the xiphocaridids in this superfamily. In addition, recent molecular and morphological analyses found a close phylogenetic relationship between xiphocaridids and atyids (Page et al., 2008), as have other studies that have examined the relationships between the xiphocaridids, atyids, and nematocarcinids (Bracken et al., 2009; De Grave & Goulding, 2011).

Larval morphology provides phylogenetically informative characters in Crustacea (Marques & Pohle, 1995; Waloßek, 1995; Clark, 2009; Hultgren et al., 2009) and has been used to support taxonomic changes in Brachyura (see Schubart et al., 2002; Cuesta & Schubart, 2007; Schubart & Cuesta, 2010).

Larval development in freshwater atyid shrimps has been described for many species (Benzie, 1982; Benzie & Silva, 1983; Salman, 1987; Walsh, 1993; Shy et al., 2001; Lai & Shy, 2009), including a number of those from the neotropics: *Atya lanipes* by Hunte (1975), *A. innocous* by Hunte (1979a) and *Micratya poeyi* by Hunte (1979b). All these species undergo an extended larval development in the plankton (Hunte, 1977, 1979a, b).

The present study describes the morphology of the first larval stage of *Xiphocaris elongata* (Guerin-Meneville, 1855) from Jamaica, and discusses the implications of these findings for its relationships with other freshwater shrimp taxa.

MATERIAL AND METHODS

Ovigerous females of *Xiphocaris elongata* were collected in the Great River in western Jamaica (Westmoreland) at the bridge in Marchmont (18°15.527′N 77°52.712′W, 203 m a.s.l.) on 23 October 2005 by SR and CDS. Other freshwater decapods found at the same locality were *Sesarma fossarum* (Brachyura: Sesarmidae), *Potimirim potimirim*, *Atya innocous* (both Caridea: Atyidae) and the introduced Australian crayfish *Cherax quadricarinatus* (Astacidea: Parastacidae), which is a first-time record for this species in any Jamaican river draining to the north. Ovigerous shrimp were transported live to the University of Regensburg, and then to the Biologische Anstalt Helgoland where they were maintained in 5 l freshwater aquaria at 24°C until the larvae hatched. A sample of larvae in the zoea I stage were fixed in ethanol, but the rest of the larvae died before reaching the next stage.

Five specimens of zoea I were dissected for morphological descriptions using a Nikon SMZ800 stereo microscope equipped with an image analyzing system (AnalySIS, SIS, Münster, Germany). An Olympus BH-2 microscope was used for detailed morphological examination of appendages, after mounting samples in a polyvinyl medium.

The following measurements were taken: total length (TL): distance from the tip of the rostrum to the posterior dorsal margin of the telson; cephalothorax length (CL): distance from the tip of the rostrum to the posterior margin of the carapace.

TAXONOMY

Family XIPHOCARIDIDAE Ortmann, 1895
Genus **Xiphocaris** von Martens, 1872
Xiphocaris elongata (Guérin-Méneville, 1855)
(figs. 1, 2)

Description. — Zoea I. TL = 2.2 ± 0.1 mm; CL = 0.68 ± 0.03 mm; n = 5.

Carapace (fig. 1A, B). Rostrum short, directed slightly downward, exceeding anterior margin of sessile eyes. Pterygostomial spine well-developed.

Antennule (fig. 1C). Peduncle incipiently 2-segmented, without setae. Inner flagellum represented by a long plumose seta; outer flagellum unsegmented with 3 terminal aesthetascs and one seta plus one subterminal plumose seta.

Antenna (fig. 1D). Biramous, peduncle (protopod) with strong distal spine, about 1/2 length of endopod. Endopod slender, slightly more than 1/2 length of exopod and with 2 terminal setae (one long and plumose). Exopod (scaphocerite) 6-segmented with 3, 2, 1, 1, 2, 3 setae respectively.

Mandible (fig. 1E). Incisor and molar process without teeth. One minute spine (lacinia mobilis) between incisor and molar processes.

Maxillule (fig. 1F). Coxal and basial endites with 5 and 4 setae respectively. Endopod partially 2-segmented, with 2, 2 setae. Exopod minute with 3 long terminal plumose setae.

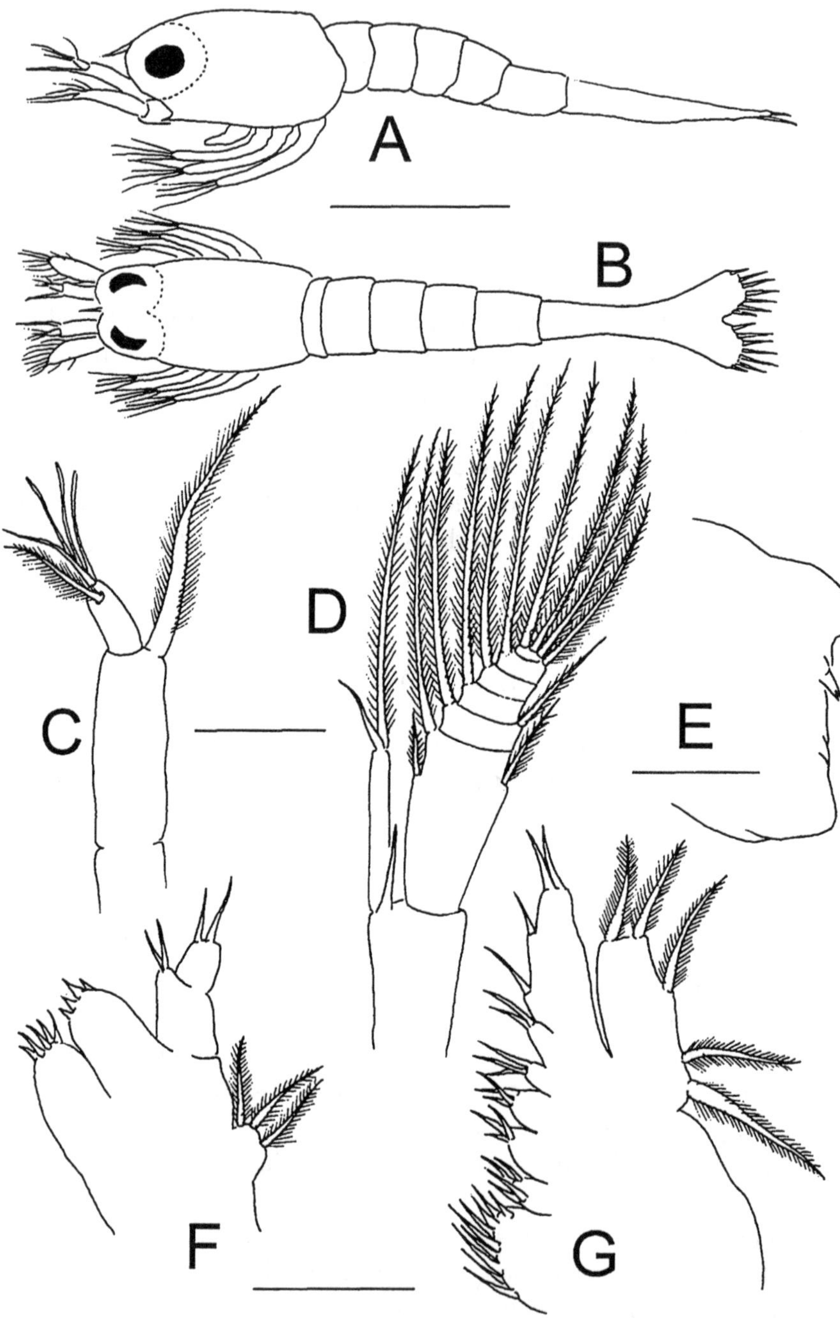

Fig. 1. *Xiphocaris elongata.* Zoea I. A, lateral view; B, dorsal view; C, antennule; D, antenna; E, mandible; F, maxillule; G, maxilla. Scale bars of A and B = 500 μm; of C and D = 100 μm; of E = 20 μm; of F and G = 50 μm.

Maxilla (fig. 1G). Coxa bilobed, with 9 + 4 setae. Basial endite bilobed with 3 + 3 setae. Endopod unsegmented, with 3 + 2 + 1 + 1 + 2 setae. Exopod (scaphognathite) with 5 plumose marginal setae.

First maxilliped (fig. 2A). Coxa and basis with 2 and 8 setae on inner margin respectively. Endopod 4-segmented, with 3, 1, 2, 1 + 3 setae. Exopod incompletely segmented with 4 terminal long plumose natatory setae.

Second maxilliped (fig. 2B). Coxa and basis with 1 and 6 setae on inner margin respectively. Endopod 4-segmented, with 2, 1, 2, 1 + 4 setae. Exopod incompletely segmented with 4 terminal long plumose natatory setae.

Third maxilliped (fig. 2C). Protopod with 2 inner setae. Endopod 4-segmented, with 2, 0-1, 2, 1 + 3 setae. Exopod incompletely segmented with 4 terminal long plumose natatory setae and 1-2 subterminal short plumose setae.

Pereiopods. Absent.

Pleon (fig. 1A, B). Five somites plus telson, without pleopods and uropods.

Telson (fig. 2D, E). Broad posteriorly, with a strong median indentation and 7 + 7 setae, outermost smaller. Posterior margin with minute teeth (fig. 2E).

DISCUSSION

Newly hatched first zoeae of *X. elongata* have the following morphological characteristics, which have also been observed in larvae of Atyidae that have non-abbreviated larval development (table I): (1) antennal exopod (scaphocerite) segmented; (2) maxillular exopod present, with 3 long terminal plumose setae; (3) maxillular endopod 2-segmented with 2, 2 setae; (4) maxilla with coxal endite bilobed; (5) exopod of the maxillipeds with 4 long terminal plumose setae; (6) posterior margin of the telson with a median indentation (telson bilobed). None of these characters is exclusively found in atyid larvae (Hunte, 1979a, b; Hayashi & Hamano, 1984; Walsh, 1993) and thus cannot be considered diagnostic for the family. Therefore, it is only a combination of features rather than a particular character set (synapomorphies) that defines the shared features of atyid larvae.

Especially relevant is the presence of the outer lobe in the maxillula in *X. elongata* (which represents the exopod) (Gurney, 1942). The exopod is absent in most caridean species except Atyidae, Rhynchocinetidae and Nematocarcinidae (table I). First zoeae of Rhynchocinetidae species are similar to those of Atyidae, but present three terminal plumose setae on the exopods of the maxillipeds (Matoba & Shokita, 1998; Maihara & Kyoya, 2001; Maihara,

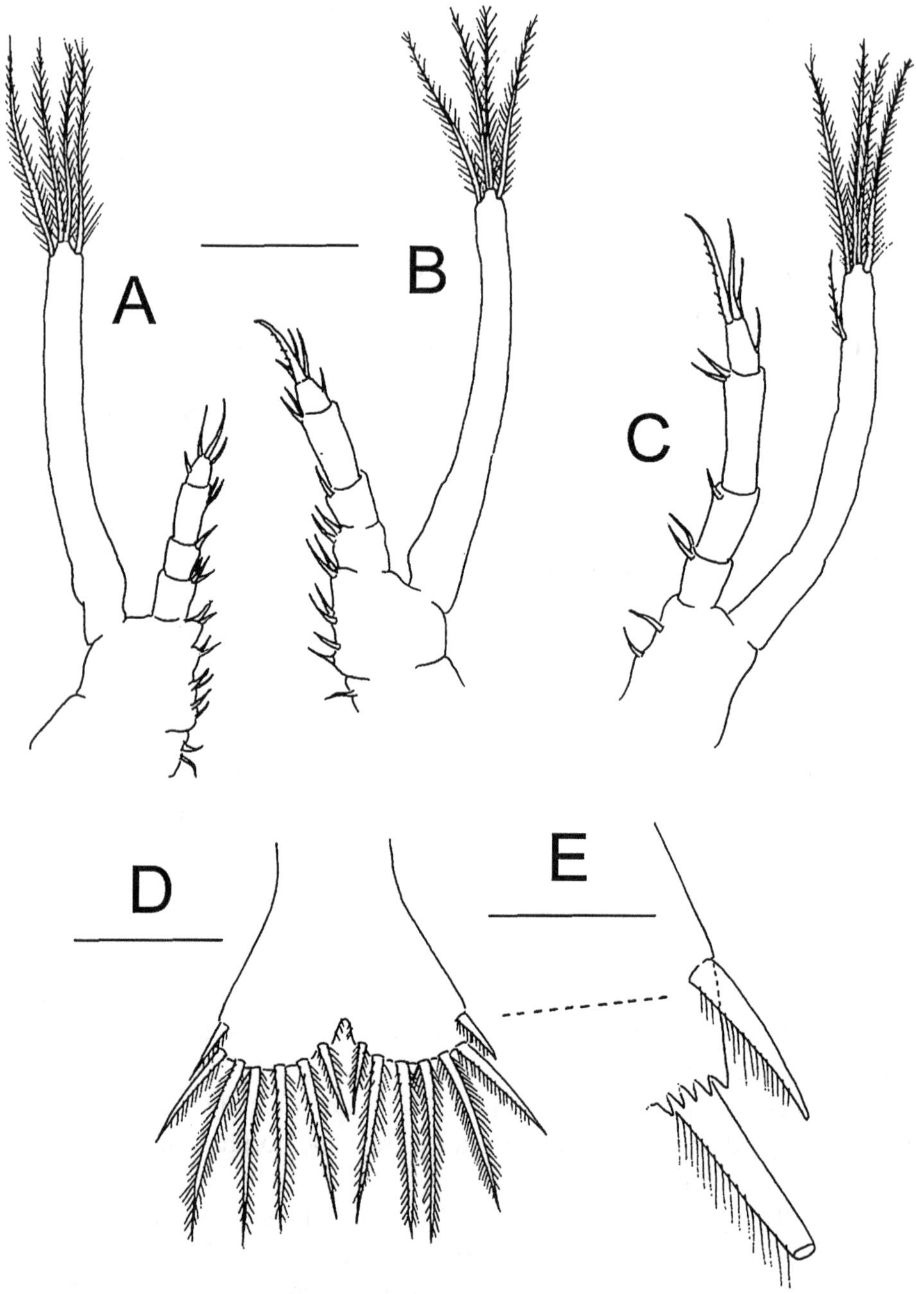

Fig. 2. *Xiphocaris elongata*. Zoea I. A, first maxilliped; B, second maxilliped; C, third maxilliped; D, telson; E, detail of the telson. Scale bars of A-C = 100 μm; of D = 200 μm; of E = 50 μm.

TABLE I

Morphological differences between the first zoeal stage of: *Paratya australiensis* Kemp, 1917 (Atyidae); unidentified Atyidae from Jamaica (unpublished data); *Rhynchocinetes conspiciocellus* Okuno & Takeda, 1992 (Rhynchocinetidae); *Nematocarcinus longirostris* Bate, 1888 (Nematocarcinidae); *Hippolyte leptocerus* (Heller, 1863) (Hippolytidae); *Austropandalus grayi* (Cunningham, 1871) (Pandalidae); *Ogyrides limicola* Williams, 1955 (Ogyrididae); *Processa macrodactyla* Holthuis, 1952 (Processidae); *Philocheras fasciatus* (Risso, 1816) (Crangonidae); *Palaemon adspersus* Rathke, 1837 (Palaemonidae); *Athanas nitescens* (Leach, 1814) (Alpheidae); and *Xiphocaris elongata*. Abbreviations: Exo, exopod; lps, long plumose terminal setae; mi, strong median indentation; mxlpd, maxilliped; s, setae

Species	Antenna Exopod	Maxillule Endopod(s)	Maxillule Exopod	Maxilla Coxal endite	Maxilla Endopod(s)	Mxlpd 1 Exo (lps)	Mxlpd 2 Exo (lps)	Mxlpd 3 Exo (lps)	Telson (mi)	Reference
X. elongata	Segmented	2, 2	Present	Bilobed	3, 2, 1, 1, 2	4	4	4	Present	Present study
P. australiensis	Segmented	2, 2	Present	Bilobed	3, 2, 1, 1, 2	4	4	4	Present	Walsh (1993)
Unidentified Atyidae	Segmented	2, 2	Present	Bilobed	3, 2, 1, 1, 2	4	4	4	Present	Unpublished data
R. conspiciocellus	Segmented	2, 3	Present	Bilobed	4, 2	3	3	3	Present	Matoba & Shokita (1998)
N. longirostris	Segmented	2, 3	Present	Bilobed	3, 2, 1, 2, 2	3	3	3	Absent	Thatje et al. (2005)
H. leptoceros	Segmented	2, 3	Absent	Bilobed	3, 2, 1, 2, 2	3	3	3	Absent	Guerao et al. (2011)
A. grayi	Segmented	2, 3	Absent	Bilobed	3, 2, 1, 1, 2	3	3	3	Present	Thatje & Bacardit (2000)
O. limicola	Segmented	2, 2	Absent	Bilobed	3, 2, 1, 1, 2	4	4	4	Absent	Sandifer (1974)
P. macrodactyla	Unsegmeted	2, 3	Absent	Bilobed	3, 2, 1, 1, 2	3	3	3	Present	Ortega et al. (2005)
P. fasciatus	Unsegmented	2, 3	Absent	Bilobed	3, 2, 1, 2	3	3	3	Present	González-Gordillo & Rodríguez (2000)
P. adspersus	Segmented	1	Absent	Unilobed	2, 1	4	4	4	Absent	Fincham (1985)
A. nitescens	Segmented	2	Absent	Unilobed	4	4	4	4	Absent	Bartilotti et al. (2005)

2002; Dupré et al., 2008) (table I). Nematocarcinidae larvae do not possess a posterior telson margin with a median indentation (Thatje et al., 2005).

This zoeal evidence supports the assignment of the Xiphocarididae as a sister group to the Atyidae. This is in agreement with the findings of Page et al. (2008), Bracken et al. (2009), De Grave & Goulding (2011), and unpublished 16S mtDNA data by CDS.

An abbreviation of larval development is common in freshwater carideans (Gurney, 1942; Anger, 2001). Atyid shrimps exhibit a great deal of variability in the number of zoeal stages, from just one or two stages with the complete suppression of all planktonic life (Benzie & De Silva, 1983; Shy et al., 2001), to between 9-12 zoeal stages (Hunte, 1979a, b). Hayashi & Hamano (1984) divided the atyids into three groups based on their egg size, clutch size, and larval development: (1) species with large eggs, a small clutch size, and abbreviated larval development; (2) species with intermediate numbers of medium-sized eggs, a medium clutch size, and several larval stages; (3) species with small eggs, a large clutch size, and an extended larval development. The morphology of the zoeal stages is related to the duration of zoeal development. In atyids with abbreviated larval development, the first larva hatches in a morphologically more or less advanced condition, with a set of characters that appears only in later stages of species with extended development (Benzie, 1982; Benzie & Silva, 1983; Shy et al., 2001; Lai & Shy, 2009). In contrast, the morphology of the first zoea of *X. elongata* is typical for those species that have an extended larval development and a long planktonic period (Hunte, 1977, 1979a, b; Hayashi & Hamano, 1984; Walsh, 1993; Shy et al., 2001; Lai & Shy, 2009): (1) antennular peduncle unsegmented or 2-segmented, without setae; (2) antennular endopod and exopod unsegmented; (3) antennal endopod unsegmented; (4) antennal exopod segmented; (5) absence of pereiopods (in some species a first pereiopod bud is present); (6) absence of pleopods; (7) telson broad posteriorly, with 7+7 setae. Complete larval development in *X. elongata* requires a long period of time in the marine plankton. This would explain the genetic homogeneity within and among the freshwater shrimp taxa found in the Caribbean islands (Cook et al., 2008; Page et al., 2013).

ACKNOWLEDGEMENTS

This study was financially supported by a six year research grant to Christoph D. Schubart (Schu 1460/3) through the Deutsche Forschungsgemeinschaft within the priority program 1127: "Adaptive Radiation — Origin of Biological Diversity". The work of Guillermo Guerao was supported by the Ministry of Science and Research to GG (post-doctoral fellowship; INIA).

REFERENCES

ALMEIDA, A. O., P. A. COELHO, J. R. LIZ, J. T. ALMEIDA DOS SANTOS & N. R. FERRAZ, 2008. Decapod crustaceans in fresh waters of southeastern Bahia, Brazil. Revista de Biología Tropical, **56**(3): 1225-1254.

ANGER, K., 2001. The biology of decapod crustacean larvae. Crustacean Issues, **14**: i-xii, 1-419. (Balkema Publishers, Lisse, Abingdon, Exton, Tokyo).

BARTILOTTI, C., R. CALADO & A. DOS SANTOS, 2005. Correct diagnosis of early zoeal stages of *Athanas nitescens* (Leach, 1814) (Decapoda, Caridea, Alpheidae) using laboratory-raised larvae. Journal of Plankton Research, **27**(11): 1189-1194.

BENZIE, J. A. H., 1982. The complete larval development of *Caridina mccullochi* Roux, 1926 (Decapoda, Atyidae) reared in the laboratory. Journal of Crustacean Biology, **2**(4): 493-513.

BENZIE, J. A. H. & P. K. DE SILVA, 1983. The abbreviated larval development of *Caridina singhalenssi* Ortmann, 1894 (Decapoda, Atyidae). Journal of Crustacean Biology, **3**(1): 117-126.

BOUVIER, E. L., 1925. Recherches sur la morphologie, les variations et la distribution systématique des crevettes d'eau douce de la famille des atyidés. Encyclopédie Entomologique A, **4**: 1-370.

BRACKEN, H. D., S. DE GRAVE & D. L. FELDER, 2009. Phylogeny of the Infraorder Caridea based on mitochondrial and nuclear genes (Crustacea: Decapoda). In: J. W. MARTIN, K. A. CRANDALL & D. L. FELDER (eds.), Decapod crustacean phylogenetics. Crustacean Issues, **18**: 281-305. (Taylor & Francis/CRC Press, Boca Raton, Florida).

CHACE, F. A. JR., 1992. On the classification of the Caridea (Decapoda). Crustaceana, **63**: 71-80.

CHACE, F. A. JR. & H. H. HOBBS, 1969. The freshwater and terrestrial decapod crustaceans of the West Indies with special reference to Dominica. United States National Museum Bulletin, **292**: 1-298.

CHRISTOFFERSEN, M. L., 1990. A new superfamily classification of the Caridea (Crustacea: Pleocyemata) based on phylogenetic pattern. Journal of Zoological Systematics and Evolutionary Research, **28**(2): 94-106.

CLARK, P., 2009. The bearing of larval morphology on brachyuran phylogeny. In: J. W. MARTIN, K. A. CRANDALL & D. L. FELDER (eds.), Decapod crustacean phylogenetics. Crustacean Issues, **18**: 221-241. (Taylor & Francis/CRC Press, Boca Raton, Florida).

COOK, B. D., C. M. PRINGLE & J. M. HUGHES, 2008. Molecular evidence for sequential colonization and taxon cycling in freshwater decapod shrimps on a Caribbean island. Molecular Ecology, **17**: 1066-1075.

CUESTA, J. A. & C. D. SCHUBART, 1997. The first zoeal stage of *Glyptograpsus impressus*, with comments on the subfamilial arrangement of Grapsidae (Crustacea: Brachyura). Cahiers de Biologie Marine, **38**: 291-299.

De Grave, S. & C. H. J. M. Fransen, 2011. Carideorum Catalogus: the recent species of the dendrobranchiate, stenopodidean, procarididean and caridean shrimps (Crustacea: Decapoda). Zoologische Mededelingen, **85**: 195-588.

De Grave, S. & L. Y. D. Goulding, 2011. Comparative morphology of the pereiopod 1 carpo-propodal (P1-CP) antennal flagellar grooming brush in caridean shrimps (Crustacea, Decapoda). Zoologischer Anzeiger, **250**: 280-301.

De Grave, S., N. D. Pentcheff, S. T. Ahyong, T.-Y. Chan, K. A. Crandall, P. C. Dworschak, D. L. Felder, R. M. Feldmann, C. H. J. M. Fransen, L. Y. D. Goulding, R. Lemaitre, M. E. Y. Low, J. W. Martin, P. K. L. Ng, C. E. Schweitzer, S. H. Tan, D. Tshudy & R. Wetzer, 2009. A classification of living and fossil genera of decapod crustaceans. Raffles Bulletin of Zoology, (Supplement) **21**: 1-109.

Fincham, A. A., 1985. Larval development of British prawns and shrimps (Crustacea: Decapoda: Natantia) 5. *Palaemon* (*Palaemon*) *adspersus* Rathke, 1837. Bulletin of the British Museum of Natural History (Zoology), London, **48**(4): 215-231.

González-Gordillo, J. I. & A. Rodríguez, 2000. Larval development of *Philocheras fasciatus* (Risso, 1816) (Decapoda, Caridea) reared in the laboratory, comparison with plankton larvae and ocurrence of accelerated development. Journal of Plankton Research, **22**(10): 1909-1923.

Guerao, G., E. Hernández & A. Úrzua, 2011. Early zoeal development of the shrimp *Hippolyte leptocerus* (Decapoda, Caridea, Hippolytidae). Zootaxa, **2988**: 53-65.

Gurney, R., 1942. Larvae of decapod Crustacea: 1-306. (The Ray Society, London).

Hayashi, K.-I. & T. Hamano, 1984. The complete larval development of *Caridina japonica* De Man (Decapoda, Caridea, Atydae) reared in the laboratory. Zoological Science, **1**: 571-589.

Hobbs, H. H. Jr. & C. W. Hart Jr., 1982. The shrimp genus *Atya* (Decapoda: Atyidae). Smithsonian Contributions to Zoology, **364**: 1-143.

Holthuis, L. B., 1993. The recent genera of the caridean and stenopodidean shrimps (Crustacea, Decapoda) with an appendix on the order Amphionidacea: 1-328. (Nationaal Natuurhistorisch Museum, Leiden).

Hultgren, K. M., F. Palero, G. Guerao & F. P. L. Marques, 2009. Phylogeny of the brachyuran crab superfamily Majoidea: a comparison of molecular and morphological characters. In: J. W. Martin, K. A. Crandall & D. L. Felder (eds.), Decapod crustacean phylogenetics. Crustacean Issues, **18**: 797-808. (Taylor & Francis/CRC Press, Boca Raton, Florida).

Hunte, W., 1975. *Atya lanipes* Holthuis, 1963, in Jamaica, including taxonomic notes and description of the first larval stage. Crustaceana, **28**: 66-72.

— —, 1977. Laboratory rearing of freshwater atyid shrimps *Atya innocous* Herbst and *Micratya poeyi* Guerin-Meneville (Decapoda, Atyidae). Aquaculture, **11**: 373-378.

— —, 1978. The distribution of freshwater shrimps (Atyidae and Palaemonidae) in Jamaica. Zoological Journal of the Linnean Society, **64**: 135-150.

— —, 1979a. The complete larval development of the freshwater shrimp *Atya innocous* (Herbst) reared in the laboratory (Decapoda, Atyidae). Crustaceana, (Supplement) **5**: 231-242.

— —, 1979b. The complete larval development of the freshwater shrimp *Micratya poeyi* (Guérin-Méneville) reared in the laboratory (Decapoda, Atyidae). Crustaceana, (Supplement) **5**: 153-166.

Lai, H.-T. & J.-Y. Shy, 2009. The larval development of *Caridina pseudodenticulata* (Crustacea: Decapoda: Atyidae) reared in the laboratory, with a discussion of larval metamorphosis types. Raffles Bulletin of Zoology, (Supplement) **20**: 97-107.

LI, C. P., S. DE GRAVE, T.-Y. CHAN, H. C. LEI & K. H. CHU, 2011. Molecular systematic of caridean shrimps based on five nuclear genes: implications for superfamily classification. Zoologischer Anzeiger, **250**: 270-279.

MAIHARA, Y., 2002. Larval stages of the rhynchocinetid shrimp, *Rhynchocinetes uritai* Kubo, 1942 (Decapoda, Caridea, Rhynchocinetidae) reared under laboratory conditions. Science Reports of the Museum Tokai University, **4**: 59-77.

MAIHARA, Y. & N. KYOYA, 2001. Larval stages of the rhynchocinetid shrimp, *Cinetorhynchus striatus* (Nomura & Hayashi, 1992) (Decapoda, Caridea, Rhynchocinetidae) reared in the laboratory. Science Reports of the Museum Tokai University, **3**: 43-62.

MARQUES, F. & G. POHLE, 1995. Phylogenetic analysis of the Pinnotheridae (Crustacea, Brachyura) based on the larval morphology, with emphasis on the *Dissodactylus* species complex. Zoologica Scripta, **24**(4): 347-364.

MARTIN, J. W. & G. E. DAVIS, 2001. An updated classification of the recent Crustacea. Natural History Museum of Los Angeles Country, Science Series, **39**: 1-124.

MATOBA, H. & S. SHOKITA, 1998. Larval development of the rhynchocinetid shrimp, *Rhynchocinetes conspiciocellus* Okuno & Takeda (Decapoda: Caridea: Rhynchocinetidae) reared under laboratory conditions. Crustacean Research, **27**: 40-69.

ORTEGA, A., H. QUEIROGA & J. I. GONZÁLEZ-GORDILLO, 2005. Planktonic stages of *Processa macrodactyla* (Decapoda: Caridea: Processidae) reared in the laboratory. Journal of Marine Biology Association of the United Kingdom, **85**: 1449-1460.

PAGE, T. J., B. D. COOK, T. VON RINTELEN, K. VON RINTELEN & J. M. HUGHES, 2008. Evolutionary relationships of atyids shrimps imply both ancient Caribbean radiations and common marine dispersals. Journal of the North American Benthological Society, **27**(1): 68-83.

PAGE, T. J., L. S. TORATI, B. D. COOK, A. BINDERUP, C. M. PRINGLE, S. REUSCHEL, C. D. SCHUBART & J. M. HUGHES, 2013. Invertébrés Sans Frontières: large scales of connectivity of selected freshwater species among Caribbean islands. Biotropica, **45**: 236-244.

SALMAN, S. D., 1987. Larval development of *Caridina babaulti basrensis* Al-Adhub & Hamzah (Decapoda, Caridea, Atyidae) reared in the laboratory. Crustaceana, **52**(3): 229-244.

SANDIFER, P. A., 1974. Larval stages of the shrimp, *Ogyrides limicola* Williams, 1955 (Decapoda, Caridea) obtained in the laboratory. Crustaceana, **26**(1): 37-60.

SCHUBART, C. D. & J. A. CUESTA, 2010. Phylogenetic relationships of the Plagusiidae Dana, 1851 (Brachyura), with description of a new genus and recognition of Percnidae Števčić, 2005, as an independent family. In: P. CASTRO, P. J. F. DAVIE, P. K. L. NG & B. RICHER DE FORGES (eds.), Studies on Brachyura: a homage to Danièle Guinot. Crustaceana Monographs, **11**: 279-299.

SCHUBART, C. D., J. A. CUESTA & D. L. FELDER, 2002. Glyptograpsidae, a new brachyuran family from Central America: larval and adult morphology, and a molecular phylogeny of the Grapsoidea. Journal of Crustacean Biology, **22**(1): 28-44.

SHY, J.-Y., H.-T. LAI & H.-P. YU, 2001. On the larval development of *Caridina formosae* (Decapoda, Atyidae) reared in the laboratory. Crustaceana, **74**(10): 1159-1168.

THATJE, S. & R. BACARDIT, 2000. Larval development of *Austrapandalus grayi* (Cunningham, 1871) (Decapoda, Caridea, Pandalidae) from the southwestern Atlantic ocean. Crustaceana, **73**(5): 609-628.

THATJE, S., R. BACARDIT & W. E. ARNTZ, 2005. Larvae of the deep-sea Nematocarcinidae (Crustacea: Decapoda: Caridea) from the Southern Ocean. Polar Biology, **28**: 290-302.

WALOSSEK, D., 1995. The Upper Cambrian *Rehbachiella*, its larval development, morphology and significance for the phylogeny of Branchiopoda and Crustacea. Hydrobiologia, **298**(1-3): 1-13.

WALSH, C. J., 1993. Larval development of *Paratya australiensis* Kemp, 1917 (Decapoda: Caridea: Atyidae), reared in the laboratory, with comparisons of fecundity and egg and larval size between estuarine and riverine environments. Journal of Crustacean Biology, **13**(3): 456-480.

First received 4 October 2011.
Final version accepted 29 July 2013.

DIVERSITY AND DISTRIBUTION OF AUSTRALIAN FRESHWATER CRAYFISH WITH A CHECK-LIST OF THE WORLD PARASTACIDAE AND A KEY TO THE GENERA (DECAPODA, ASTACIDEA, PARASTACOIDEA)

BY

SHANE T. AHYONG[1])

Australian Musuem, 6 College St., Sydney, NSW 2010, Australia, and School of Biological, Earth and Environmental Sciences, University of New South Wales, Kensington, NSW 2052, Australia

ABSTRACT

The Southern Hemisphere freshwater crayfish family, Parastacidae, currently includes 178 species and 15 genera. Australia is the centre of parastacid diversity, with 10 genera and 137 species, or 78% of species and 67% of genera. An overview of the diversity, distribution, and conservation status of Australian freshwater crayfishes is presented, including a global key to the genera of the Parastacidae and checklist of the world species.

INTRODUCTION

Freshwater crayfish are distributed worldwide in two superfamilies and three families containing more than 600 species (De Grave et al., 2009). Astacoidea includes the northern hemisphere crayfish, Astacidae and Cambaridae. Parastacoidea contains the Southern Hemisphere crayfish, Parastacidae, that occur in South America, Madagascar, New Zealand, Australia, and New Guinea. The parastacids exhibit a classic Gondwanan distribution and have a limited fossil record dating back to the early Cretaceous of Australia (Martin et al., 2008). The taxonomy of the parastacids has been extensively studied over the last four decades with 15 genera and 178 species known in total (table I). Australia has long been recognised as a centre of diversity of freshwater crayfish and hosts the most diverse parastacid fauna in the world, including the three largest and second smallest known species of freshwater crayfish.

[1]) e-mail: shane.ahyong@austmus.gov.au

 Advances in freshwater decapod systematics and biology: 245-271

TABLE I

World species of the Parastacidae. Abbreviations: AUST, Australia; NSW, New South Wales; NT, Northern Territory; QLD, Queensland; SA, South Australia; TAS, Tasmania; VIC, Victoria; WA, Western Australia

Genus and species	Type locality
Astacopsis Huxley, 1879	
A. franklinii (Gray, 1845)	AUST, TAS, no other data
A. gouldi Clark, 1936	AUST, TAS, Circular Head
A. tricornis Clark, 1936	AUST, TAS, Lake St. Clair
Astacoides Guérin-Méneville, 1839	
A. betsileoensis Petit, 1923	MADAGASCAR, Vicinity of Fianarantosa
A. caldwelli (Bate, 1865)	MADAGASCAR, Ampamaherana, 25 km from Fianarantosa
A. crosnieri Hobbs, 1987	MADAGASCAR, no other data
A. granulimanus Monod & Petit, 1929	MAGAGASCAR, Ikongo, Vinanitelo
A. hobbsi Boyko, Ravoahangimalala, Randriamasimanana & Razafindrazaka, 2007	MADAGASCAR, Ranomadio River
A. madagascarensis (H. Milne Edwards & Audouin, 1839)	MADAGASCAR, no other data
A. petiti Hobbs, 1987	MADAGASCAR, Isaka
Cherax Erichson, 1846	
C. albertisii (Nobili, 1899)	PAPUA NEW GUINEA, Katau River, near Fly River mouth
C. albidus Clark, 1936	AUST, VIC, Nurrabiel, south-west of Horsham
C. aruanus Roux, 1911	WEST PAPUA, Ngaiguli, Trangan Island, Aru Islands
C. barretti Clark, 1941	AUST, NT, Japanese Creek, Wessell Islands
C. bicarinatus (Gray, 1845)	AUST, NT, Port Essington
C. boschmai Holthuis, 1949	PAPUA, Paniai Lake
C. boesemani Lukhaup & Pekny, 2008	WEST PAPUA, Kais River Drainage, shoreline of Ajamaru Lake
C. buitendijkae Holthuis, 1949	PAPUA, Paniai Lake
C. cainii Austin & Ryan, 2002	AUST, WA, Scott River, 12 km northeast of Augusta
C. cairnsensis Riek, 1969	AUST, QLD, Near Anserson Street, Cairns
C. cartalacoolah Short, 1993	AUST, QLD, Cape Flattery
C. communis Holthuis, 1949	PAPUA, Paniai Lake
C. crassimanus Riek, 1967	AUST, WA, Beedelup Halls, Pemberton area
C. cuspidatus Riek, 1969	AUST, NSW, 32 km south of Port Macquarie
C. depressus Riek, 1951	AUST, QLD, Mt Coot-tha, Brisbane
C. destructor Clark, 1936	AUST, VIC, Melbourne University pond, Melbourne

TABLE I
(Continued)

Genus and species	Type locality
C. dispar Riek, 1951	AUST, QLD, Sandy Creek, Moorooka, Brisbane
C. glaber Riek, 1967	AUST, WA, 12 miles north of Augusta
C. holthuisi Lukhaup & Pekny, 2006	WEST PAPUA, Aitinjo Lake, Kais River Drainage
C. leckii Coughran, 2005	AUST, NSW, tributary of Koreelah Creek, Koreelah National Park
C. longipes Holthuis, 1949	PAPUA, Tigi Lake
C. lorentzi Roux, 1911	WEST PAPUA, Near Mapar, west of Geelvink Bay, Manikion district
C. minor Holthuis, 1996	PAPUA, Sungai Kurao, Wurigelebur
C. misolicus Holthuis, 1949	WEST PAPUA, Misool Island
C. monticola Holthuis, 1950	PAPUA, Ibele River, 15 km north-east of Habbema Lake
C. murido Holthuis, 1949	PAPUA, Paniai Lake
C. nucrifraga Short, 1991	AUST, NT, Palm Springs, near Channel Point
C. pallidus Holthuis, 1949	PAPUA, Paniai Lake
C. paniaicus Holthuis, 1949	PAPUA, Paniai Lake
C. papuanus Holthuis, 1949	PAPUA NEW GUINEA, Lake Kutubu
C. parvus Short & Davie, 1993	AUST, QLD, O'Leary Creek, Tully River, above Koombooloomba Dam
C. peknyi Lukhaup & Herbert, 2008	PAPUA NEW GUINEA, Tamu Creek, Fly River Drainage
C. preissii (Erichson, 1846)	AUST, WA, Near Albany
C. punctatus Clark, 1936	AUST, QLD, Cooran
C. quadricarinatus (von Martens, 1868)	AUST, QLD, Cape York
C. quinquecarinatus (Gray, 1845)	AUST, WA, Swan River
C. rhynchotus Riek, 1951	AUST, QLD, Mapoon, Gulf of Carpentaria
C. robustus Riek, 1951	AUST, QLD, Lake Birrabeen, Fraser Island
C. rotundus Clark, 1941	AUST, QLD, Severnlea of Severn River, south of Stanthorpe
C. setosus Riek, 1951	AUST, NSW, Booral, Karuah River, Port Stephens
C. solus Holthuis, 1949	PAPUA, Tigi Lake
C. tenuimanus (Smith, 1912)	AUST, WA, Margaret River
C. urospinosus Riek, 1969	AUST, QLD, Indooroopilly, Brisbane
C. wasselli Riek, 1969	AUST, QLD, Bridge Spring, between Rocky River and Scrubby Creek, Cape York

TABLE I
(Continued)

Genus and species	Type locality
Engaeus Erichson, 1846	
E. affinis Smith & Schuster, 1913	AUST, VIC, Badger Creek, Warburton, Mount Dandenong, upper, Yarra
E. australis Riek, 1969	AUST, VIC, Lilly Pilly Gully, Wilson's Promontory
E. cisternarius Suter, 1977	AUST, TAS, Tributary of the Dip River
E. cunicularis (Erichson, 1846)	AUST, TAS, no other data
E. curvisuturus Horwitz, 1990	AUST, VIC, Yarra River Plains, near bridge over Yarra River at Warburton
E. cymus (Clark, 1936)	AUST, VIC, Buffalo River, Dondangadale
E. disjuncticus Horwitz, 1990	AUST, TAS, Rubicon River, east of Elizabeth Town on Weetah Road
E. fossor (Erichson, 1846)	AUST, TAS, no other data
E. fultoni Smith & Schuster, 1913	AUST, VIC, Cape Otway forest and Ferntree Gully
E. granulatus Horwitz, 1990	AUST, TAS, Mouth of Browns Creek, east of Port Sorell
E. hemicirratulatus Smith & Schuster, 1913	AUST, VIC, Gippsland (Thorpdale, Warrugal, Moyarra, Trafalgar)
E. karnanga Horwitz, 1990	AUST, VIC, Lilly Pilly Gully, Wilson's Promontory
E. laevis (Clark, 1941)	VIC, AUST, Bunyip
E. lengana Horwitz, 1990	AUST, TAS, East coast of Rocky Cape, National Park
E. leptorhynchus Clark, 1939	AUST, TAS, Pioneer Mine, Derby
E. lyelli (Clark, 1936)	AUST, VIC, Gisborne
E. mairener Horwitz, 1990	AUST, TAS, Off Big Hill Rd, 1 km east of Lefroy
E. mallacoota Horwitz, 1990	AUST, VIC, Double Creek, Croajingolong National Park
E. martigener Horwitz, 1990	AUST, SA, Fotheringate Creek, Strzelecki Peaks, Flinders Island
E. merosetosus Horwitz, 1990	AUST, VIC, Waurn Ponds Creek, 2 km north of Mt Moriac
E. nulloporius Horwitz, 1990	AUST, TAS, 1.5 km south of 3-way corner (Frankford, Glengarry, Birralee) of road
E. orientalis Clark, 1941	AUST, VIC, 11 miles north of Cann River, Cann River Valley
E. orramakunna Horwitz, 1990	AUST, TAS, About 3 km south of Lilydale, tributary of Pipers River
E. phylloceros Smith & Schuster, 1913	AUST, VIC, Gippsland (Narracan River, Thorpdale, Trafalgar)
E. quadrimanus Clark, 1936	AUST, VIC, Warrugal, Gippsland

TABLE I
(Continued)

Genus and species	Type locality
E. rostrogaleatus Horwitz, 1990	AUST, VIC, Ryton Junction on Midland Highway, Eastern Strzelecki Ranges
E. sericatus Clark, 1936	AUST, VIC, Croydon, Mortlake and Warburton
E. spinicaudatus Horwitz, 1990	AUST, TAS, Surveyors Creek after Scottsdale
E. sternalis (Clark, 1936)	AUST, VIC, Warrugal, Gippsland
E. strictfrons (Clark, 1936)	AUST, VIC, Portland
E. tayatea Horwitz, 1990	AUST, TAS, Pearly Brook, 8 km west of Mount Horror
E. tuberculatus Clark, 1936	AUST, VIC, Sherbrook Fall
E. urostrictus Riek, 1969	AUST, VIC, Dandenong Creek, at Alpine Road, east of Melbourne
E. victoriensis Smith & Schuster, 1913	AUST, VIC, Dandenong ranges
E. yabbimunna Horwitz, 1994	AUST, TAS, Burnie park, gully of Shorewell Creek below Oldacre Falls
Engaewa Riek, 1967	
E. pseudoreducta Horwitz & Adams, 2000	AUST, WA, near Osmington, north-east of Margaret River
E. reducta Riek, 1967	AUST, WA, north of Quindalup Common, near Dunsborough
E. similis Riek, 1967	AUST, WA, Augusta
E. subcoerulea Riek, 1967	AUST, WA, Inlet River, 24 km north-west of Walpole
E. walpolea Horwitz & Adams, 2000	AUST, WA, The Knoll Scenic Drive, Walpole
Euastacus Clark, 1936	
E. armatus (von Martens, 1886)	AUST, Murray River
E. australasiensis (H. Milne Edwards, 1837)	AUST, NSW, Sydney area
E. balanensis Morgan, 1988	AUST, QLD, Tributary of Davies Creek, Lamb Range, Atherton Tableland
E. bidawalus Morgan, 1986	AUST, VIC, Chandlers Creek, near junction with Cann River
E. bindal Morgan, 1989	AUST, QLD, Upper North Creek, Mount Elliott
E. bispinosus Clark, 1941	AUST, VIC, Glenelg River
E. brachythorax Riek, 1969	AUST, NSW, Brown Mountain
E. clarkae Morgan, 1997	AUST, NSW, Tributary of Cockerawombeeba Creek, north of Birdwood
E. claytoni Riek, 1969	AUST, NSW, Maclaughlin River
E. crassus (Riek, 1969)	AUST, ACT, Bendora

TABLE I
(Continued)

Genus and species	Type locality
E. dalagarbe Coughran, 2005	AUST, NSW, Brindle Creek, Border Ranges National Park
E. dangadi Morgan, 1997	AUST, NSW, Stockyard Creek, Ingalba State Forest, north of Kempsey
E. dharawhalus Morgan, 1997	AUST, NSW, Stream above Fitzroy Falls
E. diversus Riek, 1969	AUST, VIC, 48 km north of Orbost
E. eungella Morgan, 1988	AUST, QLD, Tributary of Cattle Creek, near Mount Dalrymple, Eungella National Park
E. fleckeri (Watson, 1935)	AUST, QLD, Roots Creek near Mount Carbine, 130 km west of Cairns
E. gamilaroi Morgan, 1997	AUST, NSW, Hanging Rock, near Nundle
E. girumulayn Coughran, 2005	AUST, NSW, Tuntable Creek, Nightcap National Park
E. gumar Morgan, 1997	AUST, NSW, Gorge Creek, Richmond Range State Forest
E. gurughi Coughran, 2005	AUST, NSW, Korrumbyn Creek, Mount Warning National Park
E. guwinus Morgan, 1997	AUST, NSW, Tianjara Creek, tributary of Shoalhaven River
E. hirsutus (McCulloch, 1917)	AUST, NSW, Belmore Falls Creek above falls, Kangaroo River
E. hystricosus Riek, 1951	AUST, QLD, Yabba Creek, Maleny
E. jagabar Coughran, 2005	AUST, NSW, Sheepstation Creek, Border Ranges National Park
E. jagara Morgan, 1988	AUST, QLD, Flaggy Creek, Mistake Mountain
E. kershawi (Smith, 1912)	AUST, VIC, Moe River, Gippsland West
E. maccai McCormack & Coughran, 2008	AUST, NSW, Back Creek Road, Riamukka State Forest
E. maidae (Riek, 1956)	AUST, QLD, Upper Currumbin Creek
E. mirangudjin Coughran, 2002	AUST, NSW, Iron Pot Creek, Toonumbar National Park
E. monteithorum Morgan, 1989	AUST, QLD, Headwaters of Kroombit Creek, Kroombit Tops
E. morgani Coughran & McCormack, 2011	AUST, NSW, Tributary of Little Nymboida River, near Lowanna, near Bindarri National Park
E. neodiversus Riek, 1969	AUST, VIC, National Park, Wilson's Promontory
E. neohirsutus Riek, 1969	AUST, NSW, 32 km west of Dorrigo
E. pilosus Coughran & Leckie, 2007	AUST, NSW, Flaggy Creek, upper Clarence River
E. polysetosus Riek, 1951	AUST, NSW, Tubrabucca Creek, Barrington Tops

TABLE I
(Continued)

Genus and species	Type locality
E. reductus Riek, 1969	AUST, NSW, Upper Allyn River, below south end of Barrington Tops
E. rieki Morgan, 1997	AUST, NSW, Snowy River Bridge, Mount Kosciusko
E. robertsi Monroe, 1977	AUST, QLD, Horan's Creek, Mount Finnigan National Park
E. setosus (Riek, 1956)	AUST, QLD, Mount Glorious
E. simplex Riek, 1956	AUST, NSW, 24 km north of Armidale
E. spinichelatus Morgan, 1997	AUST, NSW, Fenwick's Creek, Hastings River, south of Yarrowitch
E. spinifer (Heller, 1865)	AUST, NSW, New Holland (Sydney region)
E. sulcatus Riek, 1951	AUST, QLD, Binna Burra, Lamington National Park
E. suttoni Clark, 1941	AUST, QLD, Wyberba
E. urospinosus (Riek, 1956)	AUST, QLD, Obi Obi Creek, Maleny
E. valentulus Riek, 1951	AUST, QLD, Upper Currumbin Creek
E. woiwuru Morgan, 1986	AUST, VIC, Dobson's Creek, Dandenong
E. yanga Morgan, 1997	AUST, NSW, Double Creek, off Boyne Creek near Pidgeon House
E. yarraensis (McCoy, 1888)	AUST, VIC, Yarra River
E. yigara Short & Davie, 1993	AUST, QLD, O'Leary Creek, Cardwell Range
Geocharax Clark, 1936	
G. falcata Clark, 1941	AUST, VIC, Fyans Creek (south of divide), Grampians
G. gracilis Clark, 1936	AUST, VIC, Portland and Gellibrand River, south of Colac
Gramastacus Riek, 1972	
G. insolitus Riek, 1972	AUST, VIC, 8 km southwest of Moyston
G. lacus McCormack, 2014	AUST, NSW, Boomeri Swamp, Myall Lakes National Park
Ombrastacoides Hansen & Richardson, 2006	
O. asperranus Hansen & Richardson, 2006	AUST, TAS, Birches Inlet, Landing Creek quarry
O. brevirostris Hansen & Richardson, 2006	AUST, TAS, Birches Inlet, Landing Creek quarry
O. decemdentatus Hansen & Richardson, 2006	AUST, TAS, The Needles, Strathgordon Road
O. denisoni Hansen & Richardson, 2006	AUST, TAS, Little Denison River
O. dissitus Hansen & Richardson, 2006	AUST, TAS, Lune River
O. huonensis Hansen & Richardson, 2006	AUST, TAS, west of Scotts Peak Road near Harlequin Hill

TABLE I
(Continued)

Genus and species	Type locality
O. ingressus Hansen & Richardson, 2006	AUST, TAS, East side of Victoria Pass, Lyell Highway
O. leptomerus (Riek, 1951)	AUST, TAS, Lake Lilla, Cradle Mountain
O. parvicaudatus Hansen & Richardson, 2006	AUST, TAS, near King River, Lyell Highway
O. professorum Hansen & Richardson, 2006	AUST, TAS, Allens Creek, Crotty Road
O. pulcher (Riek, 1967)	AUST, TAS, Lake Pedder
Paranephrops White, 1842	
P. planifrons White, 1842	NEW ZEALAND, Waihou River, North Island
P. zealandicus (White, 1847)	NEW ZEALAND, no other data
Parastacus White, 1842	
P. brasiliensis (von Martens, 1869)	BRAZIL, Porto Alegre, Rio Grande do Sul
P. defossus Faxon, 1898	URUGUAY, Montevideo
P. laevigatus Buckup & Rossi, 1980	BRAZIL, Joinville (Estrada da Cidra, Chácara dos ipes), Santa Catarina
P. nicoleti (Philippi, 1882)	CHILE, Valdivia
P. pilimanus (von Martens, 1869)	BRAZIL, Porto Alegre, Rio Grande do Sul
P. pugnax (Poeppig, 1835)	CHILE, Talcahuano
P. saffordi Faxon, 1898	URUGUAY, Montevideo
P. varicosus Faxon, 1898	BRAZIL or URUGUAY
Samastacus Riek, 1971	
S. spinifrons (Philippi, 1882)	CHILE, Valdivia
Spinastacoides Hansen & Richardson, 2006	
S. inermis (Clark, 1939)	AUST, TAS, Adamsons Peak
S. insignis (Clark, 1939)	AUST, TAS, Bramble Cove, Port Davey
S. catinipalmus Hansen & Richardson, 2006	AUST, TAS, Indiana Creek, Lower Gordon River
Tenuibranchiurus Riek, 1951	
T. glypticus Riek, 1951	AUST, QLD, Caloundra
Virilastacus Hobbs, 1991	
V. araucanius (Faxon, 1914)	CHILE, Corral
V. retamali Rudolph & Crandall, 2007	CHILE, "Los Kakanes", Rucapihuel, Coastal Cordillera, Osorno Province
V. rucapihuelensis Rudolph & Crandall, 2005	CHILE, Rucapihuel, Coastal Cordillera, Osorno Province
V. jarai Rudolph & Crandall, 2012	CHILE, "Quinta El Porvenir", Intermediate Depression, Biobío Region

Shaw (1794) described the first freshwater crayfish from Australia under the name *Cancer serratus*, now known as *Euastacus spinifer*, but the modern foundations of Australian astacology were laid by Ellen Clark (e.g., 1936a, 1941a, b) and Edgar Riek (e.g., 1951, 1956, 1967a, b, 1969, 1972), who dealt comprehensively with all known Australian genera and species. Clark's and Riek's works have been revised and extended by subsequent workers, e.g., Morgan (1986, 1988, 1997) for *Euastacus*; Horwitz (1990c) for *Engaeus*; Zeidler & Adams (1990) for *Gramastacus*; Horwitz & Adams (2000) for *Engaewa*; Hansen & Richardson (2006) for *Spinastacoides* and *Ombrastacoides*. The taxonomy of *Cherax* has been studied primarily by Austin and colleagues (e.g., Austin, 1996; Austin & Knott, 1996; Austin & Ryan, 2002). Multiple new species of *Cherax* and *Euastacus* have also been described in recent years and discovery of new species is ongoing. The Australian Parastacidae now includes at least 138 species and 10 genera accounting for 78% of world parastacid species and 67% of genera. The present paper presents a general overview of the diversity and distribution of Australian freshwater crayfish, a key to the genera of the Parastacidae and a check-list of the world species (table I).

DIVERSITY, HABITATS AND DISTRIBUTION

The Australian freshwater crayfish live in a wide range of habitats, but broadly segregate ecologically into largely aquatic and largely terrestrial forms. The largely aquatic forms live in creeks, streams, swamps and rivers, and include *Astacopsis*, *Euastacus* and *Cherax*. Although they may forage on land, especially at night, most time is spent in water, whether in the open or in the burrow. Riek (1972) regarded these three genera as 'moderate burrowers' because they do not burrow extensively and lack the burrowing specialisations seen in the largely terrestrial crayfish. Most of the remaining genera are largely terrestrial ('primary burrowers') that are highly specialised for extensive burrowing, typified most strongly by *Engaeus* and *Engaewa*. These genera usually burrow in marginal areas of streams, seepages, swamps or other areas that have either minimal, intermittent, or no surface water. Particular adaptations for burrowing include a compact, short and slender abdomen, chelipeds held directly forward with vertically opening dactyls, a shortened rostrum, small eyes, and minimal surface ornamentation. *Engaeus* and *Engaewa* seldom leave their burrows and their presence is often indicated at the surface only by a mud mound or chimney at the burrow entrance.

Horwitz & Richardson (1986) recognised three types of burrows among Australian freshwater crayfish — Type 1 burrows, that are constructed in the bed or banks of permanent waters, which may have other terrestrial entrances but are always connected to permanent surface water; Type 2 burrows, that have a terrestrial entrance, and which reach down to the water table but are not connected to permanent surface water; and Type 3 burrows, that are not associated with either permanent surface water or groundwater and are typically built in hillsides and filled by surface runoff water. Type 3 burrows are unique to species of Australian crayfish. Burrow type correlates broadly with generic groupings. Species of *Astacopsis*, *Euastacus*, *Cherax* and *Gramastacus* typically construct Type 1 burrows, but a few build Type 2 burrows, and at least one species (*Cherax punctatus* Clark, 1936), builds a Type 3 burrow. *Spinastacoides* and *Ombrastacoides* dig both Type 1 and 2 burrows. Most primary burrowers such as *Engaeus*, *Engaewa*, *Geocharax*, and *Tenuibranchiurus* dig Type 2 burrows and some species of *Engaeus* may also construct Type 1 and 3 burrows.

Riek (1972) examined the interrelationships of all the parastacid genera based on general morphology, and recognised two broad lineages separated primarily by the vertical or horizontal orientation of the major chelipeds (reflecting their type of burrowing behaviour), and secondarily by the proximity of the cervical, postcervical and branchiocardiac grooves on the carapace. Genera from each of the Gondwanan landmasses were interspersed with each other suggesting major diversification had occurred prior to the break-up of Gondwana. The 'primary burrowers' were principally among the lineage with vertically oriented major chelipeds and separate, unmerged cervical and branchiocardiac grooves, whereas the 'moderate' burrowers correspond to the second lineage, with horizontal chelipeds and merged carapace grooves. In the last decade or so, most phylogenetic studies of freshwater crayfish have focussed on molecular data (e.g., Crandall et al., 1999, 2000; Breinholdt et al., 2009; Schultz et al., 2009). Results corroborate several aspects of Riek's (1972) hypotheses, such as the general differentiation of clades of 'primary' and 'moderate' burrowers and the phylogenetic utility of the carapace groove arrangements, but present a different picture of early cladogenesis. The most recent overall phylogenetic estimate, based on nucleotide sequences from multiple loci, indicates that the parastacoids diverged from the astacoids around 185 Ma after Pangaea split and the South American parastacids diverged from other parastacids around 158 Ma (Toon et al., 2010). The two spiny Australian parastacid genera, *Euastacus* and *Astacopsis*, plus *Gramastacus* form a clade with a group of primary burrowers, *Engaeus*, *Engaewa*, *Tenuibranchiurus*, and *Geocharax*. The relationship between these two clades and *Cherax* was less clear, however. Additionally, there was only weak support for the monophyly of the Australian parastacids, thus only partially corroborating Riek's (1972) hypotheses. The New Zealand (*Paranephrops*) and Malagasy (*Astacoides*) genera were found to be nested among the Australian genera forming a clade with the Tasmanian *Spinastacoides* and *Ombrastacoides*, albeit with equivocal nodal support, reprising Riek's (1972) hypothesis in relation to the New Zealand and two Tasmanian genera. Thus, the molecular and morphological studies are generally complementary. Although the interrelationships of the eastern Gondwanan freshwater crayfish (i.e., Australia, New Guinea, New Zealand, and Madagascar) have not been robustly resolved, the overall phylogenetic patterns emerging from current studies indicate an east-west pattern of Gondwanan divergence, with radiation of the eastern Gondwanan crayfish occurring around 152 Ma (Toon et al., 2010).

The modern distribution of Australian freshwater crayfish is in many respects relictual, reflecting the increasing aridity of Australia since the mid-Miocene (Martin, 2006). Most species occur in the moist south-east of the continent, with some in the south-western corner, and some in the wet tropics of the north-east. Crayfish are absent from the arid zone of central Australia, with the exception of *Cherax destructor*.

The ten parastacid genera represented in Australia are all endemic, except for *Cherax*, which ranges into southern and western New Guinea. The highest generic richness is among the strongly burrowing forms (*Engaeus*, *Engaewa*, *Geocharax*, *Ombrastacoides*, *Spinastacoides*, *Tenuibranchiurus*), while the highest species richness is among the moderate burrowers (*Euastacus* and *Cherax*). Overall generic and specific richness follows a south-easterly trend, peaking in Victoria and Tasmania.

Astacopsis (fig. 1) includes three species endemic to Tasmania (Hamr, 1992a), of which *A. gouldi* is the largest freshwater crayfish in the world, attaining a body length of 76 cm and weight of 6 kg (Horwitz, 1990; Threatened Species Section, 2006). *Astacopsis* occurs in vegetated and shaded flowing creeks, streams and rivers from sea-level to about 400 m a.s.l. Distributions of the species of *Astacopsis* largely reflect an east-west divide observed for other Tasmanian invertebrates, resulting from post-Holocene periglacial activity in central Tasmania (Mesibov, 1994). *Astacopsis gouldi* occurs in northern Tasmania on either side of the Tamar catchment, with a potentially cryptic lineage in the north-east (Sinclair et al., 2011). *Astacopsis*

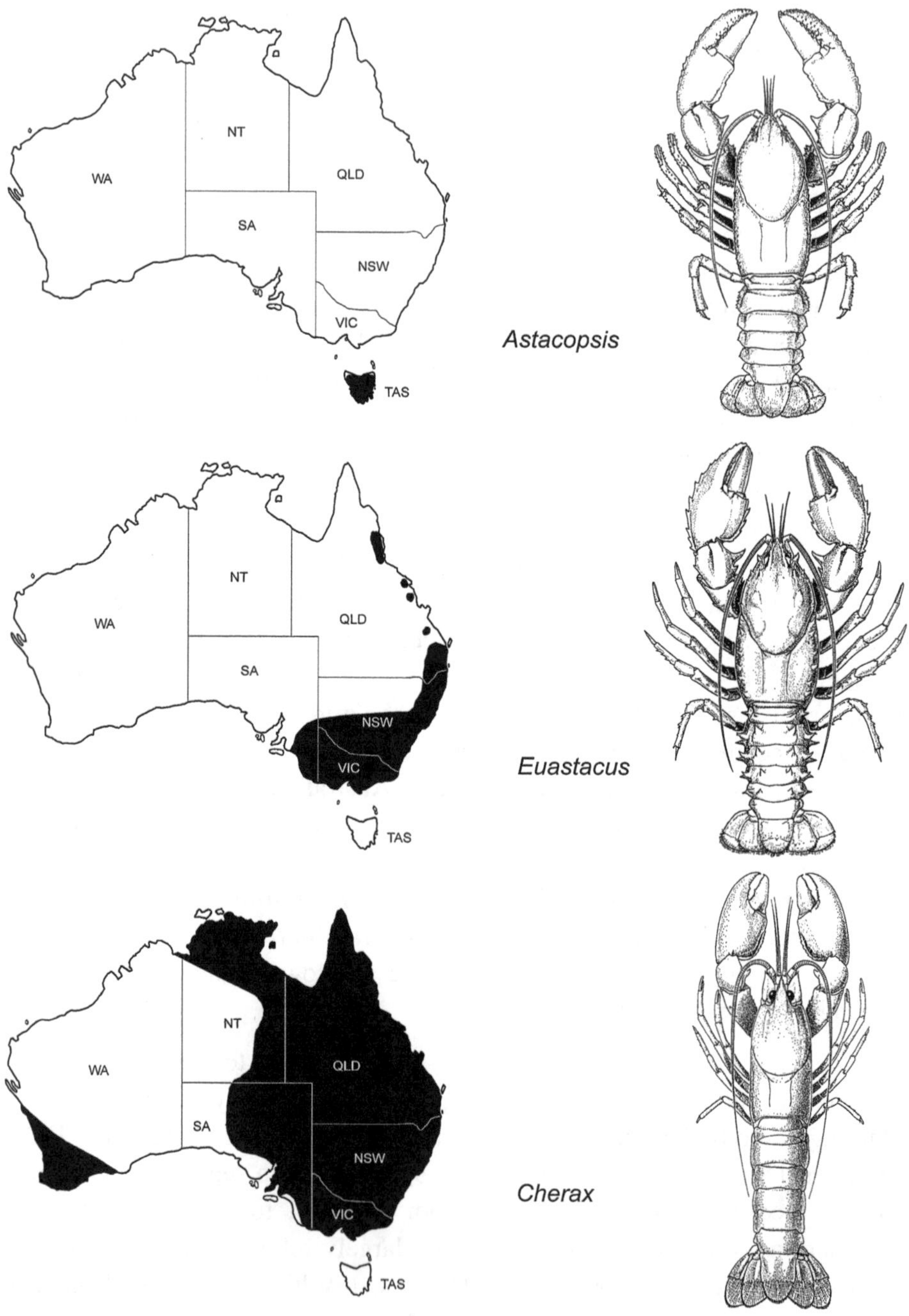

Fig. 1. Distribution of *Astacopsis*, *Euastacus* and *Cherax* (Australian only). Habitus figures depict *Astacopsis gouldi*, *Euastacus armatus* and *Cherax preissii* (modified after Riek, 1972).

tricornis and *A. franklinii* have largely discrete ranges in western and eastern Tasmania respectively.

Initial molecular phylogenetic studies indicated that *Astacopsis* might be nested within, and perhaps even paraphyletic with respect to *Euastacus* (Lawler & Crandall, 1998; Crandall et al., 1999). More extensive recent analyses, however, showed that both genera are reciprocally monophyletic sister taxa (Shull et al., 2005).

Of the three species of the genus, *A. gouldi* has been studied most extensively, principally because of its high conservation value and intrinsic scientific interest as the largest known freshwater crayfish. Studies of various aspects of the biology of *A. gouldi* include embryonic and postembryonic development (Hamr, 1992b), distribution and population biology (Horwitz, 1994; Sinclair et al., 2011), habitat characteristics (Davies et al., 2005), ecology and behaviour (Growns, 1995; Webb & Richardson, 2004; Davies et al., 2005). Species of *Astacopsis* are slow growing and long lived. *Astacopsis gouldi*, for instance, matures at about nine years in males and 14 years in females (Hamr, 1996).

Cherax (fig. 1) includes at least 44 described species found in Australia and southern/western New Guinea. Twenty-six species occur in Australia including one that also occurs in New Guinea, the Red Claw (*C. quadricarinatus*), and the remainder occur only in New Guinea. Most species do not exceed about 100 mm body length, but the largest species, *C. cainii*, may reach a length of 380 mm (Merrick & Lambert, 1991) making it the third largest freshwater crayfish in the world. *Cherax* has a disjunct distribution in Australia; most species occur in northern and eastern Australia, but eight species are restricted to the south-western corner of Western Australia. *Cherax* occurs in rivers, creeks and still waters such as ponds, lakes and swamps. Most species occur in areas of high or relatively high rainfall but an important exception is the Yabby (*Cherax destructor*), which occurs throughout the Murray-Darling system in eastern-central Australia. In addition, both the Red Claw and Yabby have artificially extended ranges having been translocated to most or all Australian states and mainland territories for aquaculture (Horwitz, 1990b).

Phylogenetic relationships of Australian *Cherax* were studied extensively by Austin and colleagues (e.g., Austin et al., 2003; Munasinghe et al., 2003, 2004a, b; Nguyen & Austin, 2005). Uncertainty remains, however, regarding the identities and taxonomic status of many of the named species from eastern Australia, which require redescription based on type material. The contemporary native distribution of *Cherax* broadly reflects divergences since the Miocene with an initial east-west separation, followed by a north-south divergence in eastern Australia (Munasinghe et al., 2004a).

The biology of most species of *Cherax* is poorly known, apart from the Gilgie (*C. quinquecarinatus*) (Beatty et al., 2005) and the three major commercial species (the Yabby, *C. destructor*; Red-Claw, *C. quadricarinatus*; and Smooth Marron, *C. cainii*), which have been extensively studied (see Merrick & Lambert, 1991, and references therein).

Species of *Cherax* have a shorter lifespan and mature more quickly than the spiny crayfish (*Astacopsis*, *Euastacus*). The Yabby and Marron mature at 1-2 years and the Red Claw may mature by six months, none living for more than 10 years (Merrick & Lambert, 1991).

Euastacus (fig. 1) is the largest parastacid genus, currently containing 50 species, all endemic to Australia. It includes the second largest freshwater crayfish globally and the largest Australian mainland crayfish, *E. armatus*, attaining about 50 cm in body length and three kilograms in weight (Horwitz, 1990a). Many other species, however, reach only relatively small sizes of less than 10 cm body length (McCormack, 2012). Most species of *Euastacus* have some degree of abdominal spination, the largest species having the most prominent spines. Morgan (1988) showed that Riek's (1956) *Euastacoides* was a junior synonym of *Euastacus*. Morphologically, Morgan (1997) recognised two major groups within the genus, those having a male cuticle partition on the gonopore (regarded as plesiomorphic), and those without the partition (regarded as derived). Phylogenetic analyses based on multiple molecular loci corroborated Morgan's (1997) division, but also recognised several lineages corresponding to northern, central and southern species of which the northern lineages were possibly the least 'derived' (Shull et al., 2005).

Euastacus ranges from northern Queensland south to Victoria and eastern South Australia with the highest diversity in the eastern drainages of southern New South Wales. As far as is known, most species have a narrow range, usually in the vegetated upper reaches of single catchments. *Euastacus* favours cool, aerated, flowing water in lowland and high altitude habitats, with different species occurring at different altitudes. In general, the extent of lowland occurrence decreases with latitude, correlating with clines in water temperature. In the south-east, lowland species include *Euastacus armatus*, *E. yanga*, *E. australasiensis* and *E. spinifer*, of which the latter two range from virtually sea-level to about 500 m altitude (Morgan, 1997). In south-east Queensland *Euastacus* occurs above 250 m altitude, above 750 m in central

Queensland and above 900 m in north Queensland, restricted to relictual montane habitats (Morgan, 1988; Ponniah & Hughes, 1998).

The biology of most species of *Euastacus* is little known, although it appears that females typically mate in autumn/winter, over-winter in berry, and release juveniles in summer (McCormack, 2012). Greatest knowledge exists for *E. armatus*, extensively summarised by Gilligan et al. (2007), and for *E. spinifer*, from a series of studies detailing reproduction, population structure, diet, growth, aging, conservation and field management (Merrick, 1997, 1998; Turvey & Merrick, 1997a-e). Other studies examined development in *E. bispinosus* (see Honan, 1998); general biology of *E. australasiensis* (see Merrick, 1998); growth and catch rates of *E. kershawi* (see Morey, 1998); population changes, growth and abundance of *E. hystricosus* (Smith et al., 1998); and general biology of *E. urospinosus* (see Borsboom, 1998). Like *Astacopsis*, *Euastacus* species are slow growing and long lived. *Euastacus armatus* matures at about four years in males, and 8-10 years in females and lives for up to 50 years (Gilligan et al., 2007). *Euastacus spinifer* typically matures at about 5 years in males and 8 years for females, and has a lifespan of up to 39 years (Turvey & Merrick, 1997e).

Engaeus (fig. 2) is found in southern New South Wales, Victoria, and Tasmania. Thirty-five species of *Engaeus* are currently recognised, with the majority occurring in Victoria (Horwitz, 1990c). Most species are small, seldom exceeding 80 mm body length. All are strong burrowers living in swamps, seepages, creek banks or pastures. Burrows are usually extensive, with one or more water-filled chambers and multiple secondary tunnels. Burrows are occupied by family groups of crayfish as well as a suite of other small invertebrates (Suter & Richardson, 1977; Horwitz, 1990; Fitzsimons & Antos, 2011). In rural areas where population densities are high and burrows numerous, *Engaeus* is considered to be either a nuisance or a pest (Clark, 1936b).

Studies of the phylogenetic relationships among *Engaeus* species using allozyme and nucleotide data (Horwitz et al., 1990; Schultz et al., 2007, 2009) indicate that this genus is paraphyletic. Ecological studies of *Engaeus* species have focussed on aspects of life history, burrow structure, associated fauna and ectosymbionts (e.g., Suter, 1977; Suter & Richardson, 1977; Richardson & Swain, 1980; Horwitz et al., 1985; Fitzsimons & Antos, 2011).

Engaewa (fig. 2) includes five species from south-western Australia, all of small size, not exceeding about 50 mm body length. Like *Engaeus*, species of

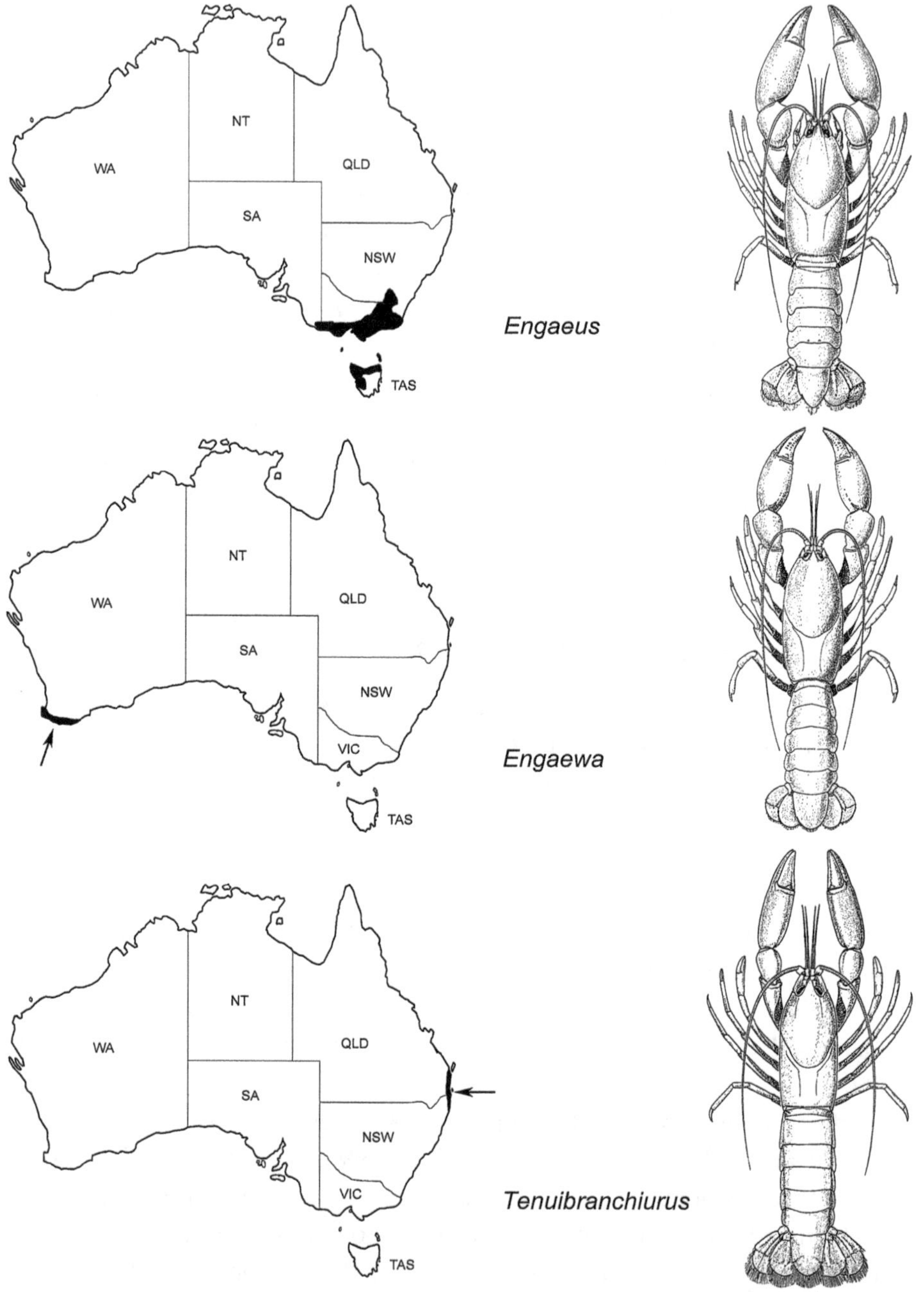

Fig. 2. Distribution of *Engaeus*, *Engaewa* and *Tenuibranchiurus*. Habitus figures depict *Engaeus lyelli*, *Engaewa subcoerulea* and *Tenuibranchiurus glypticus* (modified after Riek, 1972).

Engaewa are strong burrowers in clay, loam or sandy soils, and the two genera have similar morphological adaptations for burrowing. The interrelationships between the five known species of *Engaewa* were studied by Horwitz & Adams (2000) based on allozyme data.

Geocharax (fig. 3) includes two described (and at least two undescribed) species that range from south-western Victoria to northwestern Tasmania including King Island, Bass Strait (Schultz et al., 2007). Adults reach about 60 mm in length. Like *Engaeus*, species of *Geocharax* are strong burrowers whose burrows sometimes exceed two metres in depth (Zeidler & Adams, 1990).

Gramastacus (fig. 3) includes, *G. insolitus*, that occurs from south-eastern South Australia to south-western Victoria, and *G. lacus*, found in the coastal region of central New South Wales. Species of *Gramastacus* reach about 50 mm in body length and live in lowland swamps and seepages, and are only weak burrowers (Zeidler & Adams, 1990).

Ombrastacoides (fig. 3) contains 11 species from western and southern Tasmania that rarely exceed 80 mm body length. This species occurs in swamps, seeps, lakes and creeks and burrows into clay, mud, peat or gravel (Hansen & Richardson, 2006). *Parastacoides* was recently revised by Hansen & Richardson (2006), who evaluated the phylogenetic relationships within *Ombrastacoides* and *Spinastacoides* based on mitochondrial 16S sequences and reassigned species formerly placed in *Parastacoides* to the genera *Ombrastacoides* and *Spinastacoides* (except for *P. tasmanicus*, the type species). *Parastacoides tasmanicus*, itself proved to be conspecific with *Geocharax gracilis*, making *Parastacoides* a junior synonym of *Geocharax*.

Hamr (1992b) reported on embryonic and post-embryonic development of a crayfish referred to as *Parastacoides tasmanicus tasmanicus*, collected from burrows in a wet buttongrass plain near Harlequin Hill, south-western Tasmania. The identity of Hamr's (1992b) *Parastacoides tasmanicus tasmanicus* is not clear; it may be either *Ombrastacoides huonensis* or *Spinastacoides inermis*, both of which occur at that locality (Hansen & Richardson, 2006).

Spinastacoides (fig. 3) contains three species from southern to south-western Tasmania. Species of *Spinastacoides* attain a similar size and occupy similar habitats to *Ombrastacoides*. Lake & Newcombe (1975) and Richardson & Swain (1980) examined general ecology and habitat requirements of species of *Spinastacoides* (under the name *Parastacoides*).

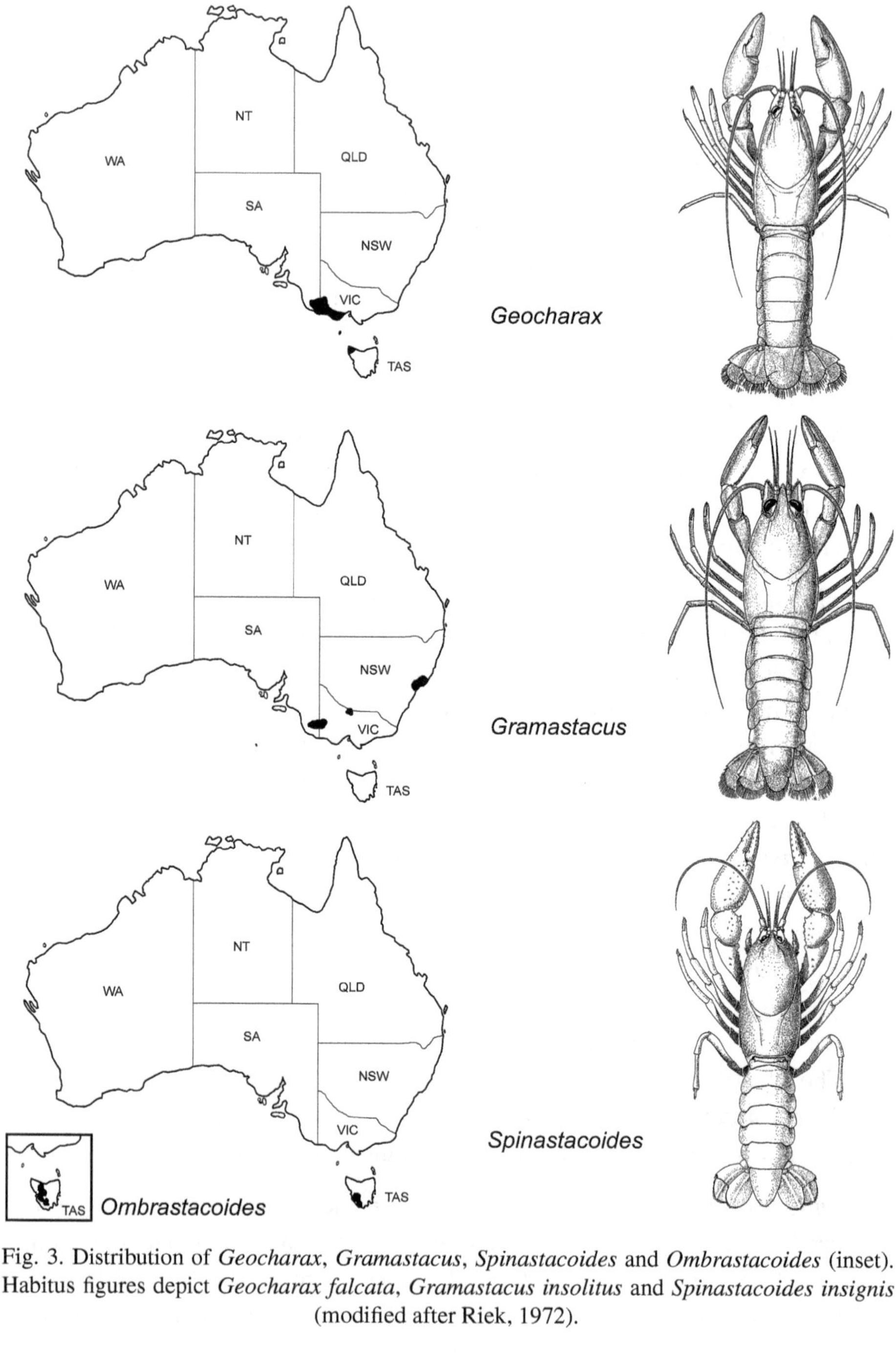

Fig. 3. Distribution of *Geocharax*, *Gramastacus*, *Spinastacoides* and *Ombrastacoides* (inset). Habitus figures depict *Geocharax falcata*, *Gramastacus insolitus* and *Spinastacoides insignis* (modified after Riek, 1972).

Tenuibranchiurus (fig. 2) ranges from south-eastern Queensland to north-eastern New South Wales, occurring in lowland coastal swamps, creeks and ephemeral wetlands. There is only one described species of *Tenuibranchiurus* (but there may be more undescribed species) which appears to have diverged during the Miocene or Pliocene (Horwitz, 1995; Dawkins et al., 2010). The type species of *Tenuibranchiurus*, *T. glypticus*, which has an adult body length of about 25 mm, is the second smallest known species of freshwater crayfish.

CONSERVATION ISSUES

The conservation status of many Australian freshwater crayfish has been addressed by several studies with many species considered to be under a significant level of conservation concern (e.g., Horwitz, 1990a, 1995; McCormack et al., 2010; Furse & Coughran, 2011a-c). Of the 125 (of 138) Australian crayfish assessed by the IUCN, 39% are considered either Endangered or Critically Endangered (IUCN, 2011). Several species are subject to strong recreational fishing pressure, primarily the largest species: *Astacopsis gouldi*, *Euastacus armatus*, *E. bispinosus*, *E. kershawi* and *E. spinifer* (Merrick, 1995; Morey, 1998).

Species translocated for aquaculture or deliberately released by private individuals have significant potential to impact other species (Lodge et al., 2012). The Hairy Marron (*Cherax tenuimanus*) has a limited range in the Warren River catchment, southwestern Australia, and could easily be displaced by the more widespread Smooth Marron (*Cherax cainii*), which is widely cultured in Western Australia (Austin & Ryan, 2002). *Cherax destructor*, and to a lesser extent *C. quadricarinatus*, have been widely translocated around Australia, especially along the eastern seaboard where they are now invasive in many waterways, actively competing with other native crayfish (Coughran & Leckie, 2007; Coughran et al., 2009). For example, *Euastacus dharawalus*, that has a limited range in the Fitzroy Falls area, southern New South Wales, is currently being displaced by introduced *C. destructor* and in November 2011 was declared Critically Endangered under the NSW Fisheries Management Act (McCormack, 2013). The major overall threat to Australian freshwater crayfish is probably habitat loss (Dawkins et al., 2010; Furse & Coughran, 2011b). Species that live close to urban and rural centres are subject to agricultural activities and are affected by habitat fragmentation and detrimental water quality through water harvesting, catchment modification, pollution, and runoff. This is particularly relevant to species with small populations and restricted distributions where small scale, localised impacts can have disproportionate consequences ("local disasters" of Merrick, 1995).

In Australia, conservation measures have been developed for various crayfish species. These include recovery plans for *Engaeus* and *Astacopsis gouldi* in Tasmania (Doran, 2005; Threatened Species Section, 2006) and *Engaewa* in Western Australia (Mantle, 2008), and legislative protections for *Euastacus armatus* (ACT Government, 1999) and *Euastacus bispinosus* in Victoria (Beeton, 2011). The long term effectiveness of these measures remains to be seen.

PROSPECT

Important challenges remain in our understanding of freshwater crayfish diversity in Australia. The fauna is generally well characterised, but important taxonomic challenges remain with respect to the identities of *Cherax* species from southern Queensland and New South Wales, most of which require revision based on type material. Although *Euastacus* was treated extensively by Morgan (1986, 1988, 1997), the status of many species, especially purportedly wide ranging species, requires closer scrutiny. Moreover, new species of *Cherax*, *Euastacus*, *Engaeus*, and *Gramastacus* continue to be discovered (e.g., Coughran, 2005a, b; Schultz et al., 2007; Coughran & McCormack, 2011). Overlaid on this are important conservation issues. Many species are known from only a few sites or individuals. Whereas many species and their habitats enjoy the relative protection of isolation deep inside National Parks or conservation reserves, a number of species occur close to human population centres or activities, and are subject to significant recreational fishing pressure or illegal poaching (including some species in National Parks). This highlights the pressing need for ongoing taxonomic and ecological research to inform appropriate management plans.

Key to genera of the Parastacidae

1. Palm and margins of major cheliped covered with spines .... *Paranephrops* (New Zealand)
– Palm and margins of major cheliped smooth or with tubercles but not spines 2

2. Branchiocardiac groove with long anterolateral extension running subparallel to, or converging on, cervical groove for some distance 5
– Branchiocardiac groove meeting cervical groove high up on carapace, without long anterolateral extension 3

3. Telson entirely calcified, not soft and membranous posteriorly, without trace of transverse suture. Stem of podobranchs without wing-like expansion .... *Astacopsis* (Australia)
– Telson with posterior portion soft, membranous, without or without distinct of transverse suture. Stem of podobranchs with wing-like expansion 4
4. Outer (flexor) margin of major cheliped palm bluntly serrated. Uropodal endopod fully calcified, not membranous distally .... *Euastacus* (Australia)
– Outer (flexor) margin of major cheliped palm smooth. Uropodal endopod fully membranous distally .... *Cherax* (Australia and southern New Guinea)
5. Branchiocardiac groove anterolaterally running extremely close to and then fusing with, or remaining subparallel to, dorsolateral portion of cervical groove. Reproductive females without subcalcified anteroventral extension on abdominal pleuron 2 6
– Branchiocardiac groove anterolaterally running distinctly separate from cervical groove, never fusing with cervical groove and only converging towards cervical groove at extreme anterolateral portion where it may curve slightly anteriorly. Reproductive females with subcalcified anteroventral extension on abdominal pleuron 2 11
6. Rostrum anterior margin truncate to subtruncate; dorsolateral carina strongly tuberculate. Outer margin of scaphocerite dentate to spinous. Telson with trace of transverse suture indicated laterally at base of posterolateral spines ..... *Astacoides* (Madagascar)
– Rostrum anterior margin triangular; dorsolateral carina smooth. Outer margin of scaphocerite straight, unarmed. Telson without trace of transverse suture 7
7. Pleurobranchs absent 8
– Pleurobranchs present 9
8. Posterior margins of uropodal endopod and exopod with spine .... *Spinastacoides* (Australia)
– Posterior margins of uropodal endopod and exopod rounded, unarmed .... *Ombrastacoides* (Australia)
9. Branchiocardiac groove not fusing anteriorly with cervical groove.................. .... *Virilastacus* (Chile)
– Branchiocardiac groove fusing anteriorly with cervical groove 10
10. Dactylus of major cheliped opening sub-vertically. Cervical groove broadly V-shaped in dorsal view .... *Parastacus* (Chile, Argentina, Uruguay, southern Brazil)
– Dactylus of major cheliped opening sub-horizontally. Cervical groove U-shaped in dorsal view .... *Samastacus* (Chile)
11. Dactylus of major cheliped opening sub-vertically 12
– Dactylus of major cheliped opening sub-horizontally or obliquely 14
12. Major chelipeds equal. Uropodal protopod produced to acute point at articulation with one or both rami. Antennal peduncle ischium with centrolateral spine................ .... *Tenuibranchiurus* (Australia)
– Major chelipeds equal or unequal. Uropodal protopod produced to blunt lobe at articulation with both rami. Antennal peduncle ischium without centrolateral spine.... 13
13. Stem of podobranchs without wing-like expansion .... *Engaeus* (Australia)
– Stem of podobranchs with wing-like expansion .... *Engaewa* (Australia)
14. Major cheliped palm of females densely setose. Mesial margin of major cheliped carpus smooth. Penes of male prominent, elongated, longer than coxa .... *Gramastacus* (Australia)

– Major cheliped palm of females not densely setose. Mesial margin of major cheliped carpus serrated. Penes of male a small papilla, much smaller than coxa. . . . . *Geocharax* (Australia)

ACKNOWLEDGEMENTS

I wish to thank Darren Yeo for the invitation to contribute to this volume and two anonymous reviewers for their constructive comments on the draft. This is a contribution from the Australian Museum Research Institute.

REFERENCES

ACT GOVERNMENT, 1999. Murray River Crayfish (*Euastacus armatus*): a vulnerable species. Action Plan, **14**: 1-12. (Environment ACT, Canberra).

AUSTIN, C. M., 1996. Systematics of the freshwater crayfish genus *Cherax* Erichson (Decapoda: Parastacidae) in northern and eastern Australia: electrophoretic and morphological variation. Australian Journal of Zoology, **44**: 259-296.

AUSTIN, C. M. & B. KNOTT, 1996. Systematics of the freshwater crayfish genus *Cherax* (Decapoda: Parastacidae) in southwestern Australia: electrophoretic, morphological and habitat variation. Australian Journal of Zoology, **44**: 223-258.

AUSTIN, C. M., T. T. T. NGUYEN, M. M. MEEWAN & D. R. JERRY, 2003. The taxonomy and phylogeny of the '*Cherax destructor*' complex (Decapoda: Parastacidae) examined using mitochondrial 16S sequences. Australian Journal of Zoology, **51**: 99-110.

AUSTIN, C. M. & S. G. RYAN, 2002. Allozyme evidence for a new species of freshwater crayfish of the genus *Cherax* Erichson (Decapoda: Parastacidae) from the south-west of Western Australia. Invertebrate Systematics, **16**: 357-367.

BEATTY, S. J., D. L. MORGAN & H. S. GILL, 2005. Life history and reproductive biology of the Gilgie, *Cherax quinquecarinatus*, a freshwater crayfish endemic to southwestern Australia. Journal of Crustacean Biology, **25**(2): 251-262.

BEETON, R. J., 2011. Advice to the Minister for Sustainability, Environment, Water, Population and Communities from the Threatened Species Scientific Committee (the Committee) on amendment to the list of Threatened Species under the Environment Protection and Biodiversity Conservation Act 1999 (EPBC Act).

BORSBOOM, A., 1998. Aspects of the biology and ecology of the Australian freshwater crayfish, *Euastacus urospinosus* (Decapoda: Parastacidae). Proceedings of the Linnean Society of New South Wales, **119**: 87-100.

BREINHOLT, J., M. PÉREZ-LOSADA & K. A. CRANDALL, 2009. The timing of the diversification of the freshwater crayfishes. In: J. W. MARTIN, K. A. CRANDALL & D. L. FELDER (eds.), Decapod crustacean phylogenetics. Crustacean Issues, **18**: 343-355. (CRC Press, Boca Raton, Florida).

CLARK, E., 1936a. The freshwater crayfishes of Australia. Memoirs of the National Museum of Victoria, **10**: 5-58.

——, 1936b. Notes on the habits of land crayfishes. Victorian Naturalist, **53**: 65-68.

——, 1941a. Revision of the genus *Euastacus* (crayfishes, family Parastacidae), with notes on the distribution of certain species. Memoirs of the National Museum of Victoria, **12**: 7-30, pls. 1-8.

— —, 1941b. New species of Australian freshwater and land crayfishes (family: Parastacidae). Memoirs of the National Museum of Victoria, **12**: 31-40.

COUGHRAN, J., 2005a. *Cherax leckii* n. sp. (Decapoda: Parastacidae): a new crayfish from coastal, northeastern New South Wales. Fishes of Sahul, **19**: 191-196.

— —, 2005b. New crayfishes (Decapoda: Parastacidae: *Euastacus*) from northeastern New South Wales, Australia. Records of the Australian Museum, **57**: 361-374.

COUGHRAN, J. & S. LECKIE, 2007. Invasion of a New South Wales stream by the tropical crayfish, *Cherax quadricarinatus* (von Martens). In: D. LUNNEY, P. EBY, P. HUTCHINGS & S. BURGIN (eds.), Pest or guest: the zoology of overabundance: 40-46. (Royal Zoological Society of New South Wales, Mosman).

COUGHRAN, J. & R. B. MCCORMACK, 2011. *Euastacus morgani* sp. n., a new spiny crayfish (Crustacea, Decapoda, Parastacidae) from the highland rainforests of eastern New South Wales, Australia. ZooKeys, **85**: 17-26.

COUGHRAN, J., R. B. MCCORMACK & G. DALY, 2009. Translocation of the Yabby *Cherax destructor* into eastern drainages of New South Wales, Australia. Australian Zoologist, **35**(1): 100-103.

CRANDALL, K. A., J. W. FETZNER JR., S. H. LAWLER, M. KINNERSLEY & C. M. AUSTIN, 1999. Phylogenetic relationships among the Australian and New Zealand genera of freshwater crayfishes (Decapoda: Parastacidae). Australian Journal of Zoology, **47**: 199-214.

CRANDALL, K. A., D. J. HARRIS & J. W. FETZNER JR., 2000. The monophyletic origin of freshwater crayfish estimated from nuclear and mitochondrial DNA sequences. Proceedings of the Royal Society of London B, **267**: 1679-1686.

DAVIES, P. E., L. S. J. COOK, S. A. MUNKS & J. MEGGS, 2005. *Astacopsis gouldi* Clark: habitat characteristics and relative abundance of juveniles. Tasforests, **16**: 1-17.

DAWKINS, K. L., J. M. FURSE, C. H. WILD & J. M. HUGHES, 2010. Distribution and population genetics of the threatened freshwater crayfish genus *Tenuibranchiurus* (Decapoda: Parastacidae). Marine and Freshwater Research, **61**: 1048-1055.

DE GRAVE, S., N. D. PENTCHEFF, S. T. AHYONG, T.-Y. CHAN, K. A. CRANDALL, P. C. DWORSCHAK, D. L. FELDER, R. M. FELDMANN, C. H. J. M. FRANSEN, L. Y. D. GOULDING, R. LEMAITRE, M. E. Y. LOW, J. W. MARTIN, P. K. L. NG, C. E. SCHWEITZER, S. H. TAN, D. TSHUDY & R. WETZER, 2009. A classification of living and fossil genera of decapod Crustaceans. Raffles Bulletin of Zoology, (Supplement) **21**: 1-109.

DORAN, N. E., 1999. Burrowing Crayfish (*Engaeus*) Group Recovery Plan 2001-2005. (Threatened Species Unit, Parks and Wildlife Service, Hobart). http://www.environment.gov.au/biodiversity/threatened/publications/recovery/burrowing-crayfish/index.html.

FITZSIMONS, J. A. & M. J. ANTOS, 2011. Ecological notes on the East Gippsland burrowing crayfish *Engaeus orientalis*, including burrow structure and associated fauna. Australian Zoologist, **35**(3): 853-857.

FURSE, J. M. & J. COUGHRAN, 2011a. An assessment of the distribution, biology, threatening processes and conservation status of the freshwater crayfish, genus *Euastacus* (Decapoda, Parastacidae), in continental Australia. I. Biological background and current status. In: A. ASAKURA (ed.), New frontiers in crustacean biology. Crustaceana Monographs, **15**: 241-252.

— — & — —, 2011b. An assessment of the distribution, biology, threatening processes and conservation status of the freshwater crayfish, genus *Euastacus* (Decapoda, Parastacidae), in continental Australia. II. Threats, conservation assessments and key findings. In: A. ASAKURA (ed.), New frontiers in crustacean biology. Crustaceana Monographs, **15**: 253-263.

— — & — —, 2011c. An assessment of the distribution, biology, threatening processes and conservation status of the freshwater crayfish, genus *Euastacus* (Decapoda, Parastacidae), in continental Australia. III. Case studies and recommendations. In: A. ASAKURA (ed.), New frontiers in crustacean biology. Crustaceana Monographs, **15**: 265-274.

GILLIGAN, D., R. ROLLS, J. MERRICK, M. LINTERMANS, P. DUNCAN & J. KOHEN, 2007. Scoping the knowledge requirements for Murray crayfish (*Euastacus armatus*). NSW Department of Primary Industries — Fisheries Final Report Series, **89**: 1-103.

GROWNS, I. O., 1995. *Astacopsis gouldi* Clark in streams of the Gog Range, northern Tasmania: the effects of catchment disturbance. Papers and Proceedings of the Royal Society of Tasmania, **129**: 1-6.

HAMR, P., 1992a. A revision of the Tasmanian freshwater crayfish genus *Astacopsis* Huxley (Decapoda: Parastacidae). Papers and Proceedings of the Royal Society of Tasmania, **126**: 91-94.

— —, 1992b. Embryonic and postembryonic development in the Tasmanian freshwater crayfishes *Astacopsis gouldi*, *Astacopsis franklinii* and *Parastacoides tasmanicus tasmanicus* (Decapoda: Parastacidae). Australian Journal of Marine and Freshwater Research, **43**: 861-878.

— —, 1996. A Giant's Tale: the life history of *Astacopsis gouldi* (Decapoda: Parastacidae). Freshwater Crayfish, **11**: 13-33.

HANSEN, B. & A. M. M. RICHARDSON, 2006. A revision of the Tasmanian endemic freshwater crayfish genus *Parastacoides* (Crustacea: Decapoda: Parastacidae). Invertebrate Systematics, **20**: 713-769.

HONAN, J. A., 1998. Egg and juvenile development of the Australian freshwater crayfish, *Euastacus bispinosus* Clark (Decapoda; Parastacidae). Proceedings of the Linnean Society of New South Wales, **119**: 37-54.

HORWITZ, P., 1990a. The conservation status of Australian freshwater Crustacea with a provisional list of threatened species, habitats and potentially threatening processes. Australian National Parks and Wildlife Service, Report Series, **14**: 1-121.

— —, 1990b. The translocation of freshwater crayfish in Australia: potential impact, the need for control and global relevance. Biological Conservation, **54**: 291-305.

— —, 1990c. A taxonomic revision of species in the freshwater crayfish genus *Engaeus* Erichson (Decapoda: Parastacidae). Invertebrate Taxonomy, **4**: 427-614.

— —, 1994. Distribution and conservation status of the Tasmanian giant freshwater lobster *Astacopsis gouldi* (Decapoda: Parastacidae). Biological Conservation, **69**: 199-206.

— —, 1995. The conservation status of Australian freshwater crayfish: review and update. Freshwater Crayfish, **10**: 70-80.

HORWITZ, P. & M. ADAMS, 2000. The systematics, biogeography and conservation status of species in the freshwater crayfish genus *Engaewa* Riek (Decapoda: Parastacidae) from south-western Australia. Invertebrate Systematics, **14**: 655-680.

HORWITZ, P., M. ADAMS & P. BAVERSTOCK, 1990. Electrophoretic contributions to the systematics of the freshwater crayfish genus *Engaeus* Erichson (Decapoda: Parastacidae). Invertebrate Taxonomy, **4**: 615-641.

HORWITZ, P. & A. M. M. RICHARDSON, 1986. An ecological classification of the burrows of Australian freshwater crayfish. Australian Journal of Marine and Freshwater Research, **37**: 237-242.

HORWITZ, P., A. M. M. RICHARDSON & P. M. CRAMP, 1985. Aspects of the life history of the burrowing freshwater crayfish *Engaeus leptorhyncus* at Rattrays Marsh, North East Tasmania. Tasmanian Naturalist, **82**: 1-5.

IUCN, 2011. IUCN Red List of Threatened Species. Version 2011.2. (Accessed 17 September 2012; www.iucnredlist.org.)

LAKE, P. S. & K. J. NEWCOMBE, 1975. Observations on the ecology of the crayfish *Parastacoides tasmanicus* (Decapoda; Parastacidae) from south-western Tasmania. Australian Zoologist, **18**(3): 197-213.

LAWLER, S. H. & K. A. CRANDALL, 1998. The relationship of the Australian freshwater crayfish genera *Euastacus* and *Astacopsis*. Proceedings of the Linnean Society of New South Wales, **119**: 1-8.

LODGE, D. M., A. DEINES, F. GHERARDI, D. C. J. YEO, T. ARCELLA, A. K. BALDRIDGE, M. A. BARNES, W. L. CHADDERTON, J. L. FEDER, C. A. GANTZ, G. W. HOWARD, C. L. JERDE, B. W. PETERS, J. A. PETERS, L. W. SARGENT, C. R. TURNER, M. E. WITTMANN & Y. ZENG, 2012. Global introductions of crayfishes: evaluating the impact of species invasions on ecosystem services. Annual Review of Ecology and Systematics, **43**: 449-472.

MANTLE, K., 2008. Dunsborough Burrowing Crayfish (*Engaewa reducta*), Margaret River Burrowing Crayfish (*Engaewa pseudoreducta*) and Walpole Burrowing Crayfish (*Engaewa walpolea*) Recovery Plan 2007-2016. Western Australian Wildlife Management Program, **41**: i-vi, 1-33.

MARTIN, A. J., T. H. RICH, G. C. B. POORE, M. B. SCHULTZ, C. M. AUSTIN, L. KOOL & P. VICKERS-RICH, 2008. Fossil evidence in Australia for the oldest known freshwater crayfish of Gondwana. Gondwana Research, **14**: 287-566.

MARTIN, H. A., 2006. Cenozoic climatic change and the development of the arid vegetation in Australia. Journal of Arid Environments, **66**(3): 533-563.

MCCORMACK, R. B., 2012. A guide to Australia's spiny freshwater crayfish: i-viii, 1-235. (CSIRO Publishing, Melbourne).

— —, 2013. Conservation status of imperiled freshwater crayfish, *Euastacus dharawalus* (Decapoda: Astacidea: Parastacidae), from the Southern Highlands of New South Wales, Australia. Journal of Crustacean Biology, **33**(2): 432-439.

— —, 2014. The eastern swamp crayfish *Gramastacus lacus* sp. n. (Decapoda, Parastacidae) a new species of freshwater crayfish from coastal New South Wales, Australia. Zookeys, **398**: 53-67.

MCCORMACK, R. B., J. COUGHRAN, J. M. FURSE & P. VAN-DER WERF, 2010. Conservation of imperiled crayfish — *Euastacus jagara* (Decapoda: Parastacidae), a highland crayfish from the Main Range, South-Eastern Queensland, Australia. Journal of Crustacean Biology, **30**(3): 531-535.

MERRICK, J. R., 1995. Diversity, distribution and conservation of freshwater crayfishes in the eastern highlands of New South Wales. Proceedings of the Linnean Society of New South Wales, **118**: 217-255.

— —, 1997. Conservation and field management of the freshwater crayfish, *Euastacus spinifer* (Decapoda: Parastacidae), in the Sydney Region, Australia. Proceedings of the Linnean Society of New South Wales, **118**: 217-225.

— —, 1998. Endemic crayfishes of the Sydney region: distribution, biology and management options. Macquarie University Graduate School of the Environment, Research Report, **9812**: 1-118.

MERRICK, J. R. & C. N. LAMBERT, 1991. The Yabby, Marron and Red Claw: production and marketing: 1-184. (J.R. Merrick Publications, Artarmon).

MESIBOV, R., 1994. Faunal breaks in Tasmania and their significance for invertebrate conservation. Memoirs of the Queensland Museum, **36**: 133-136.

MOREY, J. L., 1998. Growth, catch rates and notes on the biology of the Gippsland spiny freshwater crayfish, *Euastacus kershawi* (Decapoda: Parastacidae), in West Gippsland, Victoria. Proceedings of the Linnean Society of New South Wales, **119**: 55-69.

MORGAN, G. J., 1986. Freshwater crayfish of the genus *Euastacus* Clark (Decapoda: Parastacidae) from Victoria. Memoirs of the Museum of Victoria, **47**: 1-57.

— —, 1988. Freshwater crayfish of the genus *Euastacus* Clark (Decapoda: Parastacidae) from Queensland. Memoirs of the Museum of Victoria, **49**: 1-49.

— —, 1997. Freshwater crayfish of the genus *Euastacus* Clark (Decapoda: Parastacidae) from New South Wales, with a key to all species of the genus. Records of the Australian Museum, (Supplement) **23**: 1-110.

MUNASINGHE, D. H. N., C. P. BURRIDGE & C. M. AUSTIN, 2004a. Molecular phylogeny and zoogeography of the freshwater crayfish genus *Cherax* Erichson (Decapoda: Parastacidae) in Australia. Biological Journal of the Linnean Society, **81**: 553-563.

— —, — — & — —, 2004b. The systematics of freshwater crayfish of the genus *Cherax* Erichson (Decapoda: Parastacidae) in eastern Australia re-examined using nucleotide sequences from 12S rRNA and 16S rRNA genes. Invertebrate Systematics, **18**: 215-225.

MUNASINGHE, D. H. N., N. P. MURPHY & C. M. AUSTIN, 2003. Utility of mitochondrial DNA sequences from four gene regions for systematic studies of Australian freshwater crayfish of the genus *Cherax* (Decapoda: Parastacidae). Journal of Crustacean Biology, **23**: 402-417.

NGUYEN, T. T. T. & C. M. AUSTIN, 2005. Phylogeny of the Australian freshwater crayfish *Cherax destructor*-complex (Decapoda: Parastacidae) inferred from four mitochondrial gene regions. Invertebrate Systematics, **19**: 209-216.

PONNIAH, M. & J. M. HUGHES, 1998. Evolution of Queensland spiny mountain crayfish of the genus *Euastacus* Clark (Decapoda: Parastacidae): preliminary 16S mtDNA phylogeny. Proceedings of the Linnean Society of New South Wales, **119**: 9-19.

RICHARDSON, A. M. M. & R. SWAIN, 1980. Habitat requirements and distribution of *Engaeus cisternarius* and three subspecies of *Parastacoides tasmanicus* (Decapoda: Parastacidae), burrowing crayfish from an area of south-western Tasmania. Australian Journal of Marine and Freshwater Research, **31**: 475-484.

RIEK, E. F., 1951. The freshwater crayfish (family: Parastacidae) of Queensland. With an appendix describing other Australian species. Records of the Australian Museum, **22**: 368-388.

— —, 1956. Additions to the Australian freshwater crayfish. Records of the Australian Museum, **24**: 1-6.

— —, 1967a. The freshwater crayfish of Western Australia (Decapoda: Parastacidae). Australian Journal of Zoology, **15**: 103-121.

— —, 1967b. The Tasmanian freshwater crayfish genus *Parastacoides* (Decapoda: Parastacidae). Australian Journal of Zoology, **15**: 999-1006.

— —, 1969. The Australian freshwater crayfish (Crustacea: Decapoda: Parastacidae), with the descriptions of new species. Australian Journal of Zoology, **17**: 855-918.

— —, 1972. The phylogeny of the Parastacidae (Crustacea: Astacoidea), and description of a new genus of Australian freshwater crayfishes. Australian Journal of Zoology, **20**: 369-389.

SCHULTZ, M. B., S. A. SMITH, P. HORWITZ, A. M. M. RICHARDSON, K. A. CRANDALL & C. M. AUSTIN, 2009. Evolution underground: a molecular phylogenetic investigation of Australian burrowing freshwater crayfish (Decapoda: Parastacidae) with particular focus on *Engaeus* Erichson. Molecular Phylogenetics and Evolution, **50**: 580-598.

SCHULTZ, M. B., S. A. SMITH, A. M. M. RICHARDSON, P. HORWITZ, K. A. CRANDALL & C. M. AUSTIN, 2007. Cryptic diversity in *Engaeus* Erichson, *Geocharax* Clark and *Gramastacus* Riek (Decapoda: Parastacidae) revealed by mitochondrial 16S rDNA sequences. Invertebrate Systematics, **21**: 569-587.

SHAW, G., 1794. Zoology of New Holland. Volume 1: 1-33, pls. 1-12. (J. Davis, London).

SHULL, H. C., M. PÉREZ-LOSADA, D. BLAIR, K. SEWELL, E. SINCLAIR, S. LAWLER, M. PONNIAH & K. A. CRANDALL, 2005. Phylogeny and biogeography of the freshwater crayfish *Euastacus* (Decapoda: Parastacidae) based on nuclear and mitochondrial DNA. Molecular Phylogenetics and Evolution, **37**: 249-263.

SINCLAIR, E. A., A. MADSEN, T. WALSH, J. NELSON & K. A. CRANDALL, 2011. Cryptic genetic divergence in the giant Tasmanian freshwater crayfish *Astacopsis gouldi* (Decapoda: Parastacidae): implications for conservation. Animal Conservation, **14**: 87-97.

SMITH, G. C., A. BORSBOOM, R. LLOYD, N. LEES & J. KEHL, 1998. Habitat changes, growth and abundance of juvenile giant spiny crayfish, *Euastacus hystricosus* (Decapoda: Parastacidae), in the Conondale Ranges, South-east Queensland. Proceedings of the Linnean Society of New South Wales, **119**: 71-86.

SUTER, P. J., 1977. The biology of two species of *Engaeus* (Decapoda : Parastacidae) in Tasmania. II. Life history and larval development, with particular reference to *E. cisternarius*. Australian Journal of Marine and Freshwater Research, **28**: 85-93.

SUTER, P. J. & A. M. M. RICHARDSON, 1977. The biology of two species of *Engaeus* (Decapoda : Parastacidae) in Tasmania. III. Habitat, food, associated fauna and distribution. Australian Journal of Marine and Freshwater Research, **28**: 94-103.

THREATENED SPECIES SECTION, 2006. Giant Freshwater Lobster *Astacopsis gouldi* Recovery Plan 2006-2010: 1-40. (Department of Primary Industries and Water, Hobart).

TOON, A., M. PÉREZ-LOSADA, C. E. SCHWEITZER, R. M. FELDMANN, M. CARLSON & K. A. CRANDALL, 2010. Gondwanan radiation of the Southern Hemisphere crayfishes (Decapoda: Parastacidae): evidence from fossils and molecules. Journal of Biogeography, **37**: 2275-2290.

TURVEY, P. & J. R. MERRICK, 1997a. Reproductive biology of the freshwater crayfish, *Euastacus spinifer* (Decapoda: Parastacidae), from the Sydney Region, Australia. Proceedings of the Linnean Society of New South Wales, **118**: 131-155.

— — & — —, 1997b. Population structure of the freshwater crayfish, *Euastacus spinifer* (Decapoda: Parastacidae), from the Sydney Region, Australia. Proceedings of the Linnean Society of New South Wales, **118**: 157-174.

— — & — —, 1997c. Diet and feeding in the freshwater crayfish, *Euastacus spinifer* (Decapoda: Parastacidae), from the Sydney Region, Australia. Proceedings of the Linnean Society of New South Wales, **118**: 175-185.

— — & — —, 1997d. Moult increments and frequency in the freshwater crayfish, *Euastacus spinifer* (Decapoda: Parastacidae), from the Sydney Region, Australia. Proceedings of the Linnean Society of New South Wales, **118**: 187-204.

— — & — —, 1997e. Growth with age in the freshwater crayfish, *Euastacus spinifer* (Decapoda: Parastacidae), from the Sydney Region, Australia. Proceedings of the Linnean Society of New South Wales, **118**: 205-215.

WEBB, M. & A. RICHARDSON, 2004. A radio study of movement in the Giant Tasmanian Freshwater crayfish, *Astacopsis gouldi*. Freshwater Crayfish, **14**: 197-204.

ZEIDLER, W. & M. ADAMS, 1990. Revision of the Australian crustacean genus of freshwater crayfish *Gramastacus* Riek (Decapoda: Parastacidae). Invertebrate Taxonomy, **3**: 913-924.

First received 11 April 2012.
Final version accepted 29 July 2013.

THE FRESHWATER CRAYFISH FAUNA OF AUSTRALIA: UPDATE ON CONSERVATION STATUS AND THREATS

BY

JAMES M. FURSE[1])

Environmental Futures Research Institute and Griffith School of Environment, Gold Coast campus, Griffith University, Queensland 4222, Australia

ABSTRACT

The Australian Continent is home to one of the most diverse freshwater crayfish faunas in the world, and includes many of the world's largest species of freshwater crayfish (including the largest species at ~1 m overall length), but also some of the smallest. The Australian fauna is almost entirely endemic and features some of the world's rarest, slowest growing and longest lived species (over 30 years in some cases), the most highly endangered, but also the most iconic, charismatic and in some cases, most bizarre looking freshwater crayfish. Some species are among the hardiest, most ecologically aggressive, highly fecund, and fastest growing in the world. Due to traits desirable for fisheries and aquaculture, some species have been extensively translocated over large distances. With the exception of the arid zone, the freshwater crayfish of Australia occupy all habitat types and climatic zones on the Continent. A few species are widely distributed habitat generalists that can tolerate extremes in environmental conditions, while many other species have far smaller distributions ($\ll$10 km^2 in a number of cases) and are very closely associated with specific habitat types and rely on particular environmental conditions. The widely distributed nature of the Australian fauna exposes the Australian fauna to a broad array of very serious existing and emerging threats, all of which are mainly anthropogenic in nature. A 2010 conservation assessment of the world's freshwater crayfish versus IUCN Red List criteria indicated that the Australian freshwater crayfish fauna was the world's most endangered. This paper, 1) provides a review of earlier conservation reports and outlines our understanding of the situation to date, 2) presents the current conservation status of the Australian fauna, 3) provides updates on well-known and emerging threats, and 4) discusses recent discoveries and progress towards conservation of the Australian fauna.

THE FRESHWATER CRAYFISH FAUNA OF AUSTRALIA

The Australian Continent is one of the two main centres of freshwater crayfish diversity (Taylor, 2002), and with more than 140 species in 10 genera

[1]) e-mail: j.furse@griffith.edu.au

 Advances in freshwater decapod systematics and biology: 273-296

TABLE I

The Freshwater Crayfish of Australia (after Coughran & Furse, 2012)

Genus	Number of species	Distribution	Comments
Astacopsis	3	Tasmania	Includes the largest species in the world (*A. gouldi*)
Cherax	26	All States on mainland Australia	Includes some very large and ecologically aggressive species that have been widely translocated
Engaeus	35	Victoria and Tasmania	An unusual group of small, obligate burrowing crayfish
Engaewa	5	Western Australia	An unusual group of small, obligate burrowing crayfish
Euastacus	52	Southeastern mainland	Australia's largest genus, includes the second largest species in the world (*E. armatus*)
Geocharax	2	Victoria and Tasmania	A poorly understood group
Gramastacus	2	Southeastern mainland	A poorly understood group
Ombrastacoides	11	Tasmania	All are burrowing species, some have highly restricted distributions
Spinastacoides	3	Tasmania	All are burrowing species with widespread distributions
Tenuibranchiurus	1	Central-eastern mainland	Australia's smallest freshwater crayfish

(table I), Australia is second only to North America (including Mexico) which has more than 435 species in 13 genera (Taylor, 2002; Crandall & Buhay, 2008; Fetzner, 2012). A number of additional species from Australia are either in the process of being formally described, or are awaiting description (Coughran & Furse, 2012; Furse et al., 2013). Furthermore, there are large areas of suitable but extremely remote habitat on the Continent that have not yet been surveyed, and it is likely the Australian species' count will rise as survey expeditions to these areas are completed (Furse & Coughran, 2011a).

In addition to being noted for its diversity, the Australian Continent is also home to the iconic and "giant forms" (Provenzano, 1985) of the world's freshwater crayfish fauna. The Australian fauna includes at least five of the world's largest species of freshwater crayfish (in genera *Astacopsis* Huxley, 1878, *Euastacus* Clark, 1936, and *Cherax* Erichson, 1846) (Coughran & Furse, 2012). The world's largest freshwater invertebrate, *Astacopsis gouldi* Clark, 1936 (from Tasmania), is documented as reaching ~1 m in overall length

Fig. 1. *Astacopsis gouldi* Clark, 1936, from Tasmania, the largest species of freshwater crayfish in the world (photo: Todd Walsh).

(i.e. from tip of chelae to end of telson), and weights of ~4.5 kg (anecdotal reports of animals weighing 6 kg exist) (Threatened Species Section, 2006; Walsh & Walsh, 2013) (fig. 1); it is by a considerable margin the worlds' largest species of freshwater crayfish (Holdich, 2002a). The worlds' second largest species, *Euastacus armatus* (von Martens, 1886) is also impressively large and reportedly reaches 500 mm in overall length and can exceed 3 kg in weight (Horwitz, 1990a; Geddes et al., 1993).

The Australian Continent is also occupied by some of the world's smallest species. *Tenuibranchiurus glypticus* Riek, 1951 (overall length of ~35 mm), from central-eastern Australia is the smallest Australian species (Crandall, 2002; Dawkins et al., 2010). A number of the small and strongly, or in some cases obligate, burrowing species in the genera *Engaeus* Erichson, 1846 (from Tasmania), and *Engaewa* Riek, 1967 (from Western Australia) are among the most bizarre looking of all freshwater crayfish due to their highly reduced abdomens (fig. 2).

Australian species are also among the world's most productive aquaculture candidates, in particular *Cherax destructor* Clark, 1936, *Cherax tenuimanus* (Smith, 1912), and *Cherax quadricarinatus* (von Martens, 1868) (Riek, 1951; Morgan, 1986, 1988, 1997; Momot, 1995; Holdich, 2002a), and for this reason as species of interest to recreational fishers (Horwitz, 1990b; Kailola et al., 1993), they have been widely translocated outside of their natural

Fig. 2. The obligate burrowing species, *Engaewa pseudoreducta* Horwitz & Adams, 2000, from Western Australia, is one of the most unusual looking of Australia's freshwater crayfish due to its highly reduced abdomen (photo: Quinton F. Burnham).

ranges within Australia over the past few decades (Coughran & Leckie, 2007; Doupé, 2007; Doupé et al., 2007; Coughran et al., 2009; Coughran & Furse, 2010b, 2012; Furse & Coughran, 2011c; Leland et al., 2012). *Cherax quadricarinatus* has been translocated to the Americas, Africa, Europe, Asia and the Middle East for aquaculture research and for use in commercial aquaculture (Lawrence & Jones, 2002), but also as an ornamental species for the aquarium trade in Europe (Peay et al., 2010).

The Australian fauna also contains some of the most conspicuous and charismatic species, and some of the best known are from genus *Euastacus*; with 52 species described, *Euastacus* is the largest of Australia's 10 genera (Coughran & Furse, 2011; Coughran & McCormack, 2011; Furse & Coughran, 2011b; Furse et al., 2013).

A number of *Euastacus* feature quite remarkable and striking colourations including vivid blues and reds (fig. 3), and fluorescent oranges (e.g., *Euastacus sulcatus* Riek, 1951, and *Euastacus fleckeri* (Watson, 1953)) (Jones & Morgan, 2002; Coughran, 2006b; Furse & Coughran, 2011a), and some species (particularly *E. sulcatus*) are well known to bushwalkers and local landowners in southeast Queensland and north-eastern New South Wales due to their tendency to move considerable distances overland through rainforests and along National Parks walking tracks in damp conditions (Riek, 1951; Morgan, 1988; Furse & Wild, 2002; Furse et al., 2004; Reynolds & Souty-Grosset, 2012).

Fig. 3. Some of Australia's best known and iconic crayfish, such as *Euastacus sulcatus* Riek, 1951, from the coastal mountains of central-eastern Australia, display striking colors and/or impressive arrays of spines (photo: James M. Furse).

MOTIVATIONS FOR THE CONSERVATION OF FRESHWATER CRAYFISH

In a Global sense, the case for conservation of freshwater crayfish has been well established for some time, and the justifications for conservation of these animals are typically associated with cultural, social, aesthetic and economic considerations (Holdich, 2002a, b; Horwitz, 2010); these socioeconomic motivations are also applicable to the Australian situation. However, some of the most compelling, yet occasionally overlooked motivations for the conservation of freshwater crayfish are biological and ecological.

After some decades of study the ecological role(s) of freshwater crayfish in the systems they occupy are reasonably well understood (but not yet complete), and freshwater crayfish have been identified as having a central and somewhat unique position in aquatic food webs (Hogger, 1988; Momot, 1995).

Freshwater crayfish are polytrophic omnivores and these detritivorous, herbivorous and carnivorous feeding habits positions them as major processors of organic material whereby they act as shredders, predators, collectors and grazers (i.e. shredders, Anderson & Sedell, 1979; Huryn & Wallace, 1987; Parkyn et al., 1997; Usio & Townsend, 2001; Furse, 2010).

The activities of freshwater crayfish facilitate the release of energy and nutrients (Momot, 1995), and due to their burrowing and foraging activities act as "geomorphic agents" through bioturbation of sediments (Statzner et al., 2000, 2003; Furse, 2010) and pedoturbation of soils (Richardson, 1983; Stone, 1993). These various and naturally occurring activities of freshwater crayfish satisfy the criteria for their designation as "ecosystem engineers" (sensu Jones et al., 1994; Reynolds & Souty-Grosset, 2012).

Freshwater crayfish are well known as animals that are often highly abundant (e.g., *Orconectes limosus* (Rafinesque, 1817) from Europe has been documented at densities of up to 77 m^{-2} (Nystrom, 2002)), and can therefore dominate consumer biomass in aquatic ecosystems (Mason, 1975; Huryn & Wallace, 1987; Momot, 1995; Nystrom, 2002; Furse, 2010).

Their inherent biology and high abundances often result in freshwater crayfish having very strong ecological effects, and playing key roles in the aquatic habitats they occupy. They are therefore quite appropriately recognised as "Keystone Species" (Hart, 1992; Krebs, 1994; Collier et al., 1997; Lodge et al., 2000; Nature, 2009).

However, unless specifically investigated, the ecological role(s), and strong effect(s) of freshwater crayfish within their native ranges may be underestimated, or even completely overlooked: but this of course is true for all species (Furse, 2010). The ecological effect(s) of freshwater crayfish, and the magnitude of these effects, may only become apparent when a species is removed, or eliminated from its natural habitat, or a particular area of habitat.

A number of clear examples are available from Europe where the introduced North American crayfish plague (*Aphanomyces astaci* Schikora, 1906) has resulted in on-going sporadic mass mortalities of susceptible native species. Abrahamsson (1966) attributed dramatic changes to the aquatic flora and fauna of ponds in Sweden to the loss of the native *Astacus astacus* (Linnaeus, 1758) crayfish population in 1964. Submerged aquatic macrophyte growth increased dramatically such that the surface of the ponds were, on occasions, covered by plant growth, similarly; mollusc, leech and tadpole abundances also increased dramatically (Abrahamsson, 1966). Since Abrahamsson's (1966) observations, mass mortalities of crayfish in other regions of Europe (Sweden, Unestam, 1973; Ireland, Matthews & Reynolds, 1992) have resulted in similar ecological responses being reported.

Similarly, when freshwater crayfish are translocated outside of their native ranges, many unanticipated consequences may become evident, and it is extensively documented that these consequences are rarely benign. Well known examples of these consequences include where translocation of *Orconectes rusticus* (Girard, 1852), *Pacifastacus leniusculus* (Stimpson, 1857), and *Procambarus clarkii* (Girard, 1852) outside of their native ranges have resulted in reproductive failure of large freshwater fish species (crayfish consumed the fish eggs), reduced abundance of aquatic macrophytes, and in the case of *P. clarkii*; ecological and economic damage to rice fields in California, Spain and Japan (Lodge et al., 1985; Hogger, 1988; Huner, 1988), and appreciable damage to riverbanks by *P. leniusculus* in England (as outlined in Peay et al., 2010).

Due to their underlying biology (especially their high fecundity and growth rates), tolerance of extreme environmental conditions and their ecologically aggressive nature, there is now ample evidence that the Australian *C. destructor* and *C. quadricarinatus* are displacing native, and in some cases, Critically Endangered species (Leland et al., 2012). A good example of the relatively small *C. destructor* displacing a somewhat larger species is provided by the case of the Critically Endangered *Euastacus dharawalus* Morgan, 1997 (as outlined in Coughran & Furse, 2010a; Furse & Coughran, 2011c). The impacts, if any, of the various feral populations of *C. quadricarinatus* that have (or may have) established overseas are not yet understood, but the potential threats to native biota have been identified by various workers in Singapore (Ahyong & Yeo, 2007), and Great Britain (Peay et al., 2010). Non-native species of freshwater crayfish are well known for their capacity to displace native crayfish, fish and amphibian species (e.g., Lodge et al., 1985; Horwitz, 1990b; Harlioğlu & Harlioğlu, 2006), and the potential consequences of translocating these very hardy Australian species anywhere beyond their natural ranges should not be underestimated.

IUCN CONSERVATION ASSESSMENTS OF THE AUSTRALIAN FRESHWATER CRAYFISH FAUNA

Australian freshwater crayfish have featured in the IUCN Red List of threatened species since publication of the IUCN Invertebrate Red Data Book by Wells et al. (1983), and at that time one Australian species, *A. gouldi* was listed as Vulnerable (Wells et al., 1983); since then, the 1996 Red List assessment evaluated 19 Australian species (IUCN, 2012). A major step forward in the conservation of freshwater crayfish was the first global assessment of the world's freshwater crayfish versus IUCN Red List Criteria in 2010, where 528 species were assessed, including 121 Australian species (tables II and III).

TABLE II

History of IUCN Red List Assessments for Australian freshwater crayfish (after Wells et al., 1983; IUCN, 2011b)

IUCN category	Number of Australian species (by year of assessment)			Type of category
	1983	1996	2010	
Critically Endangered			26	Threatened
Endangered		9	24	
Vulnerable	1	8	10	
Data Deficient*		2	9	Not Assessable
Near Threatened			7	Non-Threatened
Least Concern			45	
Total	1	19	121	

* Data Deficient (DD) assessments indicate there was insufficient data to make a direct, or indirect, assessment of a taxon (IUCN, 2011a). Species with DD status should not be regarded as "Non-Threatened", for instance the DD *Euastacus armatus* is exposed to a number of very serious threats, and may well be at considerable risk of extinction (Coughran & Furse, 2010b).

TABLE III

Results of 2010 IUCN Red List Assessments of the Australian freshwater crayfish (data in part IUCN; after Coughran & Furse, 2012)

Genus	Number of species assessed	Percent in IUCN Red List Threatened* Categories
Astacopsis	3	30%
Cherax	11**	23%
Engaeus	35	29%
Engaewa	5	60%
Euastacus	49	80%***
Geocharax	2	50%
Gramastacus	1	0%
Ombrastacoides	11	27%
Spinastacoides	3	0%
Tenuibranchiurus	1	100%

* The IUCN Red List Threatened Categories include: Critically Endangered, Endangered and Vulnerable.

** Only 11 of the 24 species of Australian *Cherax* species described at the time were included in the 2010 IUCN Red List assessments, only 3 of the 11 species assessed were in Threatened Categories.

*** The world's most endangered genus of crayfish (N. Dewhurst, pers. comm.).

The results of the 2010 assessment were not encouraging with 25% of the world's species in IUCN threat categories (or extinct), and placed freshwater crayfish in the 5 most highly threatened groups of animals (N. Dewhurst, pers. comm.). Furthermore, 22% of species were so poorly understood (i.e. Data Deficient (DD)) that insufficient data was available to make direct or indirect assessments: this provided a clear indication that freshwater crayfish had been overlooked in the past (Coughran & Furse, 2012). Data Deficient status accorded to so many species should be a matter of great concern as many DD species are exposed to numerous and well understood threats, and are therefore already of considerable conservation concern (e.g., *E. armatus*, see Coughran & Furse, 2010b), and likely at risk of extinction: yet a lack of data impedes simple conservation assessments.

As discussed with regard to genus *Euastacus*, Coughran & Furse (2010b) and Furse & Coughran (2011b, c) pointed to the fact that some fragmented populations of species may in fact be genetically distinct taxa. Cryptic diversity has been identified in Australian mayflies, freshwater shrimp and crayfish (Baker et al., 2004; Sinclair et al., 2011), and the potential for uncovering additional "cryptic diversity" in the genera *Cherax*, *Engaewa*, *Euastacus*, *Engaeus*, and *Tenuibranchiurus* is very high (Coughran, 2006a; Coughran et al., 2008b; Dawkins et al., 2010; Furse, 2010; Furse & Coughran, 2011c; Burnham et al., 2012). Investigation of cryptic diversity should be a research priority as a number of non-threatened species would qualify for Critically Endangered status and most likely warrant immediate active conservation management if identified as distinct species (as outlined for *E. sulcatus* in Furse & Coughran, 2011c).

Genus *Tenuibranchiurus* provides an excellent case study of the potential for overlooking cryptic diversity. Dawkins et al. (2010) identified there are additional species of *Tenuibranchiurus* (i.e. cryptic diversity), and these "Evolutionarily Significant Units" (ESUs, Moritz, 1994) are at considerable risk due to their coastal *Melaleuca* swamp habitat being located in the rapidly urbanising coastal corridor of the central-east of Australia, which includes the fastest growing urban regions in Australia (ABS, 2009; Dawkins et al., 2010). As outlined by (Coughran et al., 2008a), the 2 small, isolated, and geographically restricted populations of this single species (*T. glypticus*) are facing a number threats that could easily lead to the rapid extirpation of either population due to a simple accident (Merrick, 1995), such as an oil, fuel or chemical spill on any of the nearby major roadways.

Overall, the 2010 IUCN Red List assessment did provide some encouragement, in that just over half (i.e. 53%) of the world's crayfish species were not in IUCN threat categories (e.g., were Near Threatened or not of any immediate conservation concern), however this should not be interpreted as an improving conservation situation (Coughran & Furse, 2012), but more as an initial baseline evaluation of the conservation status of those species. An on-going monitoring programme should be implemented on these species, or at least some carefully selected "indicator" or "reference" species, as data will be required for future IUCN assessments and subsequent conservation status updates.

THREATS TO THE FRESHWATER CRAYFISH OF AUSTRALIA — UPDATE ON EXISTING AND EMERGING THREATS

Existing threats

Our understanding of the threats to freshwater crayfish has been developing since publication of the IUCN Invertebrate Red Data Book (Wells et al., 1983) which first highlighted the main threats to invertebrates in general (i.e. habitat destruction, pollution, exotic species, and human exploitation). After many years, and much study, this now common suite of threats is relatively well understood and impacts a good deal of the world's biota.

More recently, the 2010 IUCN Red List assessment pointed to a number of additional and very worrying threats and threatening processes regarding the freshwater crayfish of Australia, particularly species with restricted ranges, fragmented distributions and those species exploited by humans. Research conducted over the last decade has facilitated clarification of the distributions, and distributional characteristics of many species (e.g., identifying restricted ranges and fragmentation), and this in turn has allowed many new conservation assessments and the refinement of existing assessments.

As outlined by Coughran & Furse (2012), the situation of many species with highly restricted ranges, fragmented ranges, and/or those species occupying single localities is indeed precarious. Of particular concern is the fact that in each of these cases, there are many species.

For instance, in the previously discussed case of *Tenuibranchiurus*, the *Melaleuca* swamp habitat of these restricted range (and ESU) species is in very close proximity to, surrounded by, or being encroached upon by corridors of rapid urbanisation, broad-acre agricultural activities (especially sugarcane farming and forestry) or other anthropogenic activities (e.g., light industry and construction of infrastructure, especially transport links). Species that occupy such small ranges, in such situations, are exposed to highly elevated risks of extinction from small scale and direct habitat destruction.

Coughran & Furse (2012) also pointed to the indirect anthropogenic threats to restricted range species with fragmented distributions. Species in the genus *Engaewa* are primary burrowing crayfish ("primary burrowers", Horwitz & Richardson, 1986) and are regarded as completely reliant (i.e. obligate burrowers, Burnham et al., 2012), on their extensive and typically deep networks of burrows (Horwitz & Adams, 2000). While these crayfish rarely appear aboveground, and would seem to be largely isolated from the common suite of "aboveground" threats, their conservation status is of considerable concern due to reliance on subterranean (or subsurface) water and their associations with specific habitat characteristics; in particular very specific soil types and profiles (Horwitz & Adams, 2000; Burnham, 2010; Burnham et al., 2012).

Engaewa pseudoreducta Horwitz & Adams, 2000, was assessed by Burnham (2010) as Critically Endangered in the 2010 IUCN Red List assessment due to its highly restricted and fragmented distribution, and reliance on subterranean water (Horwitz et al., 2008; Burnham, 2010; Burnham et al., 2012). It appears the species has been extirpated from its type locality, and water extraction for agriculture (and other uses), and the compaction of soil by agricultural activities threatens the remaining populations (Horwitz et al., 2008; Burnham, 2010; Burnham et al., 2012).

The biology of many of the exceptionally rare and endangered species of Australian crayfish (i.e. slow growing, late maturing, low fecundity and very long lived) renders them completely unsuitable for any level of exploitation by humans (e.g., recreational fishing, aquarium trade or collection by enthusiasts) (Coughran & Furse, 2010b, 2012; Furse & Coughran, 2011a). Legal recreational over-fishing (and illegal fishing) is often cited as one of the primary reasons that *E. armatus* is of considerable conservation concern (Van-Praagh, 2003; Gilligan et al., 2007), and are also contributing factors to why *A. gouldi* is now listed as Endangered on the IUCN Red List (Doran & Walsh, 2012).

Emerging threats

Increased environmental temperature and changes to climatic and weather patterns. — While the idea of increased environmental temperature (and subsequent changes to climatic conditions and weather patterns), as threats to Australian freshwater crayfish is by no means new (e.g., Global Warming, Horwitz, 1990a), the effects of severe weather events (SWEs, Furse et al., 2012b) on freshwater crayfish have started to be documented (McKinnon, 1995; Parkyn & Collier, 2004; Lewis & Morris, 2008; Furse et al., 2012b).

The very serious threats posed by changing temperatures and altered climatic and weather patterns to many already rare and endangered species is only now becoming apparent as our understanding of these animals and their distributions improves, and in some cases existing threats may be exacerbated (e.g., further spread of exotic species, habitat loss) (Furse & Coughran, 2011b; Coughran & Furse, 2012; Furse et al., 2012b).

Australian montane crayfish species with restricted and/or fragmented ranges that occupy isolated, cool and damp pockets of habitat, without the capacity to relocate to habitat at higher and cooler attitudes, are at particular risk from the direct effect of climate change: increased environmental temperature (Horwitz, 1990a, 2010; Furse & Coughran, 2011b; Furse et al., 2012b; Bone et al., 2014). The long term survival of these poikilotherms that lack the capacity to appreciably regulate their temperature and acclimate to new thermal regimes (Lowe et al., 2010), or relocate to cooler habitats is doubtful (Furse, 2010; Bone et al., 2014).

Also at considerable risk from increased temperature are lowland species, such as *Tenuibranchiurus*, with very specific abiotic habitat requirements: it is unlikely this very small species will be able to relocate to more suitable habitat in response to temperature increases or saltwater intrusion into their low-lying coastal habitats (Coughran et al., 2008a). Similarly, *Engaewa* spp. have no prospect of relocating to new habitat (Burnham, 2010; Burnham et al., 2012) in the event of any depletion of natural groundwater reserves due to increased temperature and subsequent increases in evaporation from the soil column. The potential issue of decreasing groundwater availability due to increased temperature is likely to be compounded by additional extraction pressures on groundwater reserves due to various human activities, such as agriculture and the rapidly increasing urbanisation in that region (Horwitz et al., 2008; Burnham et al., 2012; Q. F. Burnham, pers. comm.).

In addition to the direct threat of increased environmental temperature, there are a number of worrying second-order ecological effects of climate change (e.g., changes in floral and faunal assemblages and food webs), that are more likely to arise from changes in rainfall than directly from changed temperature itself.

Many species of freshwater crayfish are associated with very specific habitat types (in particular vegetation types) and it is has been established that vegetation types will "shift" in accord with changes in temperature, but especially rainfall; one such Australian example is decreased precipitation resulting in a change from montane rainforest to drier sclerophyllous forest

(Hilbert et al., 2001; Hughes, 2003). Such changes in floral assemblages are a most serious threat to the many habitat-specialist species that exist in Australia.

Predicted changes to weather patterns include increased frequency and severity of weather events such as floods, droughts, tropical cyclones and storms (Hughes, 2003; IPCC, 2007; Specht, 2008). A number of cases of freshwater crayfish populations being impacted by SWEs, including severe flooding events, have been reported over the last few years, and the frequency of these events being reported appears to be increasing (McKinnon, 1995; Parkyn & Collier, 2004; Lewis & Morris, 2008; Furse et al., 2012b; McCarthy et al., 2014).

In this context, species with highly restricted and/or fragmented ranges that occupy areas that are already subject to episodic and heavy rainfall, or areas that are predicted to receive additional precipitation (Chiew & McMahon, 2002; Hughes, 2003) are of considerable concern. Furse et al. (2012b) discussed the very real possibility that a localised SWE, such as a single severe storm or heavy rainfall event, could conceivably eliminate a restricted range species such as *Euastacus jagabar* Coughran, 2005, which has a entire distribution of 2.5 km^2. Australia has a number of species with entire distributions <10 km^2 and SWE's are clearly an emerging and considerable threat (Burnham, 2010; Dawkins et al., 2010; Furse & Coughran, 2011b; Furse et al., 2012a).

Exotic non-crayfish species. — It is well established that exotic species are a considerable threat to freshwater crayfish, but in particular restricted range crayfish species. In the Australian context, Coughran & Furse (2010b, 2012) highlighted that highly mobile species such as foxes (*Vulpes vulpes* (Linnaeus, 1758)), wild pigs (*Sus scrofa*, Linnaeus, 1758), goats (*Capra hircus*, Linnaeus, 1758), cats (*Felis catus* (Linnaeus, 1758)) and the toxic cane toad (*Rhinella marina* (Linnaeus, 1758)) pose very serious potential threats to populations of rare and isolated populations of native crayfish species.

Also well known, but primarily from other regions of the world, are the most serious threats posed by other crayfish (see Harlioğlu & Harlioğlu, 2007; Peay et al., 2010; Reynolds & Souty-Grosset, 2012). It is well documented that losses of formerly common and widespread native crayfish species are attributable to other non-indigenous crayfish being intentionally or inadvertently translocated between waterbodies or drainages (Lodge et al., 1985, 1994, 2000; Kozubíková et al., 2008; Holdich et al., 2009; Peay et al., 2010).

Translocation of Australian crayfish species. — Concerns surrounding the translocation of crayfish within Australia are not new (Horwitz, 1990b;

Horwitz & Knott, 1995; Coughran & Leckie, 2007; Coughran et al., 2009), however, we are now seeing the effects of these translocations, and it is clear that "other" Australian species are becoming problematic. As previously outlined, the species *E. dharawalus* is at very serious risk of near-term extinction due to the more aggressive *C. destructor* (Coughran et al., 2009; Coughran & Furse, 2010b; Coughran & Daly, 2012).

More recently, an ESU population of *Tenuibranchiurus* in northern New South Wales was identified as at risk from a nearby feral population of *C. quadricarinatus* (Dawkins et al., 2010; Leland et al., 2012) at a site where *Cherax cuspidatus* Riek, 1969, was formerly well known (Coughran et al., 2008b). Despite extensive trapping efforts, the formerly common *C. cuspidatus* has not been captured at the site recently, and the species may in fact have been eliminated by the apparently large population of *C. quadricarinatus* (Leland et al., 2012).

The reasons behind some of these translocations are known, and include 1) escapees from farm dam translocations, and failed aquaculture ventures, 2) domestic aquarium escapees and discards, 3) bait-bucket discards and illegal introductions by recreational fishers, and 4) illegal translocations between waterbodies (including between States) by "Grey Nomads" (retirees on long-duration or frequent caravan/camping/off-road tours of Australia).

Of these, the aquarium trade is of particular concern as various Australian species are freely available well outside of their natural ranges, often as "feeders" for larger predatory aquarium fish, plus berried females or breeding pairs are often available for purchase. Typically, *C. destructor* (in particular the blue colour morph, the "Electric Blue Yabby") and *C. quadricarinatus* are the commonly available species. Species of crayfish that are protected (such as all species of *Euastacus* in Queensland, that cannot be legally taken from the wild without appropriate permits), have been seen on sale from time-to-time in aquarium stores.

Outcomes of purchases from the aquarium trade typically include additional movements of the species around Australia, or aquarium discards to local waterbodies (both as a consequence of pet owners moving residences), or inadvertent escapees from aquariums to local waterbodies. Escapees are not an unlikely scenario, in fact given that freshwater crayfish are well known as "consummate escape artists" (Huner, 1999) crayfish escapees from any domestic or commercial/retail facility are almost assured.

"Private" and unlicensed collectors. — The emergence of private collectors was discussed by Coughran (2007) and the author's and colleagues experiences with these people since that time suggests the situation has not improved.

These private collectors typically claim to be "hobby aquarists", or "professional wildlife photographers", or claim to be conducting "scientific research". Irrespective of monikers, it appears their main objective(s) are to collect specimens of every possible species (often in large numbers, sometime more than 40 specimens), including rare and critically Endangered Species (including berried females). In many cases these activities seem to have been conducted illegally in National Parks (i.e. without scientific collection permits).

Recently (2012), at a freshwater ecology conference in Europe, the author observed live specimens (i.e. on display in aquaria) of at least 6 species of freshwater crayfish from New Guinea and Australia (and one species from the United States that was apparently not previously known to occur in Europe): all specimens had presumably been collected and transported to Europe by a well-known "private collector" who arranged the display.

The various activities of private collectors have the capacity to transmit parasites, commensal organisms and diseases/pathogens between waterbodies within Australia (e.g., the amphibian chytrid fungus). Unregulated and illegal removal of poorly understood, rare and/or Critically Endangered Species (especially sexually reproductive animals, but in particular brooding females) has the potential to cause serious problems in wild populations: particularly reproduction, and the production and recruitment of juvenile crayfish to higher-age classes.

Unregulated International translocations of crayfish. — International translocation of crayfish for any reason (such as the *Cherax* spp. to Europe case outlined above) has the potential for catastrophic consequences; as continues to be demonstrated by the well known case of North American crayfish species carrying the crayfish plaque *A. astaci* to Europe and its devastating consequences.

The Southern Hemisphere crayfish are well known for their diverse commensal and ectosymbiont fauna (e.g., Temnocephalan flatworms) (Cannon & Jennings, 1987; Cannon, 1991, 1993; Cannon & Sewell, 1994, 1995, 2001; Sewell & Cannon, 1998a, b; Wild & Furse, 2004) but as our understanding of this area is incomplete, there are certainly undiscovered parasitic and ecto-commensal species on the crayfish of the Southern Hemisphere. As has been clearly demonstrated elsewhere, the known and unknown fauna on crayfish poses a potential and considerable threat to the freshwater biota of any country that receives these crayfish without appropriate inspections and controls.

This is of course also true for illegal importation of crayfish from the Northern Hemisphere (or elsewhere) to Australia, importation of a "healthy carrier" of some disease could be catastrophic for the unique crayfish fauna of Australia (sensu the crayfish plague). This scenario is not beyond the realms of possibility; numerous commercial websites offer species for sale, and various internet forums specialise in these matters.

The rise of the Internet, Social Media and the consequent risks. — In 2007, Coughran pointed to the various internet forums that featured extensive discussions about illegal collection and guidelines for keeping and care of protected species in domestic aquariums (Coughran, 2007). Since at least 2010, entire books (published in Australia), with sections devoted to providing detailed instructions on keeping these animals in captivity have been available. The rise of social media has seen the emergence of numerous additional online "groups" with very large International memberships (from Europe, The Middle East, Asia, Southeast Asia, the Americas, Oceania and Africa). Trade in freshwater crayfish, shrimps, fish and freshwater crabs to-and-from from all regions of the Globe are openly discussed and publicised: it appears that International sales and translocations are also openly arranged through these same groups. In one case, the live (and illegal) export of *E. armatus* for a private aquarist in Asia had been arranged through one of these groups. Obviously the reliability of these ephemeral websites and forums is highly questionable and cannot be evaluated, and they may have short lifespans, but in any case, even the most basic search of the internet will quickly reveal there are many such sites active at any one time.

This rise of Social Media and its effects on the International translocation of freshwater crayfish (and other freshwater biota) is extremely concerning as it is, 1) unregulated, 2) apparently a rapidly expanding phenomenon, 3) dealing in rare and endangered species (and in these cases, clearly in breach of CITES regulations), and 4) there are few mechanisms to control the International movements of specimens between countries. The only real barriers to International trade of this kind are the postal, customs and quarantine inspections of international mail and travellers at ports of entry and exit.

RECENT PROGRESS AND THE WAY FORWARD

Efforts to conserve the freshwater crayfish of Australia are ongoing with a small number of highly dedicated and active researchers, research teams and their volunteers, addressing the numerous topics that have been previously identified as research imperatives. Due to the work of these researchers (which is very much "enabling research"), preparation of nominations for listing of species under State conservation legislation are now becoming possible.

Nominations have been prepared, submitted, and it is anticipated that *T. glypticus* and *Euastacus bindal* Morgan, 1989 (from far North Queensland) will be listed under the Queensland Nature Conservation Act (1992) as "Endangered", which is the highest threat category under Queensland State legislation. These will be the first species of freshwater crayfish listed under Queensland State legislation.

In late 2011, *E. dharawalus* from New South Wales was listed as a "Critically Endangered Species" under New South Wales State legislation (the Fisheries Management Act (1994)). It is encouraging to see that since listing the species in 2011, the NSW State Government has prepared a factsheet that among other things identifies recovery and conservations actions for this species, and outlines the legal consequence of breaching the fisheries legislation and regulations that protect this species.

The next steps towards conservation of the Australian freshwater crayfish fauna are reasonably well understood and have been outlined and discussed at length by various workers over the past few years (for reviews see Horwitz, 2010; Furse & Coughran, 2011c; Coughran & Furse, 2012).

In short, the next steps to ensure there is a chance that the unique freshwater crayfish fauna of Australia can be conserved, previously identified research imperatives will need to be addressed, Legislation will need to be reviewed (and in some cases enacted), positive action(s) will need to be taken by Government Authorities, recovery and monitoring programmes will need to be designed and implemented, collaboration between all stakeholders will be required (especially for species that occur across State Borders), and the chronic shortage of funding for basic biological, and "enabling" scientific research (including taxonomic research) will need to be rectified (Horwitz, 2010; Furse & Coughran, 2011c; Coughran & Furse, 2012).

ACKNOWLEDGEMENTS

Many and sincere thanks to Quinton F. Burnham and Todd Walsh for kindly providing the images of *Engaewa pseudoreducta* and *Astacopsis gouldi*. Similarly, thanks are extended to the editors of this special issue of Crustaceana Monographs for the kind invitation to contribute. This paper was written, in part, in Innsbruck (Austria), and I sincerely thank my hosts for providing the ideal surroundings for such a task. Particular thanks are due to Dr. Leopold

Füereder and his team from the Institute of Ecology, University Of Innsbruck, Austria. I appreciate my anonymous reviewers for providing helpful comments on an earlier version of this paper. Special thanks are due to my colleague Jason Coughran for our discussions and collaborations on our previous papers on this topic, all of which provided the basis from which this update was constructed. This work was supported by the Environmental Futures Research Institute, and the Griffith School of Environment, Griffith University, Gold Coast campus, Queensland, Australia.

REFERENCES

ABRAHAMSSON, S. A. A., 1966. Dynamics of an isolated population of the crayfish *Astacopsis Astacopsis* Linne. Oikos, **17**: 96-107.

ABS, 2009. Regional population growth, Australia, 2007-08. Report number 3218.0. The Australian Bureau of Statistics, Canberra, ACT, Australia. Available online via: http://www.abs.gov.au/AUSSTATS/abs@.nsf/DetailsPage/3218.02007-08?OpenDocument.

AHYONG, S. T. & D. C. J. YEO, 2007. Feral populations of the Australian Red-Claw crayfish (*Cherax quadricarinatus* von Martens) in water supply catchments of Singapore. Biological Invasions, **9**(8): 943–946.

ANDERSON, N. H. & J. R. SEDELL, 1979. Detritus processing by macroinvertebrates in stream ecosystems. Annual Review of Entomology, **24**: 351-377.

BAKER, A. M., J. M. HUGHES, J. C. DEAN & S. E. BUNN, 2004. Mitochondrial DNA reveals phylogenetic structuring and cryptic diversity in Australian freshwater macroinvertebrate assemblages. Marine and Freshwater Research, **55**(6): 629-640.

BONE, J. W. P., C. H. WILD & J. M. FURSE, 2014. Thermal limit of *Euastacus sulcatus* (Decapoda: Parastacidae), a freshwater crayfish from the highlands of central eastern Australia. Marine and Freshwater Research, **65**(7): 645-651. DOI:10.1071/MF13189.

BURNHAM, Q., 2010. *Engaewa pseudoreducta*. In: IUCN Red List of Threatened Species (Version 2011.1). (The International Union for Conservation of Nature and Natural Resources, Gland, Switzerland).

BURNHAM, Q., A. KOENDERS & P. HORWITZ, 2012. The status of the critically endangered freshwater crayfish *Engaewa pseudoreducta* (Crustacea: Parastacidae) in south-western Australia. Records of the Western Australian Museum, **27**(1): 45-54.

CANNON, L. R. G., 1991. Temnocephalan symbionts of the freshwater crayfish *Cherax quadricarinatus* from northern Australia. Hydrobiologia, **227**: 341-347.

— —, 1993. New temnocephalans (Platyhelminthes): ectosymbionts of freshwater crabs and shrimp. Memoirs of the Queensland Museum, **33**: 17-44.

CANNON, L. R. G. & J. B. JENNINGS, 1987. Occurrence and nutritional relationships of four ectosymbionts of the freshwater crayfish *Cherax dispar* Riek and *Cherax punctatus* Clark (Crustacea: Decapoda) in Queensland. Australian Journal of Marine and Freshwater Research, **38**: 419-427.

CANNON, L. R. G. & K. B. SEWELL, 1994. Symbionts and biodiversity. Memoirs of the Queensland Museum, **36**: 33-40.

— — & — —, 1995. Craspedellinae Baer, 1931 (Platyhelminthes: Temnocephalida) ectosymbionts from the branchial chamber of Australian crayfish (Crustacea: Parastacidae). Memoirs of the Queensland Museum, **38**: 397-418.

— — & — —, 2001. A review of *Temnosewellia* (Platyhelminthes: Temnocephalida) ectosymbionts of *Cherax* (Crustacea: Parastacidae) in Australia. Memoirs of the Queensland Museum, **46**: 385-389.

CHIEW, F. & T. MCMAHON, 2002. Modelling the impacts of climate change on Australian streamflow. Hydrological Processes, **16**(6): 1235-1245.

COLLIER, K. J., S. M. PARKYN & C. F. RABENI, 1997. Koura: a keystone species? Water and Atmosphere, **5**(1): 18-20.

COUGHRAN, J., 2006a. Biology of the freshwater crayfishes of Northeastern New South Wales, Australia. (Ph.D. thesis, School of Environmental Science and Management, Southern Cross University, Lismore, New South Wales, Australia).

— —, 2006b. Field guide to the freshwater crayfishes of Northeastern New South Wales (Compact Disk Format). (Natureview Publishing, Bangalow, NSW, Australia. ISBN: 0 9581149 3 5).

— —, 2007. Distribution, habitat and conservation status of the freshwater crayfishes, *Euastacus dalagarbe*, *E. girurmulayn*, *E. guruhgi*, *E. jagabar* and *E. mirangudjin*. Australian Zoologist, **34**(2): 222-227.

COUGHRAN, J. & G. DALY, 2012. Potential threats posed by a translocated crayfish: the case of *Cherax destructor* in coastal drainages of New South Wales, Australia. Crustacean Research, (Special Number) **7**: 5-13.

COUGHRAN, J., K. L. DAWKINS & J. M. FURSE, 2008a. An assessment of genus *Tenuibranchiurus* (1 species) versus IUCN Red List Criteria. Unpublished report prepared for the global species conservation assessment of crayfishes for the IUCN Red List of Threatened Species. A report jointly prepared by the School of Environmental Science and Management (Southern Cross University), and Griffith School of Environment (Griffith University): 2 pp.

COUGHRAN, J. & J. FURSE, 2010a. *Euastacus dharawalus*. In: IUCN 2012. IUCN Red List of Threatened Species. Version 2012.1. www.iucnredlist.org. Downloaded on 8 August 2012.

— — & — —, 2010b. An assessment of genus *Euastacus* (49 species) versus IUCN Red List Criteria. Report prepared for the global species conservation assessment of crayfishes for the IUCN Red List of Threatened Species: 170 pp. (The International Association of *Astacopsis*, Auburn, Alabama, USA. ISBN: 978-0-9805452-1-0).

— — & — —, 2011. Chapter 5.1: Conservation of Australia's Spiny Crayfish — genus *Euastacus*. In: T. KAWAI & K. NAKATA (eds.), Seibutsu Kenkyusha, Tokyo, Japan: 398-402. [In Japanese.] (ISBN 987-4-915342-62-2).

— — & — —, 2012. Conservation of freshwater crayfish in Australia. Crustacean Research, (Special Number) **7**: 25-34.

COUGHRAN, J. & S. R. LECKIE, 2007. Invasion of a New South Wales stream by the Tropical Crayfish, *Cherax quadricarinatus* (von Martens). In: D. LUNNEY, P. EBY, P. HUTCHINGS & S. BURGIN (eds.), Pest or guest: the zoology of overabundance: 40-46. (Royal Zoological Society of New South Wales, Mosman, NSW, Australia. ISBN 978-0-9803272-1-2).

COUGHRAN, J., S. LECKIE & D. GARTSIDE, 2008b. Distribution, habitat and conservation status of the freshwater crayfishes, *Cherax cuspidatus* Riek and *Cherax leckii* Coughran (Decapoda: Parastacidae). Freshwater Crayfish, **16**: 19-26.

COUGHRAN, J. & R. B. MCCORMACK, 2011. *Euastacus morgani* sp. n., a new spiny crayfish (Crustacea, Decapoda, Parastacidae) from the highland rainforests of eastern New South Wales, Australia. ZooKeys, **85**: 17-26.

COUGHRAN, J., R. B. MCCORMACK & G. DALY, 2009. Translocation of the Yabby *Cherax destructor* into eastern drainages of New South Wales, Australia. Australian Zoologist, **35**(1): 100-103.

CRANDALL, K. A., 2002. Crayfish as model organisms. Freshwater Crayfish, **13**: 3-10.

CRANDALL, K. A. & J. E. BUHAY, 2008. Global diversity of crayfish (Astacidae, Cambaridae, and Parastacidae—Decapoda) in freshwater. Hydrobiologia, **595**(1): 295-301.

DAWKINS, K. L., J. M. FURSE, C. H. WILD & J. M. HUGHES, 2010. Distribution and population genetics of the threatened freshwater crayfish genus *Tenuibranchiurus* (Decapoda: Parastacidae). Marine and Freshwater Research, **61**(9): 1048-1055.

DORAN, N. & T. WALSH, 2012. *Astacopsis gouldi*. In: IUCN 2012. IUCN Red List of Threatened Species. Version 2012.1. (The International Union for Conservation of Nature and Natural Resources, Gland, Switzerland).

DOUPÉ, R. G., 2007. Where are they now? A baseline distributional description of introduced redclaw crayfish *Cherax quadricarinatus* (von Martens) in the east Kimberley region. Journal of the Royal Society of Western Australia, **90**(4): 215-218.

DOUPÉ, R. G., D. L. MORGAN, H. S. GILL & A. J. ROWLAND, 2007. Introduction of redclaw crayfish *Cherax quadricarinatus* (von Martens) to Lake Kununurra, Ord River, Western Australia: prospects for a 'yabby' in the Kimberley. Journal of the Royal Society of Western Australia, **87**(4): 187-191.

FETZNER, J. W. JR., 2012. The crayfish and lobster taxonomy browser. Available online via: http://iz.carnegiemnh.org/crayfish/NewAstacidea/infraorder.asp?io=Astacidea.

FURSE, J. M., 2010. Ecosystem engineering by *Euastacus sulcatus* (Decapoda: Parastacidae) in the Hinterland of the Gold Coast, Queensland, Australia. (Ph.D. thesis, Griffith School of Environment, Griffith University. Gold Coast, Queensland, Australia).

FURSE, J. M., J. W. P. BONE, S. D. APPLETON, J. C. LELAND & J. COUGHRAN, 2012a. Conservation of imperiled crayfish — *Euastacus bindal* (Decapoda: Parastacidae), a highland crayfish from far North Queensland, Australia. Journal of Crustacean Biology, **32**(4): 677-683.

FURSE, J. M. & J. COUGHRAN, 2011a. An assessment of the distribution, biology, threatening processes and conservation status of the freshwater crayfish, genus *Euastacus* (Decapoda: Parastacidae), in continental Australia. I. Biological background and current status. In: A. ASAKURA (ed.), New fronties in crustacean biology: proceedings of the TCS summer meeting, Tokyo, 20-24 September 2009. Crustaceana Monographs, **15**: 241-252.

— — & — —, 2011b. An assessment of the distribution, biology, threatening processes and conservation status of the freshwater crayfish, genus *Euastacus* (Decapoda: Parastacidae), in continental Australia. II. Threats, conservation assessments and key findings. In: A. ASAKURA (ed.), New fronties in crustacean biology: proceedings of the TCS summer meeting, Tokyo, 20-24 September 2009. Crustaceana Monographs, **15**: 253-263.

— — & — —, 2011c. An assessment of the distribution, biology, threatening processes and conservation status of the freshwater crayfish, genus *Euastacus* (Decapoda: Parastacidae), in continental Australia. III. Case studies and recommendations. In: A. ASAKURA (ed.), New fronties in crustacean biology: proceedings of the TCS summer meeting, Tokyo, 20-24 September 2009. Crustaceana Monographs, **15**: 265-274.

FURSE, J. M., J. COUGHRAN & C. H. WILD, 2012b. Report of a mass mortality of *Euastacus valentulus* (Decapoda: Parastacidae) in southeast Queensland, Australia, with a discussion of the potential impacts of climate change induced severe weather events on freshwater crayfish species. Crustacean Research, (Special Number) **7**: 15-24.

FURSE, J. M., K. L. DAWKINS & J. COUGHRAN, 2013. Two new species of *Euastacus* (Decapoda: Parastacidae) from the Gondwana Rainforests of Central-Eastern Australia. Freshwater Crayfish, **19**(1): 103-113.

FURSE, J. M. & C. H. WILD, 2002. Terrestrial activities of *Euastacus sulcatus*, the Lamington Spiny Crayfish (Decapoda: Parastacidae). Freshwater Crayfish, **13**: 604.

FURSE, J. M., C. H. WILD & N. N. VILLAMAR, 2004. In-stream and terrestrial movements of *Euastacus sulcatus* in the Gold Coast hinterland: developing and testing a method of accessing freshwater crayfish movements. Freshwater Crayfish, **14**: 213-220.

GEDDES, M. C., R. J. MUSGROVE & N. J. H. CAMPBELL, 1993. The feasibility of re-establishing the River Murray Crayfish, *Euastacus armatus*, in the Lower Murray River. Freshwater Crayfish, **9**: 368-379.

GILLIGAN, D., R. ROLLS, J. MERRICK, M. LINTERMANS, P. DUNCAN & J. KOHEN, 2007. Scoping the knowledge requirements for Murray crayfish (*Euastacus armatus*). Fisheries Final Report Series, **89**: 1-107. (New South Wales Department of Primary Industries, Narrandera, New South Wales, Australia).

HARLIOĞLU, M. M. & A. G. HARLIOĞLU, 2006. Threat of non-native crayfish introductions into Turkey: global lessons. Reviews in Fish Biology and Fisheries, **16**(2): 171-181.

HART, D. D., 1992. Community organization in streams: the importance of species interactions, physical factors, and chance. Oecologia, **91**: 220-228.

HILBERT, D. W., B. OSTENDORF & M. S. HOPKINS, 2001. Sensitivity of tropical forests to climate change in the humid tropics of north Queensland. Austral Ecology, **26**(6): 590-603.

HOGGER, J. B., 1988. Ecology, population biology and behaviour. In: D. M. HOLDICH & R. S. LOWERY (eds.), Freshwater crayfish: biology, management and exploitation: 114-144. (Croom Helm, London, England. ISBN 0-7099-3792-X).

HOLDICH, D. M., 2002a. General biology: background and functional morphology. In: D. M. HOLDICH (ed.), Biology of freshwater crayfish: 3-29. (Blackwell Science, Oxford, England. ISBN 0-632-05431-X).

— —, 2002b. Part 3: Conclusions. In: D. M. HOLDICH (ed.), Biology of freshwater crayfish: 673-682. (Blackwell Science, Oxford, England. ISBN 0-632-05431-X).

HOLDICH, D. M., J. D. REYNOLDS, C. SOUTY-GROSSET & P. J. SIBLEY, 2009. A review of the ever increasing threat to European crayfish from non-indigenous crayfish species. Knowledge and Management of Aquatic Ecosystems, **11**(394-395): 1-46.

HORWITZ, P., 1990a. The conservation status of Australian freshwater Crustacea with a provisional list of threatened species, habitats and potentially threatening processes. Report Series, **14**: 1-121. (Australian National Parks and Wildlife Service, Canberra, Australia).

— —, 1990b. The translocation of freshwater crayfish in Australia: potential impact, the need for control and global relevance. Biological Conservation, **54**(4): 291-305.

— —, 2010. The conservation status of freshwater crayfish: the basis for concern, listing and recovery processes, and community involvement. Freshwater Crayfish, **17**: 1-12.

HORWITZ, P. & M. ADAMS, 2000. The systematics, biogeography and conservation status of species in the freshwater crayfish genus *Engaewa* Riek (Decapoda: Parastacidae) from south-western Australia. Invertebrate Taxonomy, **14**(5): 655-680.

HORWITZ, P., D. BRADSHAW, S. HOPPER, P. DAVIES, R. FROEND & F. BRADSHAW, 2008. Hydrological change escalates risk of ecosystem stress in Australia's threatened biodiversity hotspot. Journal of the Royal Society of Western Australia, **91**(1): 1-11.

HORWITZ, P. & B. KNOTT, 1995. The distribution and spread of the yabby *Cherax destructor* complex in Australia: speculations, hypotheses and the need for research. Freshwater Crayfish, **10**: 81-91.

HORWITZ, P. H. J. & A. M. M. RICHARDSON, 1986. An ecological classification of the burrows of Australian crayfish. Australian Journal of Marine and Freshwater Research, **37**: 237-242.

HUGHES, L., 2003. Climate change and Australia: trends, projections and impacts. Austral Ecology, **28**(4): 423-443.

HUNER, J. V., 1988. *Procambarus* in North America and elsewhere. In: D. M. HOLDICH & R. S. LOWERY (eds.), Freshwater crayfish: biology, management and exploitation: 239-261. (Croom Helm, London, England. ISBN 0-7099-3792-X).

— —, 1999. Keeping crawfish for pets. Aquaculture Magazine, **March-April**: 54-62.

HURYN, A. D. & J. B. WALLACE, 1987. Production and litter processing by crayfish in an Appalachian mountain stream. Freshwater Biology, **18**: 277-286.

IPCC, 2007. Climate Change 2007 — Synthesis Report. An assessment of the Intergovernmental Panel on Climate Change. Fourth Assessment Report: 52 pp. (Intergovernmental Panel on Climate Change, Geneva, Switzerland).

IUCN, 2011a. Guidelines for using the IUCN Red List Categories and Criteria. Version 9.0 (September, 2011): 87 pp. (Prepared by the Standards and Petitions Subcommittee of the IUCN Species Survival Commission, Gland, Switzerland and Cambridge, UK).

— —, 2011b. IUCN Red List of Threatened Species (Version 2011.1). (The International Union for Conservation of Nature and Natural Resources, Gland, Switzerland). Available online via: http://www.iucnredlist.org/.

— —, 2012. IUCN Red List of Threatened Species (Version 2012.1). (The International Union for Conservation of Nature and Natural Resources, Gland, Switzerland). Available online via: http://www.iucnredlist.org/.

JONES, C. G., J. H. LAWTON & M. SHACHAK, 1994. Organisms as ecosystems engineers. Oikos, **69**: 373-386.

JONES, D. & G. MORGAN, 2002. Field guide to crustaceans of Australian waters: 1-224. (Reed, Chatswood, New South Wales, Australia).

KAILOLA, P., M. WILLIAMS, P. STEWART, R. REICHELT, A. MCNEE & C. GRIEVE, 1993. Australia's fisheries resources: 1-422. (Bureau of Resource Sciences and the Fisheries Research and Development Corporation, Canberra, ACT, Australia. ISBN 0642188769).

KOZUBÍKOVÁ, E., A. PETRUSEK, Z. ĎURIŠ, M. P. MARTÍN, J. DIÉGUEZ-URIBEONDO & B. OIDTMANN, 2008. The old menace is back: recent crayfish plague outbreaks in the Czech Republic. Aquaculture, **274**(2-4): 208-217.

KREBS, C. J., 1994. Ecology: the experimental analysis of distribution and abundance: 1-801. (HarperCollins, New York, USA. ISBN 0-06-500410-8).

LAWRENCE, C. & C. JONES, 2002. Crayfish of commercial importance: Cherax. In: D. M. HOLDICH (ed.), Biology of freshwater crayfish: 635-666. (Blackwell Science, Oxford, England. ISBN 0-632-05431-X).

LELAND, J. C., J. COUGHRAN & J. M. FURSE, 2012. Further translocation of the Redclaw, *Cherax quadricarinatus* (Decapoda: Parastacidae), to Lake Ainsworth in northeastern New South Wales, Australia. Crustacean Research, (Special Number) **7**: 1-4.

LEWIS, A. & F. MORRIS, 2008. Report on a major stranding of crayfish at Meldon, River Wansbeck, UK. Crayfish News, **30**(4): 7-8.

LODGE, D. M., A. L. BECKEL & J. J. MAGNUSON, 1985. Lake bottom tyrant. Natural History, **8**: 33-37.

LODGE, D. M., M. W. KERSHNER, J. E. ALOI & A. P. COVICH, 1994. Effects of an omnivorous crayfish (*Orconectes rusticus*) on a freshwater littoral food web. Ecology, **75**(5): 1265-1281.

LODGE, D. M., C. A. TAYLOR, D. M. HOLDICH & J. SKURDAL, 2000. Nonindigenous crayfishes threaten North American freshwater biodiversity: lessons from Europe. Fisheries, **25**(8): 7-20.

LOWE, K., S. FITZGIBBON, F. SEEBACHER & R. S. WILSON, 2010. Physiological and behavioural responses to seasonal changes in environmental temperature in the Australian spiny crayfish *Euastacus sulcatus*. Journal of Comparative Physiology B: Biochemical, Systemic, and Environmental Physiology, **180**(5): 653-660.

MASON, J. C., 1975. Crayfish production in a small woodland stream. Freshwater Crayfish, **2**: 449-479.

MATTHEWS, M. & J. D. REYNOLDS, 1992. Ecological impact of crayfish plague in Ireland. Hydrobiologia, **234**(1): 1-6.

MCCARTHY, B., S. ZUKOWSKI, N. WHITEROD, L. VILIZZI, L. BEESLEY & A. KING, 2014. Hypoxic blackwater event severely impacts Murray crayfish (*Euastacus armatus*) populations in the Murray River, Australia. Austral Ecology. DOI:10.1111/aec.12109.

MCKINNON, L. J., 1995. Emersion of Murray Crayfish, *Euastacus armatus* (Decapoda: Parastacidae), from The Murray River due to post-flood water quality. Proceedings — Royal Society of Victoria, **107**(1): 31-38.

MERRICK, J. R., 1995. Diversity, distribution and conservation of freshwater crayfishes in the eastern highlands of New South Wales. Proceedings of the Linnean Society of New South Wales, **151**: 247-258.

MOMOT, W. T., 1995. Redefining the role of crayfish in aquatic ecosystems. Reviews in Fisheries Science, **3**(1): 33-63.

MORGAN, G. J., 1986. Freshwater crayfish of the genus *Euastacus* Clark (Decapoda, Parastacidae) from Victoria. Memoirs of the Museum of Victoria, **47**(1): 1-57.

— —, 1988. Freshwater crayfish of the genus *Euastacus* Clark (Decapoda: Parastacidae) from Queensland. Memoirs of the Museum of Victoria, **49**(1): 1-49.

— —, 1997. Freshwater crayfish of the genus *Euastacus* Clark (Decapoda: Parastacidae) from New South Wales, with a key to all species of the genus. Records of the Australian Museum, (Supplement) **23**: 110.

MORITZ, C., 1994. Defining 'Evolutionarily Significant Units' for conservation. Trends in Ecology and Evolution, **9**(10): 373-375.

NATURE, 2009. Natural value. Nature, **457**(7231): 764.

NYSTROM, P., 2002. General biology: ecology. In: D. M. HOLDICH (ed.), Biology of freshwater crayfish: 192-235. (Blackwell Science, Oxford, England. ISBN 0-632-05431-X).

PARKYN, S. M. & K. J. COLLIER, 2004. Interaction of press and pulse disturbances on crayfish populations: flood impacts in pasture and forest streams. Hydrobiologia, **527**(1): 113-124.

PARKYN, S. M., C. F. RABENI & K. J. COLLIER, 1997. Effects of crayfish (*Paranephrops planifrons*: Parastacidae) on in-stream processes and benthic faunas: a density manipulation experiment. New Zealand Journal of Marine and Freshwater Research, **31**: 685-692.

PEAY, S., D. M. HOLDICH & J. BRICKLAND, 2010. Risk assessments of non-Indigenous crayfish in Great Britain. Freshwater Crayfish, **17**: 109-122.

PROVENZANO, A. J., 1985. Culture of crustaceans: general principles. In: D. E. BLISS & A. J. PROVENZANO (ed.), The biology of crustacea. Volume 10. Economic aspects: fisheries and culture: 249-268. (Academic Press, New York).

REYNOLDS, J. D. & C. SOUTY-GROSSET, 2012. Management of freshwater biodiversity — crayfish as bioindicators: 1-374. (Cambridge University Press, Cambridge, England. ISBN 978-0-521-51400-2).

RICHARDSON, A. M. M., 1983. The effect of the burrows of a crayfish on the respiration of the surrounding soil. Soil Biology and Biochemistry, **15**(3): 239-242.

RIEK, E. F., 1951. The freshwater crayfish (family Parastacidae) of Queensland, with an appendix describing other Australian species. Records of the Australian Museum, **22**: 368-388.

SCHIKORA, F., 1906. Die Krebspest. Fischerei-Zeitung, **9**: 549-553.

SEWELL, K. B. & L. R. G. CANNON, 1998a. New Temnocephalans from the branchial chamber of Australian *Euastacus* and *Cherax* crayfish hosts. Proceedings of the Linnean Society of New South Wales, **119**: 21-36.

— — & — —, 1998b. The taxonomic status of the ectosymbiotic flatworm *Didymorchis paranephropis* Haswell. Memoirs of the Queensland Museum, **42**: 585-595.

SINCLAIR, E. A., A. MADSEN, T. WALSH, J. NELSON & K. A. CRANDALL, 2011. Cryptic genetic divergence in the giant Tasmanian freshwater crayfish *Astacopsis gouldi* (Decapoda: Parastacidae): implications for conservation. Animal Conservation, **14**: 87-97.

SPECHT, A., 2008. Extreme natural events and effects on tourism: central eastern coast of Australia: 1-66. (The Cooperative Centre for Sustainable Tourism Pty Ltd, Griffith University, Gold Coast, Queensland, Australia).

STATZNER, B., E. FIEVET, J.-Y. CHAMPAGNE, R. MOREL & E. HEROUIN, 2000. Crayfish as geomorphic agents and ecosystem engineers: biological behavior affects sand and gravel erosion in experimental streams. Limnology and Oceanography, **45**(5): 1030-1040.

STATZNER, B., O. PELTRET & S. TOMANOVA, 2003. Crayfish as geomorphic agents and ecosystem engineers: effect of a biomass gradient on baseflow and flood-induced transport of gravel and sand in experimental streams. Freshwater Biology, **48**: 147-163.

STONE, E. L., 1993. Soil burrowing and mixing by a crayfish. Soil Science Society of America Journal, **57**(4): 1096-1099.

TAYLOR, C. A., 2002. Taxonomy and conservation of native crayfish stocks. In: D. M. HOLDICH (ed.), Biology of freshwater crayfish: 236-257. (Blackwell Science, Oxford, England. ISBN 0-632-05431-X).

THREATENED SPECIES SECTION, 2006. Giant Freshwater Lobster *Astacopsis gouldi* Recovery Plan 2006-2010: 1-41. (Department of Primary Industries and Water, Hobart, Tasmania, Australia).

UNESTAM, T., 1973. Significance of diseases on freshwater crayfish. Freshwater Crayfish, **1**: 135-150.

USIO, N. & C. R. TOWNSEND, 2001. The significance of the crayfish *Paranephrops zealandicus* as shredders in a New Zealand headwater stream. Journal of Crustacean Biology, **21**(2): 354-359.

VAN-PRAAGH, B., 2003. Action statement (Flora and Fauna Guarantee Act, 1988): Glenelg Spiny Crayfish *Euastacus bispinosus*, Murray Spiny Crayfish *Euastacus armatus*: 1-10. (Department of Sustainability and Environment, Melbourne, Victoria, Australia).

WALSH, T. & B. WALSH, 2013. A study of growth and moulting rates of *Astacopsis gouldi* Clark. Freshwater Crayfish, **19**(1): 97-101.

WELLS, S. M., R. M. PYLE & N. M. COLLINS, 1983. The IUCN Invertebrate Red Data Book: 1-632. (The IUCN, Gland, Switzerland. ISBN 2-88032-602-X).

WILD, C. H. & J. M. FURSE, 2004. The relationship between *Euastacus sulcatus* and Temnocephalan spp. (Platyhelminthes) in the Gold Coast hinterland, Queensland. Freshwater Crayfish, **14**: 236-245.

First received 20 September 2012.
Final version accepted 29 July 2013.

www.ingramcontent.com/pod-product-compliance
Lightning Source LLC
La Vergne TN
LVHW020634300926
849161LV00004B/697